Springer Advanced Texts in Chemistry

Charles R. Cantor, Editor

Springer Advanced Texts in Chemistry

Series Editor: Charles R. Cantor

Principles of Protein Structure
G.E. Schulz and *R.H. Schirmer*

Bioorganic Chemistry
H. Dugas and *C. Penney*

Protein Purification: Principles and Practice
R.K. Scopes

Principles of Nucleic Acid Structure
W. Saenger

G. E. Schulz
R. H. Schirmer

Principles of Protein Structure

Springer-Verlag
New York Berlin Heidelberg Tokyo

Dr. Georg E. Schulz

Max-Planck-Institut für Medizinische Forschung
Jahnstr. 29
6900 Heidelberg
Federal Republic of Germany

Dr. R. Heiner Schirmer

Max-Planck-Institut für Medizinische Forschung
Jahnstr. 29
6900 Heidelberg
Federal Republic of Germany

Series Editor:

Prof. Charles R. Cantor

Columbia University
College of Physicians and Surgeons
Department of Human Genetics and Development
New York, New York 10032, U.S.A.

Library of Congress Cataloging in Publication Data
Schulz, Georg E., 1939-
  Principles of protein structure.
  (Springer advanced texts in chemistry)
  Bibliography: p.
  Includes index.
  1. Proteins. I. Schirmer, R. Heiner,
1942-   joint author. II. Title. III. Series.
QP551.S4255   547'.75   78-11500

Printed and bound by Halliday Lithograph, West Hanover, Massachusetts.
Printed in the United States of America.

9 8 7 6 (Sixth printing, 1985)

Hardcover edition:
ISBN 0-387-90386-0 Springer-Verlag New York Berlin Heidelberg Tokyo
ISBN 3-540-90386-0 Springer-Verlag Berlin Heidelberg New York Tokyo

Softcover edition:
ISBN 0-387-90334-8 Springer-Verlag New York Berlin Heidelberg Tokyo
ISBN 3-540-90334-8 Springer-Verlag Berlin Heidelberg New York Tokyo

Series Preface

New textbooks at all levels of chemistry appear with great regularity. Some fields like basic biochemistry, organic reaction mechanisms, and chemical thermodynamics are well represented by many excellent texts, and new or revised editions are published sufficiently often to keep up with progress in research. However, some areas of chemistry, especially many of those taught at the graduate level, suffer from a real lack of up-to-date textbooks. The most serious needs occur in fields that are rapidly changing. Textbooks in these subjects usually have to be written by scientists actually involved in the research which is advancing the field. It is not often easy to persuade such individuals to set time aside to help spread the knowledge they have accumulated. Our goal, in this series, is to pinpoint areas of chemistry where recent progress has outpaced what is covered in any available textbooks, and then seek out and persuade experts in these fields to produce relatively concise but instructive introductions to their fields. These should serve the needs of one semester or one quarter graduate courses in chemistry and biochemistry. In some cases the availability of texts in active research areas should help stimulate the creation of new courses.

New York, New York
December, 1978

CHARLES R. CANTOR

Preface

Among the four major types of biopolymers—nucleic acids, polysaccharides, lipid assemblies, and proteins—the latter are best understood because they are structurally well defined and therefore suitable for isolation and analysis. Proteins were placed at the forefront of biopolymer research more than a century ago when Kühne (1) purified and characterized the enzyme trypsin and Hoppe-Seyler (2) obtained crystals of haemoglobin. These crystals indicated that proteins were uniform down to atomic detail; but this fact was not widely recognized until atomic detail was actually made visible by X-ray diffraction techniques.

Today more than 70 protein structures are known. From a practical viewpoint these structures provide biochemists and pharmacologists with models for the design and interpretation of experiments. In general, they put biochemical sciences on a sounder basis. The aim of this book is to consider principles which have emerged from comparisons of known protein structures. These principles are the first step for understanding the relationship between a protein's blueprint as stored in DNA and the action of this protein in an organism. In addition, establishing principles facilitates and extends the usage of the complex protein architectures for scientific and practical purposes. The very existence of structural principles indicates that the diversity of protein functions is not paralleled by structural diversity. For the biomedical sciences this could mean that a comprehensive understanding of physiological and pathological processes at the molecular level is not a utopian dream, and that it may become possible to influence such processes in pin-pointed fashion.

Each chapter of this book is self-contained. The chapters can be read in any order, but those unfamiliar with protein structures are recommended to read chapter seven first. We would like to thank all our colleagues who allowed us to use their graphic and digital material. Furthermore, we thank all those who helped us in preparing the manuscript.

Heidelberg
December, 1978

G. E. SCHULZ
R. H. SCHIRMER

Contents

Appendix

Statistical Mechanics of the Helix-Coil Transition   252

References   262

Index   303

Principles of
Protein Structure

Chapter 1
Amino Acids

1.1 The 20 Standard Amino Acids

Today, the basic organizational scheme of living material is known. Nucleic acids contain all genetic information, which is laid down in a sequence of four different nucleotide bases along a chain. There are two levels of nucleic acids. The more stable deoxyribonucleic acid (DNA) is the permanent information store. The less stable ribonucleic acid (RNA), which is "transcribed" from DNA, comes as a handy blueprint that is "translated" into the amino acid sequence of proteins using the ribosomal machinery. Proteins perform virtually all types of work in an organism.

In this chapter we are concerned with the 20 standard amino acid residues occurring in proteins. These are listed in Table 1-1 according to their relative abundance. The commonly used three-letter and one-letter abbreviations together with mnemonic aids are given. The molecular weights of the residues range from 57 to 186 daltons, the weighted average being about 110. Thus, a protein with a molecular weight of 33,000 contains approximately 300 residues. The amino acids can be classified with respect to their side chains as either polar or nonpolar. The polar side chains can be further subdivided into neutral, basic, or acidic. In Figure 1-1, which illustrates the 20 common amino acid side chains, residues with similar properties are grouped near one another.

1.2 Why Were Just These Amino Acids Selected?

Presumably, life started with autocatalytic nucleic acids and not with proteins. The question of why just these amino acids were selected cannot yet be answered. Nevertheless speculating about it can provide us with some insights. First, let us discuss the origin of the basic organizational scheme introduced above, the origin of life. It is highly probable that life

Table 1-1. Notations and Properties of the 20 Standard Amino Acid Residues of Proteins[a]

Amino acid or residue thereof	Three-letter symbol	One letter symbol	Mnemonic help for one-letter symbol	Relative abundance in E. coli proteins (19) (%)	M.W. of residue at pH7.0 (daltons)	pK value of side chain (19)	ΔG values for transfer of side chain from water to ethanol at 25°C (16) (kcal/mol)
Alanine	Ala	A	Alanine	13.0	71		−0.5
Glutamate	Glu	E	gluEtamic acid	10.8	128	4.3	
Glutamine	Gln	Q	Q-tamine		128		
Aspartate	Asp	D	asparDic acid	9.9	114	3.9	
Asparagine	Asn	N	asparagiNe		114		
Leucine	Leu	L	Leucine	7.8	113		−1.8
Glycine	Gly	G	Glycine	7.8	57		
Lysine	Lys	K	before L	7.0	129	10.5	
Serine	Ser	S	Serine	6.0	87		+0.3
Valine	Val	V	Valine	6.0	99		−1.5
Arginine	Arg	R	aRginine	5.3	157	12.5	
Threonine	Thr	T	Threonine	4.6	101		−0.4
Proline	Pro	P	Proline	4.6	97		
Isoleucine	Ile	I	Isoleucine	4.4	113		
Methionine	Met	M	Methionine	3.8	131		−1.3
Phenylalanine	Phe	F	Fenylalanine	3.3	147		−2.5
Tyrosine	Tyr	Y	tYrosine	2.2	163	10.1	−2.3
Cysteine	Cys	C	Cysteine	1.8	103		
Tryptophan	Trp	W	tWo rings	1.0	186		−3.4
Histidine	His	H	Histidine	0.7	137	6.0	−0.5
					Weighted mean 108.7		

[a]Further three-letter symbol: Asx, Glx = either acid or amide.
An amino acid residue, -HN-CHR-CO-, is the part of an amino acid (Figure 1-2a) occurring within a peptide chain; R denotes the side chain (Figure 1-1).
Further one-letter symbol: B = Asx, Z = Glx, X = undetermined or nonstandard amino acid residue.

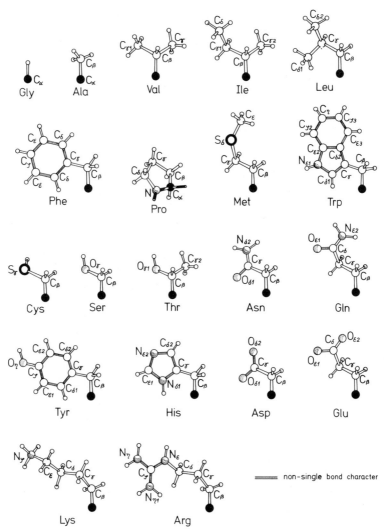

Figure 1-1. The 20 standard amino acid side chains. For proline part of the main chain is inserted. All other side chains are shown as they emerge from the C_α-atom in the main chain. The residue names are given as three-letter symbols. Atom names are those given in the IUPAC–IUB recommendations of 1969 (21). The main chain in Pro is indicated by solidly drawn bonds. All C_α-atoms are black.

started with single autocatalytic, that is, self-reproducing, molecules (3). Nucleic acids, and not proteins, are the candidates of choice for this role (4–6), because a single-stranded nucleic acid can use base pairing to align its complementary sequence of nucleotides. On linking the aligned nucleotides both strands can form the energetically favorable Watson–Crick double helix. Thus, linkage may be facilitated. In the next cycle, after reopening the double helix, the complementary strand can reproduce the original strand. With proper sequence selection by base pairing and with the double helix structure providing a free energy incentive, autocatalytic reproduction is conceivable. In contrast, starting with proteins instead of nucleic acids would be much more difficult from an organizational point of view because no mutual recognition among amino acids exists.

Our knowledge of subsequent molecular evolution is also restricted to informed guesses. Although self-reproduction must have been very inefficient at the beginning, the molecules involved showed a tendency to improve autocatalysis, because they were subjected to strong evolutionary pressure; only the most efficient systems survived. Presumably, the improvement started with "enzymes" made of nucleic acid, splinters of the autocatalytic molecule itself. It is likely that ribosomal and transfer RNA are remnants of such catalysts. Later the nucleic acid enzymes were superseded by more efficient protein enzymes.

A linear array of standard elements with standard linkage is simple. The protein enzymes developed using a very straight-forward organizational scheme. The amino acid sequence in the polypeptide chain of proteins is a colinear and unique representation of the nucleotide sequence of the underlying nucleic acid. Three adjacent nucleotides code for one amino acid (Figure 1-5b). Accordingly, the polypeptides are similar to nucleic acids in that they are *linear* chain molecules with standard elements and one standard linkage. This allows for a simple and universal nucleic acid reading and polypeptide synthesizing machinery. The simplicity of linear systems is reflected, for example, in computer technology, where information storage and retrieval is usually handled via one-dimensional arrays put in standard format on linear data carriers such as magnetic tapes.

In nucleic acids as well as in proteins not only the linkages but also the atomic groups forming the backbone of the chain are uniform; in polypeptides all 20 amino acid residues are of the α-type and have the L-configuration at the C_α-atom (Figure 1-2a). All differences and therefore all information are restricted to the rather short side chains.

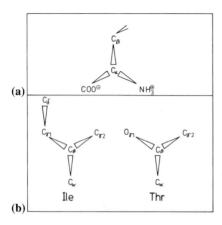

Figure 1-2. Asymmetric atoms occurring in standard amino acid residues. The view is from a hydrogen atom down to C_α or to C_β, respectively. This is opposite to the viewing direction in the Cahn-Ingold-Prelog presentation (41). (a) The standard L-isomer of an α-amino acid. The absolute configuration at C_α according to the Cahn-Ingold-Prelog notation is S. (b) The configuration at the C_β-atom of isoleucine (S) and at the C_β-atom of threonine (R).

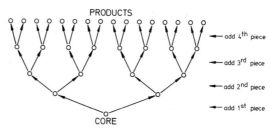

Figure 1-3. Hypothetical formation of molecular structures by adding various pieces to a core molecule, or by modifying a given precursor molecule specifically. In order to obtain different products each addition or modification must be specific, which means that it requires a specific enzymatic action. At the fourth stage $2^4 = 16$ different products are formed but $2 + 4 + 8 + 16 = 30 = 2^5 - 2$ manipulations were necessary.

Spontaneous folding is indispensable. After synthesis on the ribosome the polypeptide chain folds spontaneously to an amino acid sequence-dependent globular protein by adopting a state of lower free energy. It is possible that folding already starts during the synthesis. The resulting chain fold determines the specificity of the protein. This spontaneity simplifies the organizational scheme considerably, because otherwise "synthesizing enzymes" or "folding helper enzymes" would be required. Since very many chain folds are possible, a large number of such helping enzymes would be necessary.

Let us assume that structures were not built by spontaneous folding but, for example, by taking a small core molecule and linking subsequent pieces covalently to it until the desired form was attained. Even with the most efficient concept shown in Figure 1-3, such a process would require two enzymes for processing each intermediate and therefore an average of two enzymes per product molecule. Thus, in order to get a self-sustaining system all product molecules would have to be used as processing enzymes, each of which would have to catalyze two different synthetic reactions. However, in present day organisms generally only one enzymic action is observed for each molecule. More important, such a (super)tightly knit system would be very sensitive to errors—and it would have almost no chance to evolve from an unrefined primordial chemical soup (6). In conclusion, one appreciates the simplicity introduced by spontaneous folding.

But not all polypeptide chains fold spontaneously onto themselves. Some of the ribosomal proteins are very elongated and have very few internal contacts (7). Here, the chain assumes its final conformation by attachment to another molecule, namely the ribosomal nucleic acid core. These chains are probably very ancient and may reflect the situation which existed at the very beginning of protein production; ribosomes are highly conserved during evolution. This strong conservation is due to the ribosomes' central role in the life of an organism. Every change in the ribosome would be disruptive, giving rise to a very small chance for survival (3,4).

The standard amino acids can be produced under primordial conditions.
The set of standard links in a polypeptide chain consists of 20 amino acid
residues (Figure 1-1, Table 1-1). In order to be selected these amino acids
must have been present in the environment of the primordial autocatalytic
nucleic acid monsters. This was most likely the case, since several simulation experiments (3,8) have shown that amino acids can be produced under
primordial conditions. In these experiments electric or radiation energy
was pumped into an atmosphere considered representative of the atmosphere on early Earth. Usually, a racemic mixture of L- and D-amino acids
(Figure 1-2a) of the α-type together with β-amino acids (Figure 1-4) and
other compounds was obtained.

The L-configuration has been selected by chance. Is there any reason for
the selection of the L-configuration at the C_α-atom? We do understand that
all C_α-configurations are identical; the chain-synthesizing machinery
requires a standard configuration, otherwise synthesis would be too complex and therefore would not have a competitive chance. But there is no
obvious reason for preferring the L-configuration over its mirror image.
Although it has been shown that the intrinsic asymmetry of β-decay
expresses itself as molecular asymmetry by preferentially destroying D-
amino acids (9), the observed effect of a few percent is too small to explain
the selection. Presumably it occurred by chance and not because of a
slightly higher amino acid concentration that the L-system developed first.
Being farther advanced it then suppressed the emergence of a D-system.
Alternatively, both systems developed in parallel, and the L-system won
the ensuing competition by favorable environmental fluctuations, the
impact of which presumably far exceeded the impact of a few percent
difference in amino acid concentrations.

β-Amino acids are not suitable. Another aspect of these simulation
experiments is the production of β-amino acids (Figure 1-4) in yields
comparable to those of the α-type (8). In many respects β-amino acids
could be used in the same way as the α-type except that the resulting
polypeptide chain would contain C_α–C_β bonds which allow for additional
free rotations. As shown in section 8.1 (page 151), such flexibility would
prevent spontaneous chain folding. Thus, a reason for the selection of the
α-type seems to be at hand.

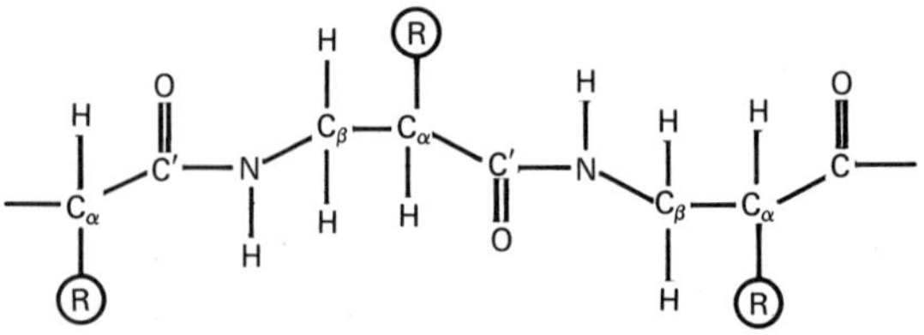

Figure 1-4. A hypothetical main chain consisting of β-amino acids. The rotation
around the C_α–C_β and the C_β–N bonds would be rather free, rendering the chain
very flexible. The argument would not change if the side chains R were attached to
C_β instead of to C_α.

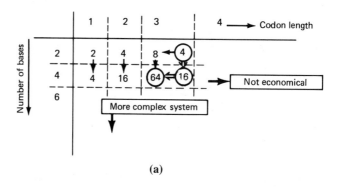

(a)

Second base

		U	C	A	G	
		Phe	Ser	Tyr	Cys	U
	U	Phe	Ser	Tyr	Cys	C
		Leu	Ser	End	End	A
		Leu	Ser	End	Trp	G
		Leu	Pro	His	Arg	U
	C	Leu	Pro	His	Arg	C
		Leu	Pro	Gln	Arg	A
		Leu	Pro	Gln	Arg	G
		Ile	Thr	Asn	Ser	U
	A	Ile	Thr	Asn	Ser	C
		Ile	Thr	Lys	Arg	A
		Met	Thr	Lys	Arg	G
		Val	Ala	Asp	Gly	U
	G	Val	Ala	Asp	Gly	C
		Val	Ala	Glu	Gly	A
		Val	Ala	Glu	Gly	G

(First base on left, Third base on right)

(b)

Figure 1-5. The genetic code for translation from nucleotide sequence to amino acid sequence. (a) The number of codable amino acids as a function of the number of different nucleotide bases and of codon length. Arrows indicate possible evolutionary directions. Judging from the present code, evolution should have followed the pathway marked by circles. (b) The genetic code table relating nucleotide triplets to amino acids.

The translation machinery fixes the set of amino acid types. Why does the entire living community on Earth use the same set of amino acids? This question is related to the translation of nucleotide sequences into amino acid sequences, applying the so-called genetic code (Figure 1-5b). The genetic code is universal because it is at the center of the organizational scheme of life. Any change in the code greatly disrupts the scheme and causes the extinction of the organism in which the change had occurred. Consequently, the code and thus the types of amino acids are extremely well conserved. There is no indication that organisms with other amino acids or other codes ever existed in reasonable numbers.

Why is the number of selected amino acid types just 20? This question is also related to the translation machinery. For the discussion some translation schemes related to the observed genetic code are given in Figure 1-5a. With a doublet code (codon length = 2 nucleotides) using four different nucleotides, $4^2 = 16$ amino acids are coded in a rather economic manner. However, 3 and not 2 was chosen as the codon length. To explain this fact it should be remembered that the codon length is related to the crucial step in translation, the recognition of the nucleotide sequence of the messenger RNA by nucleotide-base pairing to the small, amino acid-carrying transfer RNA. Conceivably, no bases were available with large enough association constants between base *doublets,* so that the codon had to be enlarged to a *triplet* in order to allow specific recognition. With four different nucleotides a triplet code can distinguish $4^3 = 64$ amino acids from each other. But only 20 amino acids are in use. This can be understood if one assumes that the genetic code did evolve and that this evolution stopped midway.

Presumably the genetic code evolved at very early stages of life. As mentioned above changes of the genetic code are extremely constrained. For instance, no change of the codon length is possible, because this would destroy the whole accumulated genetic information, leading to catastrophe. However, at very early stages of evolution, when organisms were much less refined than they are today, a change of the number of nucleotides or amino acids used is conceivable, since such a change could be rather smooth. Thus, it is quite possible that the code did evolve over a period of time.

This idea is supported by the high degeneracy of the third codon position (Figure 1-5b). The most probable starting point for the present code would have been a triplet code using two complementary nucleotides. Only the first two positions of each codon were significant, so that one triplet coded for only four different amino acids. As the next step in evolution another pair of complementary nucleotides was incorporated, leading to 16 codable amino acids. Finally, the degeneracy was partly resolved by making the third codon position significant. When organisms were so refined, that is, were competing so efficiently, that total exchange of a single amino acid type became disruptive enough to be lethal, the genetic code was frozen in and could no longer be changed. Thus, the number of amino acid residues is fixed at and will stay at 20.

Comparison of the present metabolic pathways for amino acid production with the genetic code shows that metabolically linked amino acids are also correlated with respect to their codons (10). This makes the concept of the coevolution of the genetic code and metabolism very attractive. Moreover, it suggests a historical order within the amino acids. The simpler amino acids such as Gly, Ser, Ala, Asp, and Glu were assigned as early ones, in contrast to the more complex amino acids such as Met, His, and Asn. A sequential appearance of amino acids, however, has no impact on present protein structures because amino acid residues in a protein can change easily enough so that any correlation with the early stages of life should have vanished by now.

1.3 Colinear Relation Between Nucleic Acids and Polypeptides

DNA sequencing may become a powerful method for sequencing proteins. The genetic code is degenerate; that is, most amino acids correspond to more than one codon. Therefore, a nucleotide sequence cannot be derived from the colinear amino acid sequence. But unknown amino acid sequences can be deduced from the colinear nucleotide sequences. Thus, protein sequences can be determined by analyzing the underlying nucleotide sequence. Implementation of this indirect method presents a serious drawback: experimental errors that correspond to deletions and insertions of single nucleotides along the sequence of a polynucleotide disturb the frame of nucleotide triplets in such a way that correct amino acids can no longer be deduced.

Recent work on the coat protein of tobacco mosaic virus illustrates this. A fragment of the RNA was sequenced; this nucleotide sequence can be translated into an amino acid sequence which in turn can be compared with the directly determined amino acids (11). As shown in Figure 1-6, these two peptides are by no means identical in amino acid sequence or in amino acid composition.

The possibility of determining the sequence of a polypeptide by sequencing the colinear nucleic acid is not only of academic interest. Many impor-

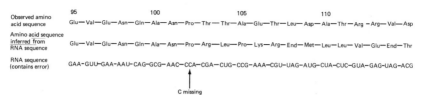

Figure 1-6. Deduction of an amino acid sequence from a corresponding nucleotide sequence. The first line shows a segment of the amino acid sequence of the coat protein of tobacco mosaic virus, and the third line shows the corresponding RNA sequence (11). One sequence error leads to a wrong inferred amino acid sequence (second line) from residue 103 onward. The error is probably a missing C at the indicated position, because this C would make most of the following amino acid sequence correct. Such frame-shift errors are often revealed by the occurrence of End codons at unexpected positions (second line).

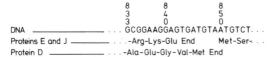

Figure 1-7. Overlapping nucleotide sequences of different genes (12). The segment 833–855 of the genome of the bacteriophage ΦX174 is shown. In this genome the sequence 850–963 codes for protein J, 392–847 codes for protein D, and 570–842 codes for protein E. This means that the nucleotide sequence coding for E is completely contained within the region coding for D but in a different reading frame (lines 2 and 3). Furthermore, the termination codon of D overlaps the initiation codon for protein J at position 850. Note, I corresponds to U in Figure 5-1b.

tant proteins can be isolated only in minute quantities; this contrasts with the progress both in amplifying genetic material and in sequencing nucleic acids. The complete sequence analysis of the polynucleotide aided by patchy sequence studies along the colinear polypeptide chain, to guarantee the correct phase when translating, might become a valuable method for determining the primary structure of proteins (12).

Some DNA segments code for more than just one protein. The colinearity of amino acid sequences and nucleotide sequences attracted renewed attention when the complete primary structures of the RNA from virus MS2 (13) and of the DNA from virus ΦX174 (12) were reported. The bacterial viruses, MS2 and ΦX174, are the first "living" particles for which the entire chemical structure of the genetic material has been elucidated.

The DNA of ΦX174 which is 5375 nucleotides in length contains overlapping gene systems (14): two proteins are read from the same DNA sequence but in different phases (Figure 1-7). This means that the total number of amino acid residues in protein sequences can be greater than only one-third of the number of nucleotides coding for them. An open question remains as to whether or not overlapping genes are present in other prokaryotic or eukaryotic organisms.

1.4 Side Chain Properties

The standard amino acids differ only with respect to their side chains. Each side chain is so specific, that it cannot be easily replaced by another one. In the following we will discuss the properties of all standard side chains.

Gly increases main chain flexibility. Glycine has no side chain but only a hydrogen. It is the exceptional amino acid without asymmetry at the C_α-atom. With no side chain hindrance Gly residues can adopt unusual dihedral angles (Figure 2-3a) giving rise to kinks in the main chain. Moreover, Gly can feed the main chain through tight places in the protein molecule. These are probably the reasons Gly is comparatively well conserved during evolution.

Alanine has one methyl group as a side chain. It is a smallish nonpolar residue that does not have much preference with respect to the inside or the surface of a protein. Ala is a very abundant amino acid, possibly because of its simplicity and easy availability. Only Gly is cheaper to produce. But the Gly frequency is restricted; too great an amount of Gly would cause the main chain to become too flexible.

Branched side chains are stiffer. The nonpolar side chains of valine, isoleucine, and leucine are branched. Branching allows for large side chains with limited internal flexibility. Val has a ramification at the C_β-atom. As a consequence the C_γ-methyl groups hinder the main chain, decreasing its flexibility. Ile is also branched at the C_β-atom. The two branches are different from each other. Thus, Ile has C_β as an additional asymmetric atom. Since all biological reactions are stereospecific, only one of the stereoisomers (Figure 1-2b) is used. Ile stiffens the main chain in a manner similar to Val. No particular main chain hindrance occurs with Leu because its ramification is as far out as C_γ. Branching stiffens the side chains. Stiff side chains are easier to fix in a certain position; the entropy reduction ΔS_{chain} on chain folding is not that large (section 3.5, page 151, and chain folding is facilitated.

All aromatic residues contain one methylene group as a spacer. Phenylalanine has the largest completely nonpolar side chain. As with the other three aromatic side chains there is a C_β methylene group between C_α and the aromatic ring. With this single methylene group the side chain flexibility is rather restricted. Without the methylene group the aromatic ring would lead to severe steric hindrance at the C_α-atom, and the main chain would become too stiff. Proline is also completely nonpolar. It is special because its side chain curls back to the main chain and seizes it. This produces an almost rigid side chain, where only different ring puckers are allowed. Furthermore, it fixes the dihedral angle between C_α and the peptide nitrogen to a small range of about $\pm 20°$.

Tryptophan has the largest side chain. It is slightly polar because the indole ring is heterocyclic. Methionine has a rather flexible side chain containing one sulfur atom in a thioether bond. This sulfur atom introduces an electrical dipole moment. All the larger nonpolar residues, Val, Ile, Leu, Phe, Pro, Trp, and, to a lesser extent, Met, are predominantly found at the inside of protein molecules.

Polar side chains form hydrogen bonds. Typical polar and neutral side chains are those of cysteine, serine, threonine, asparagine, glutamine, and tyrosine. Here, Cys plays a special role because it can form cross-bridges (cystines) between different parts of the main chain by bonding to other Cys residues (section 4.2). Ser and Thr have hydroxyl groups that can take part in hydrogen bonds. Thr has an asymmetric C_β-atom. Only one of the stereoisomers is used (Figure 1-2b). The acid amides asparagine and glutamine can also form hydrogen bonds, the amido groups functioning as hydrogen donors and the carbonyl groups functioning as acceptors. Com-

pared with asparagine, glutamine has a further methylene group, rendering
the polar group more flexible and reducing its interaction with the main
chain. With its pK value of 10.1, the polar hydroxyl group of Tyr can
dissociate at high pH values. Thus, Tyr behaves somewhat like a charged
side chain; its hydrogen bonds are rather strong. Neutral polar residues are
found at the surface as well as inside protein molecules. As internal
residues they usually form hydrogen bonds with each other or with the
polypeptide backbone (section 3.6).

His is suitable for catalytic teams. Histidine has a heterocyclic aromatic
side chain with a pK value of 6.0 (Table 1-1). In the physiological pH range
its imidazol ring can be either uncharged, or it can be charged after taking
up a H^+ ion from the solution. Since these two states are readily available,
histidine is quite suitable for catalyzing chemical reactions. As a matter of
fact, it is found in most of the active centers of enzymes.

Charged side chains are usually at the molecular surface. Aspartate and
glutamate are negatively charged residues at physiological pH. Because of
the short side chain the carboxyl group of Asp is rather rigid with respect to
the main chain. This might be the reason why the carboxyl groups of active
centers are preferentially provided by Asp and not by Glu. Generally, both
residue types are found at protein surfaces.

Most of the positively charged lysine and arginine residues are also at the
molecular surface. They are long and flexible and do not usually adopt a
defined conformation. Wobbling in the surrounding solution, they increase
the solubility of the protein globule. In several cases, however, Lys and
Arg take part in internal salt bridges or they help in catalysis. Owing to their
exposed location at the surface, Lys residues are more frequently than
others the target of enzymes. These enzymes either modify the side chain
or cleave the polypeptide chain at the carbonyl end of lysyl (and arginyl)
residues (section 4.3).

Hydrophobicity depends on surface area and dipole content. As discussed in chapters 3 and 8, the folding process of a polypeptide chain
depends on the hydrophobicity (nonpolarity) of the side chains, because the
formation of a hydrophobic core in the globule seems to be one of the
essential driving forces in folding. Kauzmann (15) and later Nozaki and
Tanford (16) have evaluated this hydrophobicity by measuring the free
energy change for a transfer of amino acids from water to organic solvents.
In order to obtain the free energy difference $\Delta G_{transfer}$ for a side chain, the
main chain contribution has to be subtracted. This was accomplished by
subtracting the value for Gly. The results are listed in Table 1-1. As
demonstrated by Chothia (17) (Figure 1-8), a linear relationship exists
between $\Delta G_{transfer}$ and the water accessible surface area for completely
nonpolar side chains. The accessible surface area is defined in Figure 1-9.
For residues containing one dipole, such as Ser, Thr, Met, Tyr, and Trp, a
similar relationship exists but with a decrease of about 1.5 kcal/mol in
$\Delta G_{transfer}$ (see section 3.5, page 40). The position of Trp in Figure 1-8

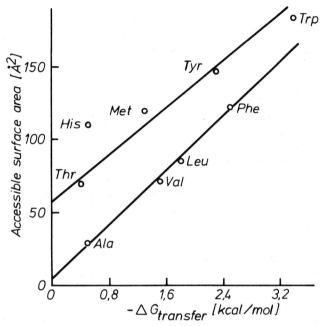

Figure 1-8. Water accessible surface areas (17) of amino acid side chains as a function of their hydrophobicity (16), that is, as a function of the free energy of transfer from water to ethanol and dioxane. These solvents are believed to resemble the protein interior. The accessible surface area is defined in Figure 1-9.

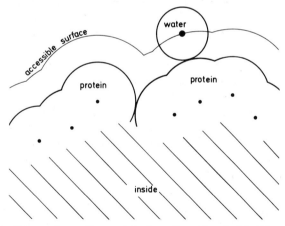

Figure 1-9. The water accessible surface of a protein (or of a side chain). The protein surface is described by van der Waals envelopes of surface atoms (dots). The protein interior is hatched. Water molecules are considered as spheres of 1.4-Å radius. The "accessible surface" is defined as the area described by the center of a water molecule that rolls over the van der Waals envelope of the protein (or of the side chain). Note that there exist inaccessible protein (or side chain) surface areas in clefts.

indicates that the indole ring is only a weak dipole. The dipole of Met seems to be rather strong. This is supported by the detection of two hydrogen bonds to Met-180 in chymotrypsin (18).

1.5 Empirical Similarities Between Amino Acid Residues

Biological diversity facilitates analyses. The present set of biological species surely can never be explained. Speciation depends so much on random events (events that we cannot but assume to be random) that we find a strong historical component in all biosciences. Although at first glance this component looks like a disastrous handicap for the elucidation of basic principles, in practice it has turned out as a major boon.

Change frequencies reveal similarities. With respect to side chains, for instance, we can derive similarities by analyzing the frequencies of amino acid changes between corresponding proteins of homologous organisms. To obtain a more accurate measure of similarity these observed frequencies are normalized, that is divided by the frequency of changes that would occur by chance. Because of differences in abundancy the latter frequencies are specific for each change; an Ala–Leu change is $13.0 \cdot 7.8/3.3 \cdot 2.2 = 14$ times more likely to be found than is a Phe–Tyr change (see abundancies in Table 1-1).

Groups of residues resemble each other. The resulting change probabilities are given in Table 1-2. Here, the amino acids are arranged in such a way that the second moment with respect to the diagonal, $\Sigma_{ij} m_{ij} \cdot (j - i)^2$, is minimized. This arrangement reveals exchange groups. Amino acids within such a group exchange preferentially with each other. Therefore, they resemble each other most with respect to their overall impact on protein structures.

The empirical residue similarities in Table 1-2 show that the aromatics Phe, Tyr, and Trp form an exchange group indicating that their roles correspond to each other. Furthermore, the positively charged residues Lys, Arg, and His form an exchange group. Note, that the negatively charged residues Glu and Asp are rather distant from the positively charged residues. Thus, changes between oppositely charged residues are rare. A high frequency would be expected if charged residues were always on the molecular surface and would do nothing but increase protein solubility.

Side chain size is important. Clearly, the large aliphatic nonpolar residues Val, Leu, and Ile (not Pro!) form an exchange group, which also contains the slightly polar Met and Cys. All small residues (Ser, Thr, Asp, Asn, Gly, Ala, as well as Glu, Gln, and Pro) are found in a single group. Thus, residues as dissimilar as the nonpolar Pro and the strongly polar Asp are lumped together. In general, the observed grouping gives the impres-

Table 1-2. Matrix of Relative Substitution Frequencies at an Evolutionary Distance of 256 Accepted Point Mutations per 100 Residues (PAM)[a]

	Gly	Pro	Asp	Glu	Ala	Asn	Gln	Ser	Thr	Lys	Arg	His	Val	Ile	Met	Cys	Leu	Phe	Tyr	Trp
Gly	29																			
Pro	12	14																		
Asp	11	10	23																	
Glu	11	11	20	20																
Ala	13	14	12	12	14															
Asn	11	10	14	12	10	17														
Gln	10	11	14	14	11	12	21													
Ser	12	11	12	11	13	13	11	13												
Thr	10	10	10	10	12	12	11	13	18											
Lys	7	8	10	10	9	12	11	12	10	24										
Arg	4	4	7	7	5	9	13	9	7	22	75									
His	6	6	8	8	8	9	12	9	6	11	20	59								
Val	7	8	7	7	10	8	9	9	11	8	5	6	22							
Ile	6	6	6	6	8	7	8	8	10	7	5	7	21	25						
Met	5	6	6	6	7	7	8	8	9	8	10	6	16	16	26					
Cys	6	4	6	6	7	6	5	5	9	4	2	3	11	9	12	166				
Leu	4	4	5	5	6	5	6	6	7	6	4	6	15	18	21	5	40			
Phe	2	2	2	3	3	3	3	4	5	3	4	7	7	11	11	2	12	70		
Tyr	1	1	1	1	2	2	2	2	3	2	3	7	3	6	6	1	6	66	137	
Trp	1	1	1	1	1	2	2	2	2	2	3	15	2	4	4	1	3	41	46	414
	Gly	Pro	Asp	Glu	Ala	Asn	Gln	Ser	Thr	Lys	Arg	His	Val	Ile	Met	Cys	Leu	Phe	Tyr	Trp

[a] The numbers are taken from Ref. (20); the unit PAM is defined in section 9.1, page 168. Each term gives a ratio; the probability that an amino acid pair will occur at a given position in proteins of common ancestry divided by the probability that it will occur by chance. All ratios are multiplied by a factor of 10. For example, in two sequences with an evolutionary distance of 256 PAM, Leu and Met would be found 2.1 times as often as would be expected by chance. The pair Asp and Tyr would be found only 0.1 times as often as by chance, that is, at a given chain fold position a change Asp → Tyr or Tyr → Asp is very unfavorable and therefore suppressed. The amino acids are ordered in such a way that $\Sigma_{ij}\, m_{ij} \cdot (i - j)^2$ becomes a minimum. Thus, the amino acids that are frequently substituted for each other at a given chain fold position are close together.

sion that for protein structures the size of a side chain is almost as important as its chemical nature.

Summary

In this chapter we have introduced the standard amino acids, the building blocks of protein. These amino acids are part of the basic organizational scheme of all organisms. Starting from the amino acids some aspects of this universal scheme can be rationalized. However, it will be quite some time before it is possible to understand why this scheme developed. Each amino acid residue has unique properties and fulfills a certain role at a given position in a given protein. Nevertheless, changes do occur to some extent during protein evolution. Analysis of change frequencies shows similarities among amino acids. Four major groups of amino acid residues can be discerned in this way.

Chapter 2
Structural Implications of the Peptide Bond

2.1 Synthesis on Ribosomes

Polypeptides are formed in a series of highly controlled reactions. Amino acids are polymerized into a polypeptide chain on ribosomes in the cell. Polymerization is based on the formation of amide bonds which are usually called "peptide bonds." The chain direction is defined as pointing from the amino end (N-terminus) to the carboxyl end (C-terminus) as shown in Figure 2-2. This definition coincides with the direction of chain synthesis *in vivo,* which in turn corresponds to the $5' \rightarrow 3'$-direction on the messenger RNA.

Free energy is spent for accuracy. For one peptide bond formed, about four acid anhydride bonds of adenosine triphosphate (ATP) and guanosine triphosphate (GTP) (22) with an overall standard free energy yield of about 25 kcal/mol amino acid (23) are split. Only part of this free energy is used to produce the peptide bond in an endergonic reaction with a standard free energy in water of about 5 kcal/mol (24). The rest is invested in translating messenger RNA to polypeptide and in making the translation as accurate as possible. The hydrolytic decay of peptide bonds is inhibited by a high barrier of activation energy, which, however, is routinely overcome by numerous protein-splitting enzymes (proteases).

Low cost bonds allow fast adaptation to environmental changes. From an organizational point of view the peptide bond is very handy because it can be synthesized and split at comparatively low expense. This gives the organism a much better chance to convert its material to the form it needs so that it can profit from more favorable and better protect itself from less favorable changes in its environment. With high cost bonds such as those in the aliphatic tails of fatty acids, the response would be slower and therefore the fitness in a Darwinian competition would be less.

2.2 Peptide Bond Dimensions

Rotation around the peptide bond is inhibited by resonance. The geometry
and the dimensions of the peptide bond are shown in Figure 2-1a. These
data have been derived by Pauling *et al.* (25) from crystal structures of
molecules containing one or few peptide bonds. Their most surprising
result was the small distance between the C'- and the N-atoms. It is 0.15 Å
or 10% shorter than normal. Moreover, the C'–O double bond is 0.02 Å
longer than that known from aldehydes and ketones (26). Pauling *et al.*
interpreted this effect as being caused by resonance between the two
extreme structures shown in Figure 2-1b. Here in structure I the C'–N
bond contains only axial symmetric σ-electrons allowing free rotation,
whereas structure II has σ- and π-electrons in the C'–N bond giving rise to
a large dipole moment and inhibiting rotation.

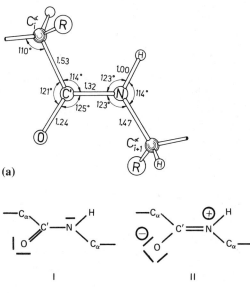

(a)

(b)   Free rotation   Planar

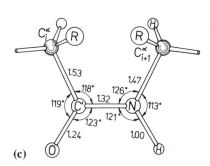

(c)

Figure 2-1. The amide or peptide bond. (a)
Angles and distances for the usual *trans*
peptide bond as given by Pauling *et al.* (25).
C' and N are planar. (b) The two limiting
electronic structures of the bond, I allowing
free rotation and II restricting rotation but
having a large dipole moment. The hybrid
structure contains 60% of I and 40% of II,
as derived from the C'–N bond length. The
resonance energy is about 20 kcal/mol. (c)
Angles and distances of the rare *cis* peptide
bond (29).

As derived from bond length differences, the resulting hybrid structure consists of structures I and II in a ratio 3:2, the π-electrons being smeared out over the C'–O and the C'–N bonds. Thermodynamic data show that the resonance energy is about 20 kcal/mol (23). Since structure II is planar, the hybrid structure, too, is planar, the six atoms, C_i^α, C_i', O_i, N_{i+1}, H_{i+1}, C_{i+1}^α, (see Figure 2-2) lying in one plane.

The backbone of a polypeptide chain is described by two dihedral angles per residue. With a stiff peptide bond and with rather rigid bond lengths and bond angles, the conformation of the polypeptide chain is essentially

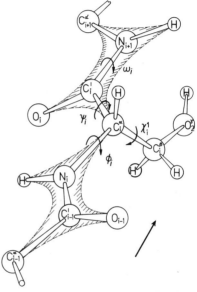

Figure 2-2. Definition of dihedral angles in a polypeptide chain. The chain direction is indicated. A Ser residue is inserted to illustrate the definition of side chain dihedral angles. The nomenclature follows the recommendations of the IUPAC–IUB commission of 1969 (21). For further side chain dihedral angles these recommendations should be consulted. The zero positions of dihedral angles are defined as follows:

$$\omega_i = 0 \text{ for } C_i^\alpha\text{–}C_i' \text{ cis to } N_{i+1}\text{–}C_{i+1}^\alpha$$
$$\psi_i = 0 \text{ for } C_i^\alpha\text{–}N_i \text{ trans to } C_i'\text{–}0_i$$
$$\phi_i = 0 \text{ for } C_i^\alpha\text{–}C_i' \text{ trans to } N_i\text{–}H_i$$
$$\chi_i^1 = 0 \text{ for } C_i^\alpha\text{–}N_i \text{ cis to } C_i^\beta\text{–}0_i^\gamma$$

Thus, the depicted main chain angles are $\phi_i = 180°$ and $\psi_i = 180°$. As indicated by hatching, usually the six atoms C_i^α, C_i', O_i, N_{i+1}, C_{i+1}^α, H_{i+1} lie in one plane, that is, ω equals 180°. H_i and H_{i+1} are the amide hydrogens. The indicated rotation directions of ω, ψ, ϕ are positive if the viewer sits on the atom at the N-terminal side of a bond and the C-terminal side of this bond is rotated according to the arrow (clockwise). A corresponding definition applies for χ with the viewer on the atom closer to the C_α-atom.

described by the dihedral angles ϕ and ψ at the C_α-atoms as defined in Figure 2-2. This definition follows the IUPAC–IUB recommendations of 1969 (21). The torsion angle ω is inserted although this rotation is inhibited. All dihedral angles in side chains are denoted by the letter χ with up to three indices. The Ser side chain illustrated in Figure 2-2 has no branches. Therefore only one index is necessary. A comprehensive description of the notation can be found in the IUPAC–IUB recommendations (21).

It should be mentioned that the main chain dihedral angles ϕ and ψ had been quasiofficially defined in 1966 with a different origin (27):

$$\phi_{1969} = \phi_{1966} - 180°$$
$$\psi_{1969} = \psi_{1966} - 180°$$

Since the time between the two publications was rather fruitful for the elucidation of protein structures, the old notation has been engraved in many memories and can still be found in the literature.

2.3 Steric Hindrance

The dihedral angles of the backbone are constrained. The stiff peptide bond restricts the polypeptide chain flexibility appreciably. In addition, the peptide unit is rather bulky and gives rise to substantial steric hindrance. While working on possible structural models for collagen, Ramachandran *et al.* (28) explored how far steric hindrance restricts the free rotation around the N–C_α and the C_α–C' bonds of the chain, that is, the range of accessible ϕ and ψ angles. For this purpose they screened the known crystal structures for van der Waals contact distances between the relevant atoms. Finally they settled on the values given in Table 2-1, a so-called "lower normal limit." In addition, they defined a "lower extreme limit," which is about 0.10 Å smaller than the lower normal limit (29). It turned out that within an error of 0.05 Å the contact distances are sums of contact

Table 2-1. Empirical Lower Limits for Nonbonded Contact Distances[a]

Type of contact	Normal limit (Å)	Extreme limit (Å)
H . . . H	2.0	1.9
H . . . O	2.4	2.2
H . . . N	2.4	2.2
H . . . C	2.4	2.2
O . . . O	2.7	2.6
O . . . N	2.7	2.6
O . . . C	2.8	2.7
N . . . N	2.7	2.6
N . . . C	2.9	2.8
C . . . C	3.0	2.9

[a]Data are from Ref. (29).

radii, that is, the contribution of a particular atom to a contact distance does not depend on its contact partner. Therefore, the values used for calculations explore space essentially as does a space-filling model using hard spheres with contact radii.

The conformational space, the range of angles ϕ and ψ, that is available for a residue without a side chain (Gly) is shown in Figure 2-3a. This $\phi\psi$-map is based on a rigid peptide bond with the dimensions of Figure 2-1a. Both lower normal and lower extreme limits of contact distances (Table 2-1) are considered. The layout of the $\phi\psi$-map has been retained from the old nomenclature (27). But the new (ϕ,ψ)-values (21) are inserted shifting the origin into the center of the map. Gly has a large and coherent allowed region comprising about 50% of the space. The most inhibitive contact arises between H_{i+1} and O_{i-1} at the center and between O_{i-1} and O_i. Less severe hindrance is found between H_i and H_{i+1}, O_{i-1} and C'_i, and O_{i-1} and N_{i+1}. The steric conflict between N_i and H_{i+1} is less serious.

Except for Gly and Pro all sterically allowed regions are essentially the same. In the presence of a C_β-atom, that is, for all residues except Gly, the allowed conformational space is reduced drastically to the areas shown in Figure 2-3b. In the hard-sphere model with normal and with extremely small contact radii, only 8% and 22% of the space, respectively, remains allowed. The C_β contacts that inhibit three large parts of the Gly area are shown in the map. The two fully allowed regions contain the right-handed α-helical and the parallel and antiparallel β-sheet conformations, respectively. The conditionally allowed region around $(\phi,\psi) = (+60°, +60°)$ contains the left-handed α_L-helix. Besides Gly only Pro is not described by Figure 2-3b. Pro behaves differently because its side chain binds to the peptide nitrogen and fixes ϕ to $-60° \pm 20°$. For Pro, the allowed ϕ-variation reflects different pucker possibilities of the pyrrolidine ring.

2.4 Conformational Energy

The hard-sphere model is a good approximation. The findings of Ramachandran *et al.* (28,29) as described by the $\phi\psi$-map (often called the "Ramachandran map") shown in Figure 2-3b are confirmed by the observed data from crystalline globular proteins. A composition of all (ϕ,ψ)-angles occurring in 13 such proteins is given in Figure 2-4. This empirical distribution shows the highest density near $(-60°, -60°)$ at the right-handed α-helix position, reflecting the high α-helix content of globular proteins. The other maximum is near $(-90°, +120°)$, corresponding to an extended chain and to residues in β-sheets. Since the density is also rather high near $(-90°, 0°)$, the hindrance between N_i and H_{i+1} is not as serious as derived from the hard-sphere model.

The region for the left-handed α_L-helix is rather sparsely occupied as compared with neighboring areas. No left-handed α_L-helix has yet been

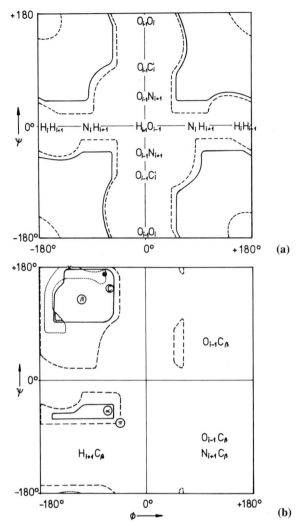

Figure 2-3. Allowed main chain dihedral angles. The peptide bond dimensions of Pauling *et al.* given in Figure 2-1a are assumed. The graph is taken from Ref. (29). These maps are often called "Ramachandran maps." (a) Map for glycine, which has no side chain, using the hard-sphere model with normal (———) contact distances and with the lower limit (−−) contact distances listed in Table 2-1. Prohibitive contacts are indicated in several areas. Atom names are given in Figure 2-2. (b) Map for amino acids with a C_β-atom, using the same hard-sphere models. The conformations of right-handed α-helix, of β-pleated sheet, and of collagen are indicated. The steric conflicts involving C_β-atoms which restrict the large glycine area to the smaller area given here are indicated. The three coherent allowed regions are sometimes denoted

α_R = right-handed α-helix, the region around label α,
α_L = left-handed α-helix, the region around $(\phi,\psi) \cong (+60°, +60°)$,
ϵ = extended chain, the region around label β.

Assuming the same hard-sphere model, the regions accessible for poly-L-*cis*-alanine (......., normal; ----, lower limit) are inserted. The conformation of Figure 2-6 is denoted by x and poly-L-proline I (all-cis) is denoted by x.

Figure 2-4. Plot of main chain dihedral angles of about 2500 residues of 13 proteins as taken from Ref. (30). This plot is sometimes named after Ramachandran.

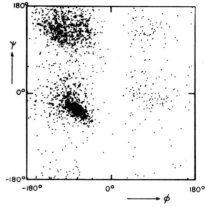

observed. About 10% of the entries fall in the regions forbidden for all residues except Gly, on the right and lower left side of the diagram. On the basis of the Gly frequency given in Table 1-1 and assuming homogeneous distribution of Gly residues over their available conformational space only half this value would be expected. Therefore, about 5% of all residues with C_β-atoms lie in a forbidden area. This fact is confirmed in more detailed plots based on eight proteins giving the (ϕ,ψ)-values separately for every type of residue (31). At this point it should be kept in mind, however, that the experimental data contain errors due to misinterpretation of electron density maps.

A potential energy map shows steric hindrance more accurately. The hard-sphere model is a simple approach to steric hindrance. Obviously, it can be improved by using potential energy functions instead of hard spheres. Such potential maps have been calculated by a number of authors (32–34) for various residues. As an example, a potential energy map for alanine is shown in Figure 2-5. There is no qualitative difference compared with the hard-sphere map. The potential energies of right-handed α-helix and left-handed α_L-helix are higher than those of the extended chain by about 0.5 and 2.5 kcal/mol, respectively. In the case of the right-handed α-helix this energy increase seems to be overcome by the advantage of efficient hydrogen bonding within the α-helix. A left-handed α-helix has no obvious structural advantage compared with a right-handed one. Both helices are rods with the same general shape. Therefore, if a rod is needed in a structure there seems to be no reason for using the energetically disfavored left-handed α_L-helix since the other one does the job just as well.

The potential energy map shows a bridge of negative energy between the extended chain region around $(\phi,\psi) = (-120°, +120°)$ and the α-helix region $= (-60°, -60°)$. No such bridge is allowed in the hard-sphere model, even when the extremely small contact radii are used. Since this so-called "bridge region" is highly populated by the observed data on crystalline proteins given in Figure 2-4, it is indeed easily accessible. Thus, the hard-sphere model errs at this position. The hindrance between N_i and H_{i+1}

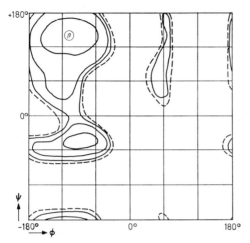

Figure 2-5. Potential energy distribution in the (ϕ,ψ)-plane for a pair of peptide units with an Ala residue in between, taken from Ref. (34). The map is calculated on the basis of the Pauling–Corey dimensions of a peptide bond (Figure 2-1a). Contours are drawn at intervals of 1 kcal/mol going negative from zero. The zero contour is dashed. The position of a twisted β-sheet is indicated. Note that this distribution is modified if hydrogen bonds to peptide units farther along the chain (for example in an α-helix) are taken into account.

(Figure 2-3a) which occurs at these angles is largely relaxed by a dipole–dipole interaction corresponding to a weak hydrogen bond. Similar interactions have been found in crystals of model compounds (35).

The peptide bond has some plasticity. The conformational maps in Figures 2-3 and 2-5 are based on the assumption of a rigid peptide bond with Pauling–Corey dimensions (Figure 2-1a). In a further refinement of the model, the potentials for changes in bond angles, bond lengths, and torsion around the peptide bond should be included. This of course makes the conformational space at one residue multidimensional and makes any exact application or comprehensive description difficult. As an estimate, the deviations corresponding to a potential energy increase of 1 kcal/mol are given:

$$\text{deviation in a bond angle} \quad \simeq 5°$$
$$\text{deviation in a bond length} \quad \simeq 0.05\text{Å}$$
$$\text{deviation in torsion angle } \omega \simeq 12°$$

Moreover, atoms moving out of the plane should also be considered (36). In principle, the plasticity can be included in the two-dimensional energy map by allowing for all plastic deformations and by always calculating the lowest energy for a given (ϕ,ψ)-angle pair. Note that the backbone conformation is not described by ϕ and ψ alone if plasticity is taken into account.

In conclusion, we have to realize that the peptide bond has appreciable plasticity. Therefore, a conformational energy map relying on fixed bond dimensions such as the one shown in Figure 2-5 tends to be inaccurate in those parts of the map where only slight steric hindrance occurs, as for example in the regions allowed only for Gly, because slight hindrance can be circumvented by small deviations in bond and torsion angles as well as in bond lengths. This might well explain the 5% of cases in Figure 2-4 in which (ϕ,ψ)-values fall in areas forbidden by steric conflict with C_β according to the hard sphere model.

2.5 *cis* versus *trans* Configuration

One of the alternatives should be universal. Up to now we have assumed that all peptide bonds are in the *trans* and not in the *cis*-configuration (Figure 2-1c). It is clear that the peptide synthesis on ribosomes is stereo-specific and therefore should yield either *cis* or *trans* bonds. In any case, one of the alternatives should be universal, otherwise the synthesizing machinery and the genetic code would become more complex, which is unfavorable from an organizational point of view. Even a post–translational isomerization would be expensive, since many specific enzymes would be necessary. A spontaneous isomerization during the folding process of the polypeptide chain is also improbable, because the energy barrier between *trans* and *cis* is high. It corresponds to the resonance energy of 20 kcal/mol between structures I and II of Figure 2-1b. This barrier is lower only for Pro. Here, concomitant changes in the pyrrolidine ring puckering decrease the barrier to 13 kcal/mol. During the folding process Pro residues undergo spontaneous isomerization (section 8.2, page 155). The energy difference between the *cis* and *trans* isomers is about 2 kcal/mol in favor of *trans*. Again, this difference is smaller for Pro because here a transition from *trans* to *cis* replaces the contact between C_i^α and C_{i+1}^α only by a similar contact between C_i^α and C_{i+1}^δ (Figure 1-1).

A couple of *cis* peptides are found in globular proteins. *cis* bonds are found in a number of small cyclic peptides, especially at the N-terminal side of (before) Pro residues (37,38). Only a few *cis* bonds have been reported from globular protein structures: for example, before Pro-93 and Pro-114 in ribonuclease-S (39), before Pro-168 in subtilisin (40), before Pro-116 in staphylococcus nuclease (E. E. Hazen, personal communication), before Pro-8 and Pro-95 of the variable chain of a Bence Jones protein (42), and between Ser-197 and Tyr-198 in carboxypeptidase-A (43). Synthetic poly-L-Pro I molecules contain only *cis* bonds. Thus, in agreement with the energy calculations, Pro seems to be *the* standard residue type which is most susceptible to *cis* bond formation. But in general, *cis* bonds are a rare exception.

For an all-*cis* protein the conformational space is too restricted. Given the necessity of one universal isomer, why was *trans* selected and not *cis*? A stringent argument favoring *trans* is the severe restriction of the available conformational space in an all-*cis* polypeptide chain. At an isolated C_α-atom the steric conflict does not differ so much between *cis* and *trans*. However, for an all-*cis* polypeptide chain, the mutual hindrance between neighboring side chains is severe. This is illustrated in Figure 2-6 for poly-L-

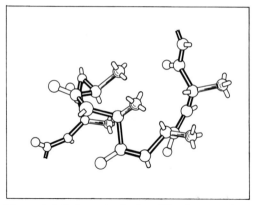

Figure 2-6 Poly-L-*cis*-alanine. The (ϕ,ψ)-angles are indicated by x in Figure 2-3b. The conformation is allowed if the hard-sphere model with the lower limit contact radii of Table 2-1 is assumed.

cis-Ala. Here, the chain looks scrambled because the *cis* bond brings consecutive C_α-atoms geometrically very close to each other.

Ramachandran and Sasisekharan (29) looked for the allowed conformations of such a chain. Assuming the hard-sphere model with normal contact radii, only a tiny area at $(-155°, +105°)$ is accessible. Even for the hard-sphere model with lower limit radii, the allowed area is very small, as indicated in Figure 2-3b. In contrast to all-*cis* polypeptide chains, the all-*trans* conformations available are diversified enough to enable the formation of a variety of globular proteins. Still, all-*trans* chains are rigid enough so that spontaneous folding does not require a very strong attractive force between different parts of the chain (see section 3.5, page 37). Note that the energy difference of 2 kcal/mol favoring a *trans* over a *cis* bond is not essential for this reasoning, because an all-*cis* chain would be more rigid and therefore would not need as much binding energy as an all-*trans* chain for assuming a unique conformation.

Summary

Peptide bonds are synthesized in a highly controlled manner on ribosomes. With the rather low cost in free energy of bond formation they can be turned over rapidly. This helps organisms to adjust quickly to environmental fluctuations. The peptide bond is rigid, which contributes significantly to protein stability because it stiffens the polypeptide chain. The rigidity of the polypeptide chain is further increased by steric hindrance within the main chain and between main and side chain atoms. The geometric constraints are best described by potential energy maps. Such maps may also include the observed main chain plasticity. When comparing *cis* and *trans* peptide bonds the *trans* bond is clearly geometrically superior, because it is geometrically not as restrictive as the *cis* bond.

Chapter 3
Noncovalent Forces Determining Protein Structure

Noncovalent forces were discovered by van der Waals (1873) in an attempt to explain the deviation of a real gas from the ideal gas law. They are of utmost importance for biological organisms. In particular, they drive the spontaneous folding of polypeptide and nucleic acid chains and the spontaneous formation of membranes. They mediate the mutual recognition of complementary molecular surfaces ("lock and key system" (44)).

A survey of noncovalent forces is given in Table 3-1. Noncovalent energies are one to three orders of magnitude smaller than covalent binding energies. They are difficult to measure and even more difficult to calculate. Moreover, a major contribution to protein stability originates from the surrounding liquid medium and is of an entropic nature. Therefore, the whole system, protein and solvent, has to be taken into account.

3.1 Dispersion Forces and Electron Shell Repulsion

Dispersion forces between any pair of atoms are attractive. Dispersion forces occur between any pair of atoms, even between totally apolar ones. Each atom behaves like an oscillating dipole generated by electrons moving in relation to the nucleus. In a pair of atoms each dipole polarizes the opposing atom. As a consequence, the oscillators are coupled, giving rise to an attractive force between the atoms. The attraction energy, to the first order, is proportional to the sixth power of the inverse distance between the atomic nuclei (45) and to the polarizabilities of the atoms (46).

When bound in a molecule the polarizability of an atom is anisotropic.

Table 3-1. Types of Noncovalent Forces Important for Protein Structures[a]

Type	Example		Binding energy (kcal/mol)	Change of free energy water→ethanol (kcal/mol)
Dispersion forces	Aliphatic hydrogen	$-\overset{\vert}{\underset{\vert}{C}}-H \cdots H-\overset{\vert}{\underset{\vert}{C}}-$	−0.03	
Electrostatic interaction	Salt bridge	$-COO^{\ominus} \cdots H_3\overset{\oplus}{N}-$	−5	−1
	2 dipoles	$\overset{\delta+ \quad \delta-}{C=O} \cdots \overset{\delta- \quad \delta+}{O=C}$	+0.3	
Hydrogen bond	Ice	$O-H \cdots O$	−4	
	Protein backbone	$N-H \cdots O=$	−3	
Hydrophobic forces	Side chain of Phe			−2.4

[a]A dielectric constant of 4 is assumed in calculating the electrostatic energies.

Therefore, the dispersion forces vary with the relative orientation of molecular groups. However, since the effects of orientation are small and difficult to measure, they are usually neglected. The dispersion forces are assumed to be isotropic.

Nonbonded electron shells repel each other. The attractive dispersion forces between a pair of nonbonded atoms are counterbalanced by the repulsion of the electronic shells. The repulsion was approximated by Lennard–Jones using a term proportional to the mth power of the inverse distance (47). Slater described the repulsion by an exponential function (48), which was modified by Buckingham (49) and now carries the latter name. For computations with proteins, the Lennard–Jones approximation with $m = 12$ is most popular. Together with London's term for dispersion forces (45), this results in the computationally simple 6-12 potential, an example of which is depicted in Figure 3-1. The two parameters needed to define this function are described in the caption to this figure. The potential of Figure 3-1 has a negative relative minimum at the distance R_m. Therefore, atoms are weakly "bound" (E_m = attractive energy) at this distance.

Crystal structures yield energy parameters. The magnitude of the Lennard–Jones 6-12 parameters can be determined from contact distances and contact energies in crystals of small molecules as derived from X-ray structure analyses and, for example, heats of sublimation. Alternatively, they can be derived from atomic and molecular beam scattering experiments. Examples of these parameters are given in Table 3-2. The differences between these results indicate the magnitude of the errors involved.

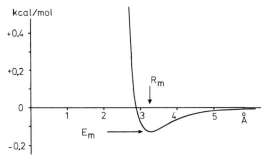

Figure 3-1. Lennard–Jones 6-12 potential for dispersion forces and electron repulsion ($R_m = 3.24$ Å, $E_m = -0.13$ kcal/mol). The two-parameter formula is given below in its computationally practical form (parameters A, B) and in its normalized form (parameters E_m, R_m). Because of repulsion between electronic shells and attraction by dispersion forces, A and B are positive, and there exists a relative minimum. With smaller B and larger A, E_m becomes smaller and the corresponding atomic distance R_m larger. The potential wall is asymmetric. Repulsion balances attraction when the atomic distance is reduced to $0.89 \cdot R_m$. On the other side the attraction energy is still one-sixth of E_m at a distance of $1.5 \cdot R_m$.

$$E = \frac{A}{R^{12}} - \frac{B}{R^6} = E_m \cdot \left[-\left(\frac{R_m}{R}\right)^{12} + 2\left(\frac{R_m}{R}\right)^6 \right]$$

$$E_m = -\frac{B^2}{4A} \qquad R_m = \sqrt[6]{\frac{2A}{B}}.$$

As shown in Table 3-2, the energy contribution per atomic contact is very small. However, the number of contacts in a protein is large. For example, in a hexagonal closest packing of identical spheres, each sphere participates in 12 contacts with its neighbors. Since contact energies add up linearly, the total attraction energy per sphere in this case is six times as large as the energy per contact.

Table 3-2. Parameters of Lennard–Jones 6–12 Potential for Electron Shell Repulsion and Dispersion Forces at a Nonbonded Contact[a]

	Momany *et al.* (69)		Lifson and Warshel (70, 71)	
	E_m (kcal/mol)	R_m (Å)	E_m (kcal/mol)	R_m (Å)
Aliphatic H . . . aliphatic H	−0.04	2.92	−0.01	2.94
Aliphatic C . . . aliphatic C	−0.04	4.12	−0.19	4.23
Carbonyl O . . . carbonyl O	−0.20	3.12	−0.23	3.00
Amide N . . . amide N	−0.11	3.51	−0.19	3.60

[a]Only contacts between identical types of atoms are considered. For a contact between nonidentical types i and j use $(E_m^{(ij)} = E_m^{(ii)} + E_m^{(jj)})^{1/2}$ and $R_m^{(ij)} = \frac{1}{2}(R_m^{(ii)} + R_m^{(jj)})$

3.2 Electrostatic Interactions

Most covalently bonded atoms carry partial charges. Since covalent bonds between different types of atoms lead to an asymmetric bond electron distribution, most atoms of a molecule carry partial charges. The partial charges of some amino acid residues are listed in Table 3-3. Since a neutral molecule has no net charge, it contains only dipoles or higher multipoles. These multipoles interact with each other according to Coulomb's law as described in Table 3-4. The interaction energy depends on the dielectric constant ϵ of the surrounding medium. We use a value $\epsilon = 4$ here, which is the macroscopic dielectric constant of amide polymers (50). The dielectric constant to be used for microscopic dimensions is difficult to calculate. The discussed values (51) range from $\epsilon = 1$ to 5.

Examples for electrostatic interactions are given in Table 3-1: An internal ($\epsilon = 4$) salt bridge between Lys and Asp has a binding energy of about 5 kcal/mol. Two carbonyl groups at contact distance R_m have an electrostatic repulsion energy of 0.3 kcal/mol.

In proteins electrostatic interactions are of a local nature. Although the calculation of electrostatic interactions is a straightforward application of Coulomb's law to all partial charges, it is very time consuming because of the large number of atoms in a protein and the long range of electrostatic interactions. But the calculation can be simplified. Since no monopoles, that is, free charges, are observed inside proteins (salt bridges are ion pairs and therefore dipoles), all partial charges form di- or multipoles. The

Table 3-3. Partial Charges of Atoms in the Polypeptide Backbone and in Three Side Chains[a]

Peptide	N	-0.36	Tyr	O_η	-0.33
Peptide	H_N	+0.18	Tyr	H_η	+0.17
Peptide	C_α	+0.06			
Peptide	C'	+0.45	Asn	C_β	-0.12
Peptide	O	-0.38	Asn	H_β	+0.06
Peptide	H_α	+0.02	Asn	C_γ	+0.46
			Asn	O_δ	-0.38
Ser	C_β	+0.13	Asn	N_δ	-0.45
Ser	H_β	+0.02	Asn	H_δ	+0.20
Ser	O_γ	-0.31			
Ser	H_γ	+0.17	Cys	S_γ	+0.01
			Cys	H_γ	+0.01

[a]Taken from the results of Momany *et al.* (52) who derived the partial charges of the 20 common amino acid residues using the CNDO/2 (complete neglect of differential overlap) method. Partial charges were also derived from *ab initio* molecular orbital calculations of small molecules (53) or by fitting observed crystal data (54). A comparison between the results shows that accuracy is low.

interaction between these decreases with at least the third power of the inverse distance (Table 3-4). Therefore, the actual range of electrostatic interactions is rather short, and the energy calculation can be restricted to the interaction between close neighbors.

It could be argued that the weak long distance interactions can add up if dipoles are aligned as found with hydrogen bond dipoles in α-helices and β-sheets. However, in β-sheets adjacent dipoles are antiparallel so that their fields cancel each other at long distance. In α-helices the dipoles form lines, cancelling each other except for charges at both ends. For this reason antiparallel helices are more favorable than parallel ones. But this electrostatic contribution is small when compared with the total binding energy between such α-helices.

Table 3-4. Electrostatic Interactions

Coulomb's law:

$$E = \frac{332}{\epsilon} \cdot \frac{q_1 \cdot q_2}{R_{12}}$$

q_i = partial charges in electron charges e_0
R_{12} = distance between partial charges in Å
ϵ = dielectric constant
E = repulsion or attraction energy in kcal/mol

Energy equivalents:

1 electron volt = 1 eV = 23 kcal/mol = 96,500 Ws/mol = $9.65 \cdot 10^{11}$ erg/mol
Unit of dipole moment: 1 debye = 1 D = 0.208 e_0Å

Interaction energy of dipoles with moments μ_1 and μ_2. Note that this formula assumes point dipoles. Hence the distance R_{12} should be large as compared to the length of the dipoles.

$$E = \frac{332}{\epsilon} \left[\frac{(\vec{\mu_1} \cdot \vec{\mu_2})}{R_{12}^3} - \frac{3(\vec{\mu_1} \cdot \vec{R_{12}}) \cdot (\vec{\mu_2} \cdot \vec{R_{12}})}{R_{12}^5} \right]$$

The interaction energy between randomly oriented dipoles is zero.

Interaction energy between two dipoles with moments of 1 D at a distance of 5 Å in a medium with $\epsilon = 4$.

$$\rightarrow \rightarrow E = -1.32 \text{ kcal/mol}$$
$$\rightarrow \leftarrow E = +1.32 \text{ kcal/mol}$$
$$\uparrow \ \uparrow \ E = +0.66 \text{ kcal/mol}$$
$$\downarrow \ \uparrow \ E = -0.66 \text{ kcal/mol}$$

3.3 Van der Waals Potentials

Van der Waals potentials comprise electron shell repulsion, dispersion forces, and electrostatic interactions. It is computationally convenient to combine all three nonbonded forces into a single, simple potential function (or force field), which is then called the "van der Waals potential" for historical reasons. To accomplish this, considerations of the electrostatic interactions have to be further simplified. First of all, only atoms in contact are taken as the close neighbors that must be considered. Moreover, electrostatic interactions are averaged over all relative orientations that are sterically allowed for two contacting groups, $>C=O$ and $H-C\leqq$, etc. Thus the resulting contribution depends only on the distance between the contacting atoms. With this simplification, the van der Waals potential is isotropic

$$E = \frac{A}{R^{12}} - \frac{B}{R^6} + \frac{q_i q_j}{R}$$

It contains three parameters: A and B as defined in Figure 3-1 and the product of the effective charges $q_i q_j$ of the contact partners.

The effective charges can be derived from partial charges of the constituent atoms (Table 3-3). Taking the electrostatic term into account, the parameters A and B describing electron shell repulsion and dispersion forces can be determined from crystal data (52). But partial charges can be calculated only with low accuracy (53). Therefore, Lifson and co-workers attempted to derive them from crystal data. This was done by fitting all three parameters of the 1-6-12 potential simultaneously to the observed crystal data (54).

A special 6-12 van der Waals potential. For computational convenience Levitt (55) discarded the R^{-1} term when refining protein structures by an energy minimization procedure. This gave rise to 6-12 van der Waals potentials that are quite distinct from the 6-12 potentials describing only electron shell repulsion and dispersion forces (Table 3-2). Examples of such potentials are given in Table 3-5. Note that in the case of contacting amide nitrogens the forces are completely repulsive. Here, the electrostatic repulsion between the negative partial charges on the nitrogens exceeds the attraction by dispersion forces given in Table 3-2.

Van der Waals radii approximately add up to contact distances. Van der Waals potentials allow the definition of "van der Waals contact distances" between given atoms. Lower limits for such distances as derived from crystals of small molecules have been given by Ramachandran and Sasisekharan (29). They amount to about 75% of the equilibrium distances R_m in Table 3-2 and correspond to a repulsion energy of about 1 kcal/mol. These contact distances have been used to evaluate the steric hindrance of the hard-sphere model at a C_α-atom in a polypeptide chain (28) (Figure 2-3).

The observed van der Waals contact distances for pairs of atoms can be converted to the more general "van der Waals radii" on the assumption

Table 3-5. Parameters of a 6-12 van der Waals Potential as Derived from Observed Protein Structures (55)[a]

	A (kcal/mol·Å^{12})	B (kcal/mol·Å^6)	E_m (kcal/mol)	R_m (Å)
Aliphatic C . . . aliphatic C	2,750,000	+1425	−0.19	3.53
Carbonyl O . . . carbonyl O	417,000	+108	−0.01	3.96
Amide N . . . amide N	417,000	0	Repulsive	
Carbonyl O . . . carbon in benzene ring	695,000	−570	Repulsive	

[a] A and B are defined in Figure 3-1.

that each distance is the sum of two atom type specific radii. This concept is only approximately valid; the resulting radii are only an average over all kinds of contact partners. Examples of van der Waals radii are given in Table 3-6. The "lower normal limits" of contact distances given in Table 2-1 are about 10% smaller than the sum of the corresponding van der Waals radii.

3.4 Hydrogen Bonds

Hydrogen bonds are predominantly electrostatic interactions. We have just seen that the atomic distances of all noncovalent atomic contacts are approximately given by the sum of the corresponding van der Waals radii. This rule is grossly violated by a number of contacts involving hydrogen atoms. The distance between an amide H and a carbonyl O, for example, is only 1.9 Å and not 2.7 Å as calculated from the van der Waals radii of Table 3-6. A closer inspection shows that this effect is always observed if the H-atom has a large positive partial charge and if its contact partner has a large negative partial charge (see Table 3-3). These charges attract each other. Since the entire electron shell of hydrogen (hydrogen has only one electron!) is appreciably shifted onto the atom to which hydrogen is covalently bound, the shell repulsion between contact partners is small, and the attracting charges can approach each other more closely. Such short distance gives rise to a high attractive Coulomb energy (Table 3-4) and also to a high dispersion energy (56). The resulting interaction energy is interme-

Table 3-6. Van der Waals Radii as Given by Bondi (73)

Type of atom	Radius (Å)
Aromatic H	1.0
Aliphatic H	1.2
O	1.5
N	1.6
C	1.7
S	1.8

diate between the energies of van der Waals contacts and covalent bonds. Accordingly, such contact has been given a name of its own, "hydrogen bond." The bonding partners are called "hydrogen bond donor atom," that is the atom to which hydrogen is covalently bound, and "hydrogen bond acceptor atom."

Biologically important hydrogen bonds do not involve much wave function overlap. There has been some controversy as to whether wave function overlap contributes significantly to the hydrogen bond (56). In the extreme case of the $[F \cdot \cdot \cdot H-F]^-$ ion this is certainly the case. Here, the F–F distance is only about 65% of the sum of H and F van der Waals radii and the H–F bond length; and with about 50 kcal/mol the binding energy approaches that of covalent bonds (57). However, in proteins distance reductions and binding energies are much smaller. Here, the contribution of wave function overlap is probably insignificant.

Bond lengths reveal bond energies. As shown in Table 3-7, the shortest length and the highest energy is found in hydrogen bonds between oxygens, in particular between phenols (Tyr–Tyr) and in water. Somewhat longer distances and smaller energies occur in hydrogen bonds between nitrogen and oxygen, and much smaller energies are to be found in hydrogen bonds between two nitrogens and in particular between nitrogen and sulfur. As compared to the sum of van der Waals radii, the reduction in distance ranges from 10% to 25%. A rough estimate for hydrogen bond energies is most easily obtained from the heat of sublimation of ice which is 13 kcal/mol. Most of this heat corresponds to hydrogen bonds, since the heats of sublimation of H_2S and H_2Se (which form much weaker hydrogen bonds) are as low as 6 kcal/mol. As shown in Figure 3-2, there are two hydrogen bonds per H_2O molecule, the energy of one hydrogen bond being somewhat more than half the difference between the heats of sublimation, namely, about 4 kcal/mol. Energies of hydrogen bonds are about 3 kcal/mol between amide and carbonyl groups, which occur frequently between main chain atoms in polypeptides, (Table 3-1).

Hydrogen bonds are linear. As mentioned above a hydrogen bond contains a large positive (H-atom) and a large negative (H-acceptor) partial charge at close distance. In general, these charges are parts of dipoles (see Table 3-3). The covalently bound neighbors (H-donor and covalent neighbor(s) of H-acceptor) carry opposite charges as for example in

$$\overset{\delta-}{\underset{}{O}}-\overset{\delta+}{H}\cdot\cdot\cdot\overset{\delta-}{O}=\overset{\delta+}{C}<$$

Thus, the positively charged H-atom is located between two negatively charged atoms.

A positive charge assumes its lowest potential energy between two negative charges when all three charges are aligned. As a consequence hydrogen bonds are linear, for example, N–H $\cdot$ $\cdot$ $\cdot$ O form a straight line. Electrostatic calculations show that a deviation of $20°$ in the angle

$$\overset{\cdot H}{\underset{N\colon \cdots \cdots O}{}}$$

Table 3-7. Hydrogen Bonds Found in Proteins (58)

Type of hydrogen bond		Distance between donor and acceptor atom (Å)	Reduction from the distance between donor and acceptor as calculated from van der Waals radii (%)	Comment
Hydroxyl–hydroxyl	$-O-H \cdots O_H$	2.8 ± 0.1	25	Bifurcated bonds are possible, for example, in ice. For phenols (Tyr); the length is 2.7 ± 0.1 Å
Hydroxyl–carbonyl	$-O-H \cdots O=C$	2.8 ± 0.1	25	
Amide–carbonyl	$>N-H \cdots O=C$	2.9 ± 0.1	20	Ubiquitous between main chain atoms in proteins; empirical length found in parvalbumin (59) is 2.9 ± 0.3 Å
Amide–hydroxyl	$>N-H \cdots O^H$	2.9 ± 0.1	20	
Amide–imidazole nitrogen	$>N-H \cdots N$	3.1 ± 0.2	15	
Amide–sulfur	$>N-H \cdots S$	3.7	10	Found with Met-180 of chymotrypsin (18) and at the FeS clusters of ferredoxin, rubredoxin, HiPiP (72) with lengths of 3.6 ± 0.3 Å

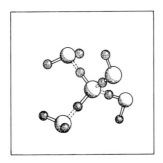

Figure 3-2. Structure of ice as revealed by X-ray and neutron diffraction analysis.

from zero reduces the binding energy by about 10% (58). Empirically, an average deviation of 20° in this angle has been found in parvalbumin (59). Obviously, a compromise has been made between maximum hydrogen bond energy and other geometrical requirements in this (as well as in other) protein structure(s).

Dipoles constituting a hydrogen bond tend to be in line. The linearity of a hydrogen bond should not be confused with the alignment of the constituting dipoles, as, for example, the alignment of a peptide amide with a carbonyl group

$$\overset{\delta-}{>N} - \overset{\delta+}{H} \cdots \overset{\delta-}{O} = \overset{\delta+}{C} <$$

As shown in Table 3-4, dipoles have the lowest energy (strongest binding) when they are aligned.[1] However, misalignment results in a much smaller energy reduction than a hydrogen bond nonlinearity of similar magnitude (58). Thus, dipole misalignment is rather common in proteins, where hydrogen bonds have to comply with various geometrical constraints. In ice, dipoles are also nonaligned (Figure 3-2). In some cases, as between an amide and a hydroxyl group, alignment is sterically impossible (Table 3-7).

3.5 Entropic Forces

Free Energy Concept

In vacuo, binding energy competes with chain entropy. Up to now we have considered only noncovalent *binding* energies, that is, van der Waals interactions and hydrogen bonds. However, protein formation and stability depend on the *free* energy and not only on the binding energy.[2] Moreover,

---

[1] At the short distances found in hydrogen bonds the dipole–dipole interactions should not be derived from two *point* dipoles as is done in Table 3-4. However, the formula of Table 3-4 applies with respect to dipole alignment.

[2] As is common in biological literature we will use the term "free energy" for "Gibbs free energy." The term "binding energy" is used for "(binding) enthalpy." In most of the biological reactions the changes of volume and volume energies are neglected, so that Gibbs free energy equals Helmholtz free energy and (binding) energy equals (binding) enthalpy.

native proteins require an aqueous environment. Therefore a rigorous treatment has to account for the free energy of the complete system, chain *and* solvent.

In order to demonstrate the relation between free energy and binding energy let us first consider a polypeptide chain in vacuo. Here, the free energy difference between the accurately folded (N = native) and the randomly extended (R = random) state is $\Delta G_{\text{chain}} = G_N - G_R$, whereas the binding energy difference is $\Delta H_{\text{chain}} = H_N - H_R$. The equilibrium constant K = (number of chains in N)/(number of chains in R) is given by the equation

$$-RT \ln K = \Delta G_{\text{chain}} = \Delta H_{\text{chain}} - T \cdot \Delta S_{\text{chain}}. \tag{3-1}$$

The right-hand part of Eq. (3-1) is displayed in Figure 3-3a. The change in entropy (disorder) $\Delta S_{\text{chain}} = S_N - S_R$ plays a decisive role. The R-state of polypeptide chains is less ordered (higher entropy) than the N-state; therefore ΔS_{chain} is negative, and the entropy term $-T \cdot \Delta S_{\text{chain}}$ is positive and favors the R-state. ΔH_{chain} is negative because hydrogen bonds and van der Waals attraction provide some binding energy in the N-state, H_N being lower (more negative) than H_R. Consequently, at a given temperature T, the sign of ΔG_{chain} (and the value of K) depends on the relative magnitudes of ΔH_{chain} and ΔS_{chain}: "only if enough binding energy is offered, the chain bothers to leave the R-state and to assume its N-state." Moreover, "the higher the temperature the more binding energy has to be offered." A more

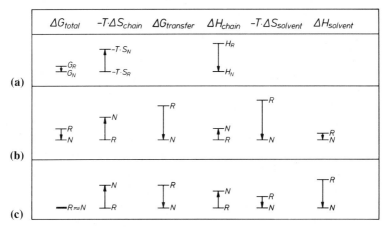

Figure 3-3. Representation of the energetic contributions to protein stability per atomic group. All magnitudes are only approximate. Arrows indicate the sign of the contributions: down favors N-state, up disfavors N-state. When separating solvent and chain contributions, one also gets a solvent–chain interaction energy term. This term is included in ΔH_{chain}. ΔG_{total} and $\Delta G_{\text{transfer}}$ are defined in Eqs. (3-2) and (3-3), respectively. (a) A polypeptide chain in a vacuum [Eq. (3-1)]. ΔG_{total} favors the N-state. (b) Nonpolar groups of the chain in an aqueous solvent [Eq. (3-2)]. For the oil–water example the index "chain" has to be replaced by "oil." (c) Polar groups of the chain in an aqueous solvent [Eq. (3-2)].

rigorous, statistical mechanics treatment of the vacuum situation is given in the Appendix and in section 8.1.

The solvent is very important. Although the vacuum situation nicely illustrates the interplay between binding energy and entropy, it is not of much biological relevance. Proteins do require a solvent. Moreover, the physical properties of the solvent are of utmost importance for protein stability. For example, almost all proteins denature in ethanol or in aqueous solutions containing sufficient amounts of sodium dodecyl sulfate (SDS) or urea. Therefore, considering the composite system of polypeptide chain *and* solvent, the total free energy difference becomes

$$-RT \ln K = \Delta G_{\text{total}} = \Delta H_{\text{chain}} + \Delta H_{\text{solvent}} - T \cdot \Delta S_{\text{chain}} - T \cdot \Delta S_{\text{solvent}}.$$
$$(3\text{-}2)$$

The interaction energy between chain and solvent is here included in the ΔH_{chain}.

A Mixture of Oil and Water

Locally, liquid water can assume an ordered, quasisolid structure. Let us discuss Eq. (3-2) for a solution of mineral oil consisting of hydrocarbons in water instead of a polypeptide chain in water. The "R-state" is assigned to the monodisperse solution of oil in water, and the "N-state" is assigned to separated phases, to an oil drop on the water surface. ΔS_{oil} is negative because a monodisperse solution is less ordered than separated phases (see Figure 3-3b). ΔH_{oil} is positive. To understand this fact one has to consider that in the N-state most oil molecules are surrounded by their own kind, whereas in the R-state all of them are surrounded by water. The van der Waals interactions between oil molecules are only dispersion forces and therefore weak. Between oil and water molecules these interactions are stronger because the polar water induces dipoles in hydrocarbons and therefore gives rise to an appreciable electrostatic term (15). Therefore, ΔH_{oil} (which is the difference between both contributions) favors the monodisperse solution. However, it is comparatively small.

It was observed experimentally (15) that $\Delta S_{\text{solvent}}$ is positive which favors the N-state. This shows that the order of the water component is higher in the monodisperse oil solution, where the total interface area between nonpolar molecules and water is larger. At such an interface water has two possibilities: either it keeps its liquid structure and forfeits hydrogen bonds, or it retains most of the hydrogen bonds but becomes locally ordered by enclosing the nonpolar molecule with a cage-like structure (15), a so-called clathrate structure or "iceberg." The former option corresponds to an increase of H_{solvent} (decrease of binding energy) whereas the latter one corresponds to a decrease of S_{solvent} (increase of order). Since the observed $\Delta S_{\text{solvent}}$ is positive, water does indeed assume a locally ordered, quasisolid structure at the interface. But even with this clathrate structure water forfeits some hydrogen bonds, which gives rise to a negative $\Delta H_{solvent}$ which favors the N-state. However, the magnitude of $\Delta H_{solvent}$ is comparatively small.

Water entropy forces oil to form droplets. Thus, we find that ΔH_{oil} and ΔS_{oil} favor the R-state whereas $\Delta H_{solvent}$ and $\Delta S_{solvent}$ favor the N-state. Since phase separation between oil and water is indeed observed, the latter contributions surpass the former ones. Furthermore, $\Delta H_{solvent}$ is small. Consequently, it is predominantly $\Delta S_{solvent}$ that drives the oil molecules into a separate phase, or into the oil drop. The effect of $\Delta S_{solvent}$ is often called "entropic forces" or "hydrophobic forces." These entropic forces also give rise to the spontaneous formation of micelles, vesicles, and membranes from amphiphilic lipids.

A Polypeptide Chain in Water

A natural polypeptide chain contains a specific pattern of polar and nonpolar groups. The situation of a polypeptide chain in water with reference to the oil–water example (see Figure 3-3) will now be discussed. The interface between a polypeptide chain and water is large for the randomly extended chain and small for the chain in its folded native conformation. In contrast to the completely nonpolar oil molecules, however, the polypeptide chain contains both polar and nonpolar groups. For example, the amide and carbonyl groups of the backbone as well as the hydroxyl (in Ser, Thr, Tyr), the carboxylate (in Asp, Glu), and the ammonium (in Lys) groups are polar and tend to remain in water (hydrophilic). The hydrocarbon side chains of Ala, Val, Leu, Ile, and Phe are nonpolar.

At first glance proteins look like oil drops. From the oil–water system it can be deduced that all nonpolar groups tend to aggregate in order to reduce the interface area between nonpolar chain surface and water. Indeed, in native proteins the majority of nonpolar side chains are removed from the water and assembled in hydrophobic cores. Accordingly, Kauzmann (15) compared proteins with oil drops.

The protein interior also contains many polar groups. Since globular proteins have diameters of about 30 Å, side chains cannot be buried in the protein interior without also burying part of the backbone, the polar amide and carbonyl groups. But polar groups coexist well with water. Burying them in the interior without loss of free energy is achieved only if they form hydrogen bonds. Indeed, such bonds are observed in as many as 90% of all internal polar groups.

Charged groups are on the surface. The removal of charged groups (in Asp, Glu, Lys, Arg) from water is energetically very unfavorable. It is found only with concomitant salt bridge formation (see page 41). But salt bridges are rare. Almost all charged side chains are at the protein surface. They increase protein solubility. Moreover, by staying in solution during and after spontaneous folding of a polypeptide chain, they constrain and therefore help direct the folding process.

Solvent entropy gives a large contribution to protein stability. The four terms of Eq. (3-2) will now be discussed. As described above ΔS_{chain} is negative for all parts of the polypeptide chain. For nonpolar parts, the magnitudes of both ΔH_{chain} and $\Delta H_{\text{solvent}}$ are small, with ΔH_{chain} positive and $\Delta H_{\text{solvent}}$ negative (Figure 3-3b). $\Delta S_{\text{solvent}}$ is positive and large for nonpolar parts.

Polar groups form hydrogen bonds with water when exposed to it and hydrogen bonds with each other in the protein interior. With our accounting system (all interactions between chain and solvent are included in H_{chain}), ΔH_{chain} is positive: one hydrogen bond with water is broken, and only half a hydrogen bond is formed in the protein interior. $\Delta H_{\text{solvent}}$ is negative: half a hydrogen bond is formed per polar group. In Figure 3-3c, the magnitude of ΔH_{chain} is given as smaller than the magnitude of $\Delta H_{\text{solvent}}$ since there are more van der Waals contacts in the tightly packed protein interior (section 3.6) than at the chain–solvent interface. This contribution favors the N-state. Since water molecules are somewhat ordered around polar groups, $\Delta S_{\text{solvent}}$ is positive but small. In all, ΔG_{total} for polar groups is around zero. Comparing all terms (Figure 3-3b, c), it can be concluded that $\Delta S_{\text{solvent}}$ from a nonpolar group gives the largest single contribution to protein stability. The large $\Delta H_{\text{solvent}}$ contribution for polar groups in Figure 3-3c is essentially cancelled by ΔH_{chain}. The magnitude of $\Delta H_{\text{solvent}}$ is only an artefact of our accounting system which includes solvent-chain interactions into H_{chain}.

Transfer free energies of side chains yield an estimate for the contributions to protein stability. It is very difficult to quantitate the contributions discussed above. In Ref. (15) one finds ΔH and ΔS values for the transfer of various small nonpolar molecules from a monodisperse solution in water into a monodisperse solution in a nonpolar hydrocarbon solvent. If these experiments are interpreted as the transfer of nonpolar side chains from water into the protein interior, it is

$$\Delta H_{\text{transfer}} = \Delta H_{\text{chain}} + \Delta H_{\text{solvent}}$$
$$\Delta S_{\text{transfer}} = \Delta S_{\text{solvent}} \qquad\qquad (3\text{-}3)$$
$$\Delta G_{\text{transfer}} = \Delta H_{\text{transfer}} - T \cdot \Delta S_{\text{transfer}}.$$

It was found that $\Delta H_{\text{transfer}}$ is small and positive and $\Delta S_{\text{transfer}}$ is large and positive as mentioned in the previously discussed oil–water example. Nozaki and Tanford (16) determined the $\Delta G_{\text{transfer}}$ values for the transfer of side chains of standard amino acids from water to ethanol or dioxane. Ethanol and dioxane simulate the polarity of the protein interior better than a hydrocarbon solvent. No separation into $\Delta H_{\text{transfer}}$ and $\Delta S_{\text{transfer}}$ was attempted. Some of the results are given in Figure 1-8. All hydrophobic side chains have negative $\Delta G_{\text{transfer}}$ values, indicating that they avoid an aqueous environment. In contrast, a glycine residue has a $\Delta G_{\text{transfer}}$ around zero. Thus, the backbone (Gly has no side chain) prefers neither the interior nor the surface of a protein.

$\Delta G_{\text{transfer}}$ **is proportional to the interface area.** The linear relationship of Figure 1-8 between the accessible surface area of a nonpolar side chain and $\Delta G_{\text{transfer}}$ is in accordance with the iceberg hypothesis (page 38), since the number of ordered water molecules is expected to be proportional to the interface area. Figure 1-8 shows further that each dipole decreases the magnitude of $\Delta G_{\text{transfer}}$ by about 1.5 kcal/mol. However, this value was measured for a transfer to ethanol or dioxane, where the hydrogen bonds are not as strong as in the protein interior. In the protein case internal hydrogen bonds partially compensate this decrease, and the values of polar side chains (for example of Thr) are nearer to the straight line at the righthand-side of Figure 1-8. Thus, the gain of $\Delta G_{\text{transfer}}$ is roughly the same for polar and for nonpolar groups, and $\Delta G_{\text{transfer}}$ is roughly proportional to the total (polar and nonpolar) accessible surface area. The proportionality constant is 0.025 kcal/mol Å².

Salt Bridges

There are only a few salt bridges in proteins. Salt bridges are observed in a number of proteins, for example, hemoglobin (60) at Lys-40 (α_2-chain)/α-carboxylate of His-146 (β_1-chain), parvalbumin (61) at Glu-81/Arg-75, and chymotrypsin (18) and trypsin (62) at α-ammonium group of Ile-16/Asp-194. Their electrostatic attraction energy can be readily calculated; it amounts to about 5 kcal/mol for adjacent carboxylate and ammonium groups (Table 3-1) in a medium with a dielectric constant $\epsilon = 4$.

The free energy contribution of a salt bridge is small. However, to assess the contribution of salt bridges to protein stability in a solvent, this attraction energy has to be compared with the attraction energies of both charges to solvent molecules, which in our accounting system are included in H_{chain}. In the solvent a charge attracts surrounding dipoles (H_2O molecules) by an electrostatic monopole–dipole interaction. The resulting energy largely offsets the electrostatic attraction energy between the ions. Therefore, the resulting ΔH_{chain} is about zero. Moreover, stability depends on free energy [Eq. (3-1)] and not only on binding energies; therefore, entropy changes have to be considered. Since the attracted dipoles are radially oriented around the charge, water shows a higher order around a charged group—just as it shows a higher order in the quasisolid structure around a nonpolar group. The $\Delta S_{\text{solvent}}$ contribution is appreciable, favoring salt bridge formation (15). The resultant net free energy $\Delta G_{\text{transfer}}$ is of the order of 1 kcal/mol in favor of salt bridge formation (60). The removal of a single charge from water without forming a salt bridge unbalances $\Delta G_{\text{transfer}}$ completely. Therefore, burying a single charge inside a protein or in other hydrophobic environments such as membranes is very unfavorable with a $\Delta G_{\text{transfer}}$ of the order of 10 kcal/mol.

3.6 Molecular Packing

Hydrogen bond energy is exploited efficiently. After discussing the rather elusive noncovalent forces which mediate the stability of a folded polypeptide chain, it is of interest to examine the efficiency with which these forces are actually applied. A survey of protein structures showed that about 90% of all internal polar groups form hydrogen bonds (17). This is consistent with the large amount of secondary structure observed in proteins (chapter 5) and indicates that almost all possible hydrogen bond energy is actually scooped in.

High packing density means wide use of dispersion forces. The efficient application of dispersion forces can be judged from the packing density in the protein interior, since packing density corresponds to the number of contacts made. The packing density is defined as the ratio of the bare molecular volume (as given by the van der Waals envelope of the molecule which in turn is defined by the van der Waals radii of the surface atoms) and the volume actually occupied in space. In crystals of small molecules the occupied volume is given by the unit cell, and the molecular volume inside the van der Waals envelope can be determined from the atomic coordinates in the crystal in conjunction with van der Waals radii.

In order to derive a global packing density for a protein molecule, the van der Waals envelope of the extended chain can be related to the envelope of the folded chain, that is, to the envelope of the native protein molecule. However, it is more promising to determine local packing densities which may vary over the protein and thus reveal features of structural significance. Integration over all local packing densities then yields the average packing density of the protein.

Voronoi polyhedra allow determination of local packing densities. Local packing densities were first determined by Richards (63). For this purpose the polypeptide chain was subdivided into small atomic groups, for example, methyl, methylene, amide, hydroxyl, etc., which contain one heavier atom and zero to three hydrogens. Aromatic rings and the guanidinium group of Arg were also considered as single groups. The atomic group centers were taken from X-ray crystallographic data. Space was then subdivided into Voronoi polyhedra as described in Figure 3-4. The local packing density at a given center is the ratio between the atomic group volume and the corresponding occupied volume, that is, the surrounding Voronoi polyhedron. To avoid surface errors, a virtual monolayer of H_2O molecules on the surface was included in the procedure. Thus, only protein atoms and no virtual H_2O molecules are enclosed by *complete* Voronoi polyhedra. This can be visualized in Figure 3-4 by considering A as the center of a protein atom and B, C, and D as the centers of virtual H_2O molecules (see also Figure 1-9). In addition this procedure defines the molecular envelope. It would be best to use the monolayer of surface water molecules that contains the maximum number of water molecules. How-

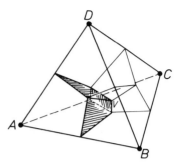

Figure 3-4. Subdivision of space into polyhedra around atoms. For a set of atoms space can be subdivided into irregular tetrahedra with one atom at each corner, one of which is depicted here. As pointed out by Voronoi (1908), the four bisecting planes between each atomic pair meet at one point, the vertex V. These planes subdivide the tetrahedron uniquely into four parts. The part belonging to atom A is indicated by the hatched surface. A given atom A may belong to any number of tetrahedra. To determine the total polyhedron around A, the illustrated construction has to be performed for all these tetrahedra, and all parts belonging to A have to be added. After repeating this procedure for each atom (atoms on the surface of the set have to be exempted), space is totally divided. The packing density of a given atom is then defined as the ratio between the volume inside its van der Waals envelope and the surrounding polyhedron. Richards (63) has adapted this procedure to protein molecules, which contain several atomic species with different van der Waals radii. Therefore the original Voronoi procedure, which is exact for equal-sized atoms, is not applicable. However, a modified form gives a reasonable approximation. Here, the bisecting planes are moved along the plane vector to a position that divides the distance between atoms, say between A and B according to the ratio of the van der Waals radii of A and B, or, if A and B are covalently bound, according to the ratio of covalent radii. With this displacement the four planes no longer meet at one point V but form a small tetrahedron near to V. The relative volume of these error tetrahedra is only about 1% and can be neglected as long as the ratios of radii do not exceed 1.5. In order to keep these ratios around 1.0, Richards lumped all (the small) H-atoms into atomic groups.

ever, since this optimal layer is difficult to find, one is usually content with a less than optimal layer. The committed error is small.

Proteins are as tightly packed as good molecular crystals. The observed local packing densities of proteins vary between about 0.68 and 0.82. Low packing densities are found in active centers (63,64), which corroborates the assumption that active centers have to be flexible. Hydrophobic cores in the protein center have high packing densities.[3] The average packing density of a protein amounts to about 0.75. In comparison, equal-sized hard spheres in closest packing have a packing density of 0.74. Crystals of small molecules that are held together by van der Waals forces have values between 0.70 and 0.78 with an average of 0.74. Glasses, oils, or exception-

---

[3]Packing density should not be confused with electron density or (mass) density. Hydrophobic cores like all hydrocarbons have much lower (mass) density than polar regions.

ally soft van der Waals crystals (or some crystals held together by directed bonds as the hydrogen bonds in normal ice, Figure 3-2) have packing densities below 0.70 and even below 0.60. Therefore proteins are indeed as densely packed as small molecules in van der Waals crystals. Accordingly, dispersion energies are efficiently activated.

The observed packing densities show that Kauzmann's oil drop metaphor for a hydrophobic core should not be taken literally. Rather, the protein interior resembles a crystal. This is also confirmed by the small compressibility of proteins when compared with oil (65,66): $\kappa_{oil}/\kappa_{protein} \approx$ 20.

Packing is conserved during evolution. Moreover, the protein manages to bring almost all interior polar groups into hydrogen bonds, although these bonds are directed and therefore pose geometric constraints (normal ice, for example, tolerates a packing density as low as 0.58 in order to form straight hydrogen bonds, Figure 3-2) which have to be reconciled with the geometric constraints of dense packing. This structural optimization is reflected in the observation that evolutionary changes in the protein interior are much rarer than changes at the surface. Moreover, internal changes tend to compensate each other (67), for example, the change Ile $\rightarrow$ Val may be accompanied by an adjacent Gly $\rightarrow$ Ala, which replaces the lost methyl group.

Packing density can serve as a criterion for structure predictions. Since the observed packing density is difficult to achieve, it can be used as a criterion to check any suggested protein structure. Thus, packing densities provide an independent check for chain folding simulation calculations (section 8.6) and for predicting structures on the basis of a known sequence by comparison with phylogenetically related proteins (chapter 9). This criterion is not restricted to a comparison of average packing densities. It turns out that the volume occupied by side chains (the sum of all occupied volumes of the constituent atomic groups) is specific for each side chain (63,68) and varies by only about 5%. Therefore, the volume occupied by each side chain can be taken as a further criterion for successful protein structure prediction.

Summary

As discussed in chapter 1, protein structure formation and stability depend heavily on noncovalent forces. These can be subdivided into (i) repulsion between nonbonded electron shells, (ii) coupling between oscillating dipoles giving rise to an attraction, (iii) electrostatic attraction and repulsion of partial charges as well as attraction of full charges as found in salt bridges, and (iv) hydrogen bonds. Since proteins exist only in aqueous environments, the whole polypeptide chain and solvent system should be taken into account. In Eq. (3-2) entropies and binding energies of the constituents are given separately. The strongest contribution to protein stability comes from hydrophobic forces. The magnitudes of the other contributions are discussed. The high packing density observed in proteins demonstrates that all noncovalent forces are efficiently used. The protein interior is as tightly packed as a good crystal.

Chapter 4
The Covalent Structure of Proteins

In section 1.1, we started out with the generalization that the functional form of a protein is obtained merely by synthesizing the polypeptide chain on ribosomes and by letting it fold spontaneously. Although this is the basic concept, there are frequent biological amendments to this scheme which are of great importance. These phenomena are discussed here.

4.1 Chain Assemblies

Functional and Structural Domains

The basic concept (74) of one structural gene[1] = one polypeptide chain = one functional unit should be refined. It no longer seems appropriate to equate polypeptide chain with functional unit. Since 1969 (75) there have been increasing indications (76) that a "domain" and not a polypeptide chain should be regarded as the basic unit and that all other entities should be referred to this basis.

---

[1] In higher organisms less than 10% of the total nucleotide sequence of a structural gene, the so-called structural zones (78, 79), code for the actual gene product. Studies on the genes coding for globin, ovalbumin, and some proteins of SV40 and polyoma virus suggest that the eukaryotic structural gene is mosaic. The nucleotide sequences of DNA which are translated into one amino acid sequence are not continuous but are interrupted by pieces of untranslated DNA. The primary RNA transcript contains internal regions which have to be excised, the final mRNA being a spliced product [for a short review see (78)].

Functional domains are functionally autonomous regions of the polypeptide chain. Originally, domains were identified in experiments that were designed to correlate gene structure and polypeptide structure (77). Domains are subregions of the polypeptide chain (encoded in subregions of a structural gene) that are autonomous in the sense that they possess all the characteristics of a complete globular protein. Many such subregions have been identified as products of limited proteolysis (76). This means that individual subregions can often be excised from a polypeptide chain without loss of their properties.

The significance of the domain concept is most obvious in certain enzymes involved in His and Trp biosynthesis (see section 9.4, page 194). Here the protein function requires a number of domains. It has been observed that this function is fulfilled independently of whether these domains are located on a single polypeptide chain, or on separate polypeptide chains associated by noncovalent forces, or on totally separate chains.

Structural domains are geometrically separate entities. Since the subregions described above are defined by observed functions (for example, ligand-binding or enzymatic activity), they are designated as "functional domains" (76). With the advent of geometric protein structure analysis it was shown that functional domains consist of one or more "structural domains." Such structural domains have been revealed by inspection of numerous three-dimensional protein structures as, for example, glutathione reductase (Figure 4-1). They are geometrically separate entities with molecular weights of around or less than 20,000 daltons. Almost all globular proteins can be subdivided into such subregions. Therefore, most functional domains with molecular weights above 20,000 presumably possess more than one structural domain. An "active site" of such a functional domain is usually located at the interface of two structural domains.

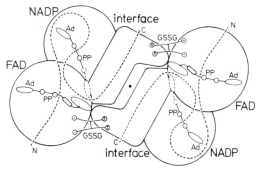

Figure 4-1. Sketch of the dimeric enzyme glutathione reductase. Each subunit consists of three domains; one of them binds FAD, another one binds NADP, and the third forms the interface. The general course of the polypeptide chain is given by a dashed line. The cofactors FAD and NADP bind in extended conformations. The substrate glutathione (GSSG) binds between subunits. Each active center is formed by four domains (124).

Proteins may be constructed using a modular system. The importance of structural domains as basic units is supported by comparisons of three-dimensional protein structures. The same structural domain identified by its characteristic chain fold can occur in different proteins. Typical examples of reoccurring structural domains are the immunoglobulin domain (Figure 4-2), the NAD-binding domain (Figure 5-17b), or the TIM-barrel (Figure 5-17e), which is present in triosephosphate isomerase and pyruvate kinase (80).

These findings, the subdivision of globular proteins into structural domains of approximately equal size and the occurrence of a given structural domain in different proteins, have led to the following hypothesis. Most proteins are constructed on the basis of a modular system with structural domains as modules (see also section 9.4). According to this hypothesis, too much importance should not be assigned to the polypeptide chain. Concentration should instead be on functional, and, whenever possible, on structural domains. The basic concept (74) should be amended as follows: "one polypeptide chain = one or more functional domains; one functional domain = one or more structural domains."

Environmental Determinants

As shown above the domain is the basic unit. Associations of these domains are of secondary importance. The following will show how the environment in which a protein operates determines the type of domain association.

Oligomers can dissociate whereas monomers cannot. Proteins are usually classified as monomers or oligomers. As defined by Klotz *et al.* (81) a protein is a "monomer" when it consists of only one polypeptide chain or when it is made of several chains, all of which are covalently linked (for example, by disulfide bonds). According to this nomenclature proteins such

---

Figure 4-2. Proteins consisting of Ig domains. In (a) the detailed structure of an Ig domain is given. In (b) and (c) each large circle represents one Ig domain. Carbohydrate moieties of the proteins are not shown. (a) A subunit of superoxide dismutase (800). The chain fold is given as C_α-backbone drawing in stereo (see section 7.2); N- and C-termini of the two chains are indicated. The subunit possesses an Ig domain and a copper-containing catalytic site with a Zn-ion close by. (b) Histocompatability antigen (human leukocyte antigen = HL-A protein). The molecule is symmetrical. Each half consists of a heavy chain H and a light chain L. The light chain which is a single Ig domain is also called β_2-microglobulin. The two heavy chains are linked by a disulfide bridge which is probably located in the membrane-bound part of the molecule. (c) Immunoglobulin G. The molecule is symmetrical. Each half consists of a heavy chain H and a light chain L. Each chain is linked to at least one other chain by at least one disulfide bridge. The standard designations for the individual domains along the L-chain and the H-chain, respectively, are given.

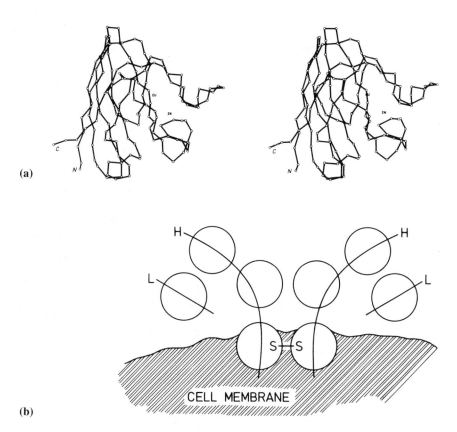

(a)

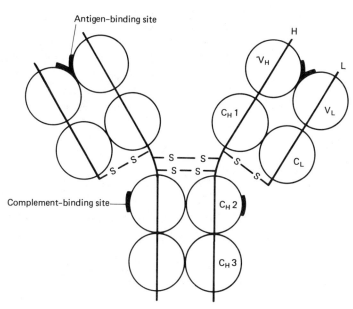

(b)

(c)

as insulin, α-chymotrypsin, and the immunoglobulins, all covalent chain-assemblies, are monomers. The characteristic of "oligomeric" proteins is that they are built up from so-called subunits, that is from noncovalently linked subordinate entities (Figures 4-1 and 5-18). As stated above, monomers can consist of several functional domains and of even more structural domains. This is also true for subunits of oligomers, although a subunit often equals a functional domain.

The Three Major Physiological Compartments

The type of bonds between functional domains (covalent, noncovalent) as well as the absence of bonds may be correlated with major physiological compartments. Most intracellular proteins are oligomers, proteins of the blood plasma are large monomers consisting of several functional domains, and proteins that operate outside[2] an organism are small monomers. To some extent this distribution can be rationalized.

Extraepithelial proteins are small monomers. In spaces outside an organism the fate of a single molecule (for example, a digestive protein or a human lysozyme) is very uncertain. Consequently, it is reasonable that as many independent entities per unit committed material as possible are produced. Thus the protein is small. Moreover, it is monomeric, because oligomers tend to dissociate by dilution.

Proteins of the blood plasma are large monomers. An essential property of true plasma proteins is that they really stay in the blood plasma. Only proteins with molecular weights around and above 60,000 are large enough so that they are not readily filtered into the large extravascular space as happens to smaller molecules (up to 20,000 daltons). Moreover, small proteins are filtered, processed, and/or excreted in the kidneys.

The large blood plasma proteins could be either oligomers or monomers. Oligomers, however, would dissociate and the subunits would be filtered in the same way as small monomers. β_2-Microglobulin (Figure 4-2b), the small soluble subunit of HL-A proteins (which are bound to the surface of certain blood cells as well as other cells), illustrates this. This subunit of an oligomeric protein is known to permeate the walls of blood capillaries; furthermore, under certain conditions (for example, in cadmium poisoning) large amounts of this subunit are excreted in the urine. Thus it is not surprising that true plasma proteins are large monomeric proteins rather than oligomeric proteins. Serum albumin (Figure 7-2b) which consists of a series of functional domains (76,82) along one polypeptide chain is a typical example. Some plasma polypeptides are appreciably enlarged by covalently bound carbohydrate moieties.

---

[2]The borders between the inside and the outside of higher animals are epithelial surfaces; the contents of the gastrointestinal tract, for example, are outside.

Intracellular proteins are as a rule oligomers. Now consider intracellular proteins. These are mostly oligomers. Small monomeric proteins such as the isoenzymes of adenylate kinase (83, 84) (22,000 daltons) are rare. When compared with large single polypeptide chains, oligomers have various advantages (section 4.1, page 52), and they are therefore preferred wherever possible. In contrast to blood plasma, cells can indeed use oligomers, because the cell membrane is impermeable even for small proteins so that oligomers are not lost in the form of subunits.—The formation of oligomers reduces the osmotic pressure of the intracellular space. Moreover, the ratio of surface:volume is smaller for the oligomer than for the monomer. Consequently, it binds less water and contributes less to the viscosity of the cell. In addition, oligomeric proteins usually show cooperativity and they can be well regulated by effectors. Tetrameric hemoglobin of mammalian erythrocytes (85) is a well-known example for the physiological relevance of these criteria. However, these criteria are also valid in less specialized cells containing a large variety of proteins.

Immunoglobulins and HL-A Proteins

The covalent linkage between immunoglobulin chains reflects a functional requirement. The plasma immunoglobulins uniquely illustrate several aspects of correlations between protein structure, function, and environment (81). In an immunoglobulin (Figure 4-2c) each of the two antigen-binding sites is constructed of two types of polypeptide chain. It is essential for the function of the molecule that two identical light chains and two identical heavy chains come and *stay* together. An uncontrolled exchange of chains among the thousands of different immunoglobulins in the blood plasma would prevent the effective interaction of these antibodies with the antigens which elicited their synthesis. The randomization of immunoglobulin structures is prevented by covalently linking the four chains which are essential for specificity (81).

HL-A proteins can be oligomeric. In contrast, the biological specificity of a given HL-A protein (Figure 4-2b) is based on only two (identical) heavy chains, the light chains of all HL-A proteins in an organism being identical (86, 87). Therefore the light chains do not have to be covalently bound to the heavy chains; only the heavy chains have to be covalently bound to each other. Dissociation and reassociation of light chains cannot scramble the molecular specificity. Indeed, the trimeric structure of HL-A proteins was predicted on the basis of their specificity requirements and of the known structure of the immunoglobulins. Thus physiological findings can lead to working hypotheses in studies on protein structure.

It can be argued that the oligomeric structure of HL-A protein is disadvantageous anyhow since it leads to the filtration and degradation of the light chain (β_2-microglobulin, Figure 4-2b) in the kidney (88). One compensating advantage could be that the light chain of HL-A proteins is produced by more or less specialized cells and that they are distributed, like hormones, throughout the organism, the partial loss in the kidney being a minor accident. It might be worthwhile to start a systematic search for light chain-producing cells. Thus results on protein structure can also lead to working hypotheses in general physiology.

Oligomeric Proteins

Oligomers are simpler than single giant polypeptide chains. There are some general advantages of oligomeric proteins vis-à-vis giant polypeptide chains with several functional domains; this may explain the fact that most large proteins are oligomers (81).

(a) *If some or all subunits are identical much less genetic material is needed to code for an oligomeric protein than for a single giant polypeptide chain.* Although the isolated subunits of many oligomeric proteins are nonfunctional (89, 90) these subunits possess all the information necessary for self-assembly to produce a functional unit. This allows a reduction of DNA (or RNA) for a given function of a protein. The effect is apparent in average-sized oligomers such as tetrameric dehydrogenases (81, 91) and is very dramatic and of great biological significance in structures such as the coat protein of tobacco mosaic virus which consists of 2130 self-assembling identical subunits (each 158 amino acid residues long).

(b) *Large polypeptide chains run a higher risk of being defective than oligomeric proteins of the same size.* If we assume that the synthetic machinery of proteins introduces 1 error per 1000 amino acid residues (81) of a protein, a single polypeptide with 1000 residues runs a high risk of being defective [the average number of correct copies amounts to $\exp(-1000 \cdot \frac{1}{1000}) = 37\%$ as derived from the Poisson distribution]. In contrast, the synthesis of four subunits each with 250 residues yields a high proportion, namely, $\exp(-250 \cdot \frac{1}{1000}) = 78\%$ of effective subunits.[3]

(c) *Oligomeric ensembles are particularly well suited for rapid molecular evolution* (92). Therefore they can more readily adjust to environmental changes than large single polypeptide chains.

(d) *Oligomers may have been a simple solution to a problem which primordial cells might have faced;* as their cell membrane was still imperfect the leakage of molecules of 20,000 daltons through membrane pores might have been appreciable. Aggregation was probably a simpler means of increasing size than the synthesis of larger chains (93).

---

[3]One may argue that the advantage is not great because the incorporation of one defective subunit into a tetrameric protein might abolish the function of the whole oligomer. However, this is usually not the case; oligomeric proteins are in a dynamic state of dissociation and reassociation; subunits between oligomers are exchanged and defective ones are eliminated. By this disproportionation inborn defects and chemical injuries of individual subunits can be corrected (81).

4.2 Disulfide Bonds

Most disulfide bonds form spontaneously *in vitro*. As mentioned above disulfide bridges between pairs of cysteine residues can cross-link different chains of a protein which gives rise to covalent chain assemblies. Disulfide bonds occur within a single polypeptide chain as well. A given cysteine residue along a polypeptide chain combines with only one other cysteine residue so that the set of disulfide bridges is unique for a given protein (94–100). Furthermore, as a rule these bonds form spontaneously *in vitro*; no external factors such as enzymes are necessary (94). Apparent exceptions, such as the disulfide bonds of insulin and chymotrypsin (101) are discussed in section 8.2. A real exception was postulated for the case of immunoglobulin A where the J-chain is essential for the formation of certain interchain disulfide bonds; the J-chain, however, is unlikely to be an external factor (102).

In vivo **S–S bridges are formed under reducing conditions.** Although cysteine residues are readily oxidized to cystine residues *in vitro,* it is not yet known how disulfide-containing proteins are formed *in vivo,* and when they originated in the course of protein evolution. These two problems are related. The ancestors of modern proteins are assumed to have evolved in a reducing atmosphere, and in modern cells a strongly reducing environment is maintained by the glutathione system (5 mM reduced glutathione and 0.1 mM oxidized glutathione). Under such conditions protein disulfide formation might not proceed spontaneously. Consequently, it has been suggested (103) that a system which involves oxidized cystamine and a membrane-bound enzyme is responsible for the introduction of S–S bonds into proteins. This enzyme is, of course, different from the protein disulfide isomerase (104–107), which reshuffles existing disulfide bonds in proteins until (*meta*)stable structures are reached.

A common function of disulfide bonds is to give extra stability to otherwise properly folded proteins [see section 8.4 and Refs. (94) and (108)]. In view of this fact it is not surprising that some S–S bridges can be broken without loss of a protein's function (Table 4-1). In the case of α-amylase, for example, all three disulfide links can be reduced without impairing the enzymic activity (109). However, there are many examples of proteins in which disulfides are involved in very specific functions.

Role of S–S Bonds in Extracellular and Intracellular Proteins

Disulfide bridges determine mechanical properties of extracellular proteins. Disulfide bridges are common in proteins that travel or operate in extracellular spaces; typical examples are snake venoms and other toxins, peptide hormones, digestive enzymes, complement proteins, immunoglobulins, lysozymes, and milk proteins. Moreover, these bridges play an important

Table 4-1. Reduction of Disulfide Bridges

Protein	Number of		S–S-Bonds split without loss of activity
	S–S-Bridges	Residues	
Bovine pancreatic trypsin inhibitor (452)	3	58	1
Ribonuclease (109)	4	124	1
Lysozyme (161, 162)	4	129	Any one of 3
Trypsin (109)	6	224	3
Chymotrypsin (109)	5	242	2
Amylase (109)	3	410	3

role in some large structures. The visco-elastic properties of various natural products are at least partly determined by disulfide bridges between structural proteins (110). The cross-links between keratin molecules contribute to the elasticity of wool and hair (110), the cohesive elastic character of wheat flour doughs is based on the disulfides of glutenin, and the three-dimensional network of disulfides in glutelin is responsible for the difficulties encountered during wet milling of corn. Thus in arts as old as milling, baking, wool processing, and hair dressing skillful disulfide engineering is essential to obtain optimal products (110).

Disulfide bridges determine chemical properties of intracellular proteins. Of the proteins which never leave their parent cell very few possess disulfide bridges. As a rule disulfides in intracellular proteins have functions other than structure stabilization. The disulfide in glutathione reductase, for example, plays an active part in catalysis (111) (Figure 11-4); the disulfides of threonine deaminase (Figure 8-3) stabilize the correct conformation of the associated protein subunits (112). Thiol-disulfide interchange (Figure 4-3) may play a functional role in the regulation of enzyme activities (113–116); this exchange mechanism has also been found to operate in the interaction of a membrane-bound contractile protein and a microtubular protein during the cleavage cycle of the sea urchin egg (117). The expectation that thiol-disulfide interchange reactions would play a role in the specific interactions between membrane proteins and extracellular proteins turned out to be incorrect for the two known systems: insulin–membrane receptor (118,119) and complement proteins–membrane receptors (120).

The enzymes tryptophanase (121) and threonine deaminase (122) (Figure 8-3) from *Salmonella* have a peculiar construction principle based on disulfide bridge formation. Here 4 identical polypeptide chains are pairwise linked with disulfide bonds so that proteins with 2 (monomeric) subunits result. This asymmetric aggregation is probably related to the fact that these enzymes possess 2 binding sites for pyridoxal phosphate. In immunoglobulins the disulfide bridge between a light and a heavy chain has a similar function, it maintains the specificity of a binding site.

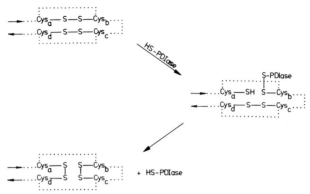

Figure 4-3. Rearrangement of disulfide bridges in a protein by thioldisulfide interchange reactions. Cys_a, Cys_b, etc. are disulfide-forming residues along a polypeptide chain, the course of which is indicated by dots. As a first step of disulfide rearrangement a thioldisulfide interchange is catalyzed by the enzyme protein-disulfide isomerase (PDIase with its catalytic thiol group HS-); in this reaction a mixed disulfide between protein and enzyme is formed (109). A series of further thiol-disulfide interchanges leads to a new pattern of disulfide bridges in the protein and to a regeneration of the catalytic thiol group of PDIase. Not only protein-disulfide isomerase (104–106) but other SH-compounds such as reduced glutathione or dithiothreitol (at pH > 7) can catalyze disulfide isomerizations in proteins.

Disulfide Bridges as Integral Parts of Structural Motifs in Proteins

Disulfide links which close tight intrachain loops have been found in several proteins [pepsin, thioredoxin, insulin A-chain, silk fibroin (145), lipoamide dehydrogenase and other pyridine nucleotide–disulfide oxidoreductases (111)]. The bridged cysteine residues are separated by two to four residues along the polypeptide chain. Model building and X-ray diffraction analysis have shown that such loops tend to be flat and rigid. For glutathione reductase and related enzymes the loop touches the isoalloxazine ring of FAD (123,124).

Another frequently encountered structural element is the sequence –Cys–Cys–, with both residues forming disulfide bridges with other cysteine residues (82); the common peptide bond does not allow a disulfide bridge between adjacent residues. Thus a –Cys–Cys– sequence forms the basis for linking three chain segments in close proximity. It is very unlikely that the presence of S–S-bonded –Cys–Cys– in 15 proteins has arisen by chance or could be due to common ancestry of these proteins. Consequently, it was proposed that the sequence –Cys–Cys– is an especially useful architectural unit in protein structure and that many proteins have independently evolved this sequence (see Figure 7-2).

4.3 Enzyme-Controlled Modifications of the Main Chain

For many aspects of protein structure it is essential to account for the physiological fate of a protein over its total life span, from synthesis to degradation. This is particularly important for those proteins which undergo post-translational (secondary) modification(s), that is, covalent modification(s) after the peptide bonds have been formed by the ribosomal synthetic machinery.

The biogenesis of collagens (85, 125–131) is the textbook example of proteins that are extensively modified and processed during and after synthesis. There are probably even more reactions involved than shown in Figure 4-4. Nevertheless, it is evident that collagens have really fascinating life stories. In the following some general aspects of protein modification *in vivo* are described.

N-terminus and C-terminus

Modification of the N-terminal amino group is frequent. In many proteins the N-terminal α-ammonium group (pK=8) undergoes secondary modification [see Ref. (132) for a review]. It is acetylated in the coat protein of tobacco mosaic virus, in most c-type cytochromes, in the muscle proteins actin, myosin, tropomyosin, parvalbumin, adenylate kinase, and lactate dehydrogenase. Other examples include wool keratin and the α-melanocyte-stimulating hormone (132). A related modification of the α-amino group, formylation, has been found in the bee venom melittin and in lamprey hemoglobin. A different principle for masking the N-terminus of a polypeptide is the conversion of N-terminal glutamate to a pyrrolidone carbonyl group. This modification has been found in many proteins that operate in the extracellular space; examples are the light and heavy chains of human immunoglobulins, hormones from various sources, and snake venoms (132–134).

The biological significance of masking α-amino groups is not clearly understood; it might protect a protein against the attack of aminopeptidases or it may be important for anchoring the N-terminal part of a polypeptide in an apolar environment, either inside of the protein itself (to prevent it from dangling in the solution) or at a receptor molecule. This argument would not hold for the methylation of the α-amino group which was found in ribosomal proteins of *Escherichia coli* (135) since methylation does not eliminate the charge. In the case of some fish hemoglobins the acetylation of the α-amino group has a defined physiological role; this modification helps to make the oxygen binding independent of pH, an adaptation which prevents excess loss of oxygen to the swim bladder (136) (section 10.3, page 219).

C-terminal modifications are rare. The only known substituent at the C-terminus of naturally occurring polypeptides is the amide group. Amidated C-terminal groups are found, for example, in hormones and bee venoms (132). The function of the amide group might be to protect the peptide chain

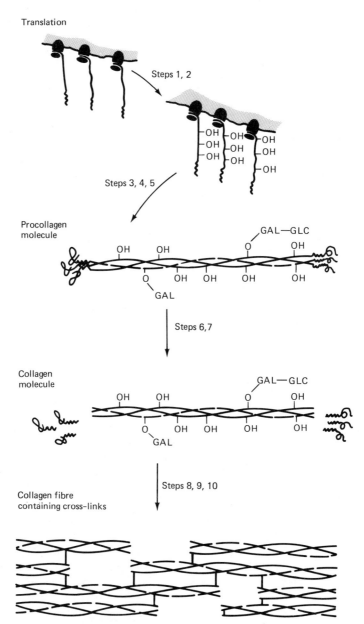

Figure 4-4. Biogenesis of collagen (131). The following stages have been identified: (Step 1) Synthesis of pro α_1-chains and pro α_2-chains in a ratio 2:1. Each chain contains 1300 residues. (Step 2) Hydroxylation of some Pro and Lys residues. (Step 3) Addition of sugars (Glc–Gal) to hydroxylysyl residues. (Step 4) Trimer formation associated with the formation of interchain disulfide bridges which are probably located in the terminal regions of the chains. (Step 5) Formation of the triple helix in the center of the procollagen molecule. (Step 6) Secretion of procollagen into the extracellular space. (Step 7) Removal of the globular extensions of the chains which gives rise to collagen. (Step 8) (Spontaneous) self-assembly of collagen molecules (fiber formation). (Step 9) Deamination of Lys- and hydroxylysyl residues to give aldehydes. (Step 10) Cross-link formation between polypeptide chains in the fiber by various reactions involving these aldehydes, and side chains of Lys and His. [For details see Refs. (127) and (131).]

against the attack by carboxypeptidases or to facilitate the accommodation of the otherwise negatively charged C-terminus (the pK value of the α-CO_2H group is around 3) in a nonpolar environment. In general, C-terminal modifications are much rarer than N-terminal ones.

Specific Cleavage of the Polypeptide Chain

Specific proteolysis is a common and convenient tool for constructing complex protein teams. In many cases proteins are modified by cleavage of one or a few peptide bonds. For this type of enzyme-catalyzed reaction which plays a dominant role in many physiological processes (137–139) the terms "limited proteolysis" or "specific proteolysis" are used (Table 4-2). Some well-known examples of specific scissions of polypeptides are the activation of digestive enzyme precursors, the morphogenetic processes in bacterial viruses, and the cascade mechanisms involved in blood coagulation and complement action (138, 140). More recently, post-translational cleavage mechanisms have been shown to be involved in the formation of proteins as different as insulin, collagen, and virus-specific proteins. In addition, proteolytic cleaving enzymes of high specificity play a role in the inactivation and activation of specific intracellular enzymes (Table 4-2).

The process opposite to specific proteolysis, namely, protein splicing (141), is also known. An example in which the C-terminus of one protein is linked to the N-terminus of a different one occurs in the assembly process of a bacterial virus (142).

Activation of Trypsinogen

Trypsinogen is converted to trypsin which in turn activates various other enzymes. In many systems the specific protease has not been identified. A representative case in which the process of limited proteolysis is known in some detail is the activation of zymogens (1, 138, 139), inactive enzyme precursors, in the digestive tract (143). The key enzyme in digestion is the protease trypsin (1), not only because of its own activity on the ingested protein but also because it is the only known activator of other zymogens such as the chymotrypsinogens, the procarboxypeptidases, proelastase, and prophospholipase A. Trypsin itself is secreted as an inactive precursor, trypsinogen, into the duodenum (Figure 4-5).

The activator of trypsinogen recognizes a specific sequence. The physiological activator of trypsinogen is enteropeptidase (also called enterokinase) which is located in the surface membrane of epithelial cells in the duodenum. The extremely narrow specificity of enteropeptidase is reflected in the fact that this enzyme hydrolyzes no bond in any tested native or denatured protein except for the one bond in trypsinogen (or in synthetic peptides) which is preceded by the sequence Asp–Asp–Asp–Asp–Lys [von Neurath, personal communication, and Ref. (138)]. Therefore, this enzyme may be specific for a certain sequence and not for a certain tertiary structure.

Table 4-2. Aspects of Specific Proteolysis

Biological process	Example of limited proteolysis	Functional aspects
Formation of virus-specific proteins in animal viruses (163, 164)	Precursor of poliovirus proteins → poliovirus proteins	Adaptation of the viruses to peculiarities of protein synthesis in animal cells
Morphogenesis of virus particles (140)	Two independent substructures of T4-virus → assembly of the two substructures	Cleavage of a polypeptide chain provides new binding sites; the irreversibility of the proteolytic reaction(s) contributes to the ordered assembly of the virus particle
Connective tissue formation (125, 127, 131)	Procollagen with N- and C-terminal extensions → triple helical collagen + large polypeptides	Release of collagen after it has been processed and transported as a soluble precursor
Digestion (138)	Trypsinogen → trypsin + a small peptide	Activation of the protease at physiologically appropriate time and space
Blood coagulation (148)	Inactive coagulation factor A → active coagulation factor A + a peptide; A then activates the inactive coagulation factor B etc.	Poised-triggered system of many proenzymes; amplification of a response through an enzyme cascade; control on each level of the cascade
Complement action (147)	Complement factor → one or two active complement proteins	Poised-triggered system of at least 18 serum proteins; a series of specific proteolytic reactions restricts the action of complement in time and space. Soluble proteins become membrane constituents
Toxin action (165)	Diphtheria toxin → enzymatically active fragment + permeation fragment	A protease of the invaded cell releases the lethal enzyme of the toxin
Protein degradation (154–156, 166)	Serine dehydratase (apoenzyme) → large fragment + small peptide	Protein inactivation and degradation is often initiated by one specific proteolytic step
Hormone formation (167)	Proglucagon → glucagon + octapeptide	The extrapeptide is the C-terminal part of the prohormone
	Proinsulin → insulin + C-peptide	Formative cleavage of a protein which can fold properly only as a precursor
Protein secretion (153)	Pretrypsinogen → trypsinogen + a hydrophobic peptide	Removal of the signal sequence which seems to carry the information "made for export"

(Met) Ala—X—X—Phe—Leu—Phe—Leu—Ala—X—Leu—Leu—Ala—Tyr—Val—Ala—Phe—Pro—Leu—Asp—Asp—Asp—Asp—Lys—Leu– ....
       1   2   3   4   5   6   7   8   9   10   11   12   13   14   15   16   17   18   19   20   21   22   23   24

├──► Pretrypsinogen   ├──► Trypsinogen   ├──► Trypsin

Figure 4-5. Sequence of pretrypsinogen from dog (preliminary data (153)). Met (which is later removed) is assumed to be the N-terminal amino acid of all nascent polypeptide chains in eukaryotes. It is as yet unclear whether residue 16 or 17 of pretrypsinogen corresponds to the N-terminal amino acid of trypsinogen in dog. The method used does not distinguish between Leu and Ile (for instance in position 24).

Disregarding correction factors accounting for the natural abundance of individual amino acids (144, 145), the sequence Asp–Asp–Asp–Asp–Lys– is a unique one out of 20^5 possibilities. This is an indication that a unique sequence may be more reliable as a specific recognition site on a protein than a unique feature of the tertiary structure; and reliability seems to be very essential for any initiator of an enzyme cascade.

Trypsinogen activation is representative for other systems. The system trypsinogen–enteropeptidase is probably representative of other systems of specific proteolysis (Table 4-2) in the following aspects:

(a) In most cases the transformation (proprotein → protein) *in vitro* can be brought about by trypsin, although at a relatively low rate and with a relatively poor specificity (146). It should be emphasized that active trypsin occurs *in vivo* only in the duodenum.

(b) The enzyme responsible for the conversion *in vivo* is more efficient and more specific; each of the reactions shown in Table 4-2 appears to require a separate enzyme. Like enteropeptidase the specific proteases are expected to recognize not only a basic residue but a broader structural unit (such as the Asp–Asp–Asp–Asp–Lys sequence in trypsinogen).

(c) The specific proteases are often bound to membranes.

Complement Action—the Consequences of Triggering a Poised System of Proproteins

"Complement" is part of the body's defense system which eliminates cells such as bacteria or tumor cells. Complement itself is composed of at least 18 distinct plasma proteins (147). In a functional sense most of these proteins are proproteins (zymogens) since they are precursors of sequentially and concertedly operating enzymes, that is, of an amplifying enzyme cascade. The whole complement system with all components present at optimal concentration is always in readiness; it is poised and can be triggered any time. With regard to this property (being a poised and triggered system of many precursors of active proteins) complement resembles the blood-clotting system (148).

Proteolytic triggering has physiological advantages. At the molecular level the complement proteins demonstrate various consequences of specific proteolytic processes. These are (138):

(a) As all proteolytic reactions are quantitative and irreversible the overall process is unidirectional.

(b) The activation and inactivation of proteins, that is the rapid decay of activated binding sites, impose rigid constraints in time and space; thus the effect of complement action is restricted to the immediate environment of the immunoglobulins which label the foreign cell.

(c) Both a given protein and its cleavage products can possess physiological functions; moreover, all these functions are abolished by specific proteolysis. The fate of the complement factor C3 dramatizes this (Figure 4-6).

(d) Complement action includes a morphogenetic aspect. A series of specific proteolytic reactions leads to an assembly process through which five different soluble proteins become membrane constituents (147).

Signal Sequences in Preproteins

N-terminal chain extensions are probably used as labels. Many newly synthesized proteins are exported via the secretory apparatus of the cell; other proteins operate in intracellular spaces, in the cytosol, in membranes,

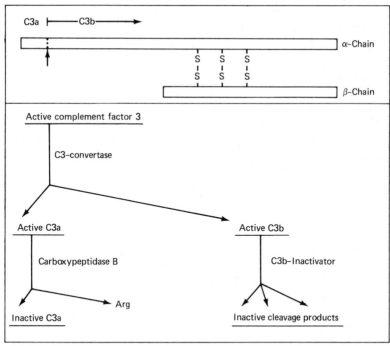

Figure 4-6. The multiple functions of complement protein C3 and its cleavage products (147). The point where C3 is cleaved by C3 convertase is indicated by an arrow pointing upwards. C3 (molecular weight 180,000) has a function in the "properdin pathway" and is the physiological precursor of two active fragments, the small C3a and the large C3b. C3a is involved in inflammatory reactions and C3b is a subunit or a modulator of four distinct complement enzymes. It should be noted that the inactivation of C3a and C3b is also based on limited proteolysis.

or in membrane-surrounded compartments. The destination of a polypeptide may be encoded in its sequence; at least for the proteins to be exported from the cell this seems likely. According to the "signal hypothesis" (149) a secretory protein is synthesized as a "preprotein," that is as a precursor with a short-lived peptide extension of the N-terminus.

The extensions, so-called signal peptides, with about 15 predominantly hydrophobic amino acid residues, have been found in the precursors of some immunoglobulin light chains (134, 150, 151) of proparathyroid hormone (152) and of various pancreatic proteins (153) (Figure 4-5). As the signal peptide emerges from a space within the ribosome it is believed to attach to the microsomal membrane; this is the first step in a sequence of events which leads to the transfer of the nascent chain across the membrane. During or immediately after this transfer the signal sequence is removed by a membrane-bound protease.

Protein Degradation

Removal is as important as synthesis. The susceptibility to inactivation and degradation is an important biological property of all protein structures. The present view (154–157) is that the first step in the degradation of a protein is the specific cleavage of one or a few peptide bonds. This is followed by nonspecific extensive proteolysis, a common final pathway for the breakdown of all proteins (156). Degradation is necessary to eliminate proteins that are defective or not desired at a given moment. Another aspect, the routine turnover of proteins, is most conspicuous for the digestive enzymes. If these proteins were not broken down to amino acids, that is, if proteases such as trypsin and chymotrypsin did not act as molecular cannibals, a person would lose 100 g of amino acids per day in the form of digestive enzymes. This amount corresponds to the human daily protein requirement in the diet.

4.4 Enzyme-Controlled Side Chain Modifications

Cross-Links on the Basis of Modified Lys Residues

The formation of ϵ-(γ-glutamyl)lysine links[4] plays a role in clotting processes, for example, in the enzymic coagulation of semen in rodents (158) or in the stabilization of fragile fibrin aggregates during blood clotting (159). The cross-links between polypeptide chains in elastin (Figure 4-7) and in collagen (Figure 4-4) are also derived from lysine side chains. The cross-

---

[4]According to Figure 1-1, ζ-(δ-glutamyl)lysine.

Figure 4-7. Desmosine, a cross-link derived from four lysyl side chains, which occurs in elastin.

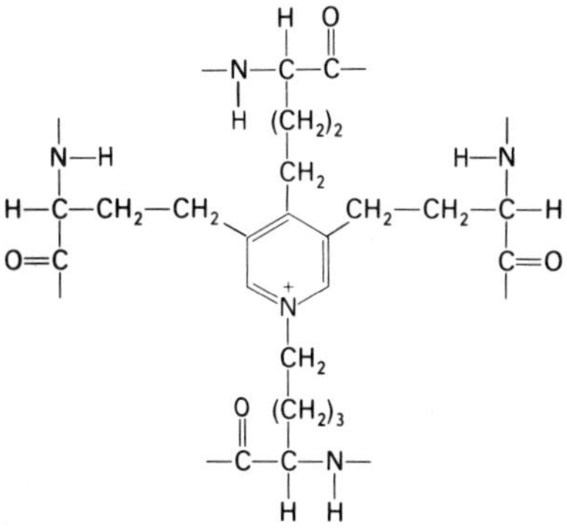

Desmosine

links in elastin, a rubberlike protein, may be important in enabling elastin fibers to return to their original size after stretching. The cross-links between collagen molecules make up for an intrinsic problem of the triple helix; this structure is thermolabile and melts into gelatine at approximately 40°C (125, 127). Thus the intra- and intermolecular cross-links between polypeptide chains in collagen prevent this most abundant protein of the human body from turning into amorphous gelatine when we run a fever. In all proteins mentioned above the covalent cross-links are involved in the stabilization, not in the formation of protein structures, a property which they share with certain disulfide bridges (section 4.2).

Other Varieties of Modification Reactions

Side chain modifications enlarge the spectrum of protein properties. Since a side chain modification endows a protein with unique properties, very few general comments can be made. In most cases only one or a limited number of side chains are modified in one protein. The structure of the protein and of the modifying enzyme as well as the nature of the side chain contribute to this specificity. Examples of side chain modifications and some functional implications are given in Table 4-3. Figures illustrating the chemistry have been compiled by Stryer (85).

Some generalizations with respect to the modifying group are:

(a) It endows a side chain with a function which in general cannot be carried out by one of the standard amino acids.

(b) It does not alter the bulk properties of a protein; the attached group has a molecular weight of a cell metabolite, that is, below 1000.

(c) The modifying group can be removed from the protein by cleaving the covalent bond. This property might distinguish a side chain modification from a ''prosthetic group'' which may be linked covalently to a side chain but which is also strongly bound by noncovalent forces to a protein. FAD (160) and haem are examples of prosthetic groups which are bound covalently to some proteins. In certain respects, of course, they are side

Table 4-3. Examples of Side Chain Modifications

Modification	Modified side chain(s)	One specific, not necessarily representative, example	Functional aspects
Adenylylation (168)	Tyr	Glutamine synthetase	By adenylylation of one Tyr per subunit the enzyme is converted from a form whose biosynthetic activity is dependent on Mg^{2+} (pH optimum = 7.6) to a form dependent on Mn^{2+} (pH optimum = 6.5)
ADP-ribosylation (169)	Lys or Arg	Elongation factor EF2	This modification catalyzed by the diphtheria enzyme (165) inhibits protein biosynthesis and leads to cell death
Carboxylation (170, 171)	Glu	Prothrombin and other blood-clotting factors (VII, IX, X)	Modified glutamyl residues are necessary for normal Ca^{2+} binding by prothrombin and thus for the Ca^{2+}-mediated binding of prothrombin to phospholipid surfaces
Glycosylation (172, 173)	Asn, Thr, Ser	Coeroloplasmin	Loss of the terminal unit of carbohydrate in side chains of serum glycoproteins is the signal for uptake and degradation of these proteins in the liver
Hydroxylation (125, 127)	Pro, Lys	Collagen	Impairment of hydroxylation prevents collagen maturation; this is the molecular basis of scurvy
Methylation (174)	Asp, Gln, His, Lys, Arg	ϵ-N-Methyllysine in flagella proteins	Despite the wide-spread occurrence of this reaction no example which establishes a solid relationship between protein methylation and a particular function is known
Phosphorylation (175)	Ser, Thr	Phosphorylase	Inactive phosphorylase b is converted to active phosphorylase a by phosphorylation of a specific serine residue in each subunit

chain modifications in these proteins; unfortunately every biological classification has to be flexible.

Summary

Scientists tend to simplify in order to create a comprehensive picture. This has been done in the preceding chapters. From time to time, however, it is advisable to face real complexity. A polypeptide chain built from 20 amino acid types, folding spontaneously, and then acting as a globular protein, is *not* the whole story! It is common, for instance, that chains assemble and function as an aggregate. The observed properties of such aggregates suggest that the basic units of proteins are not individual chains but individual domains. In proteins with unknown geometry only functional domains can be defined; structural domains cannot. However, the known protein structures show that the structural domain is the more basic concept. The observed pattern of protein aggregates is so complex that even the definition of monomers and oligomers poses problems. We use the physiologically oriented definition, according to which monomers may contain more than one polypeptide chain if the polypeptide chains are linked covalently.

The comprehensive picture of proteins given in the preceding chapters also omits cross-links and epigenetic modifications. A very common cross-link is the disulfide bridge, which serves mechanical as well as chemical purposes. Frequently, mechanically important cross-links are formed using the ϵ-amino group of Lys. Among the epigenetic modifications main chain scissures are very frequent. They are a major physiological tool, providing the right protein at the right place at the right time. Side chain modifications are also common. They endow enzymes with new properties. All these phenomena should be recalled if one tries to draw specific conclusions from the rather general principles previously introduced.

Chapter 5

Patterns of Folding and Association of Polypeptide Chains

Six levels of structural organization can be distinguished. According to Linderström-Lang (176), four levels of structural organization in proteins can be distinguished: primary, secondary, tertiary, and quaternary structure (176). These terms refer to the amino acid sequence, the regular arrangements of the polypeptide backbone, the three-dimensional structure of the globular protein, and the structures of aggregates of globular proteins, respectively. With our present knowledge, two more levels can be added: supersecondary structures denoting physically preferred aggregates of secondary structure and domains referring to those parts of the protein which form well-separated globular regions. An organizational scheme is given in Figure 5-1a. Since renaturation experiments have shown that the amino acid sequence contains the entire structural information (177), the relationship between these levels is dependent, with elements at a lower level determining the elements of higher levels.

This relationship has been schematized in an extreme simplification in Figure 5-1b. This scheme assumes that all interactions are well-segregated; there is no interference from elements nonadjacent along the polypeptide chain, or from elements of other levels. The scheme resembles a hierarchic social system (with reversed polarity). In the following we will describe all levels and finally discuss the relationships between them.

5.1 Secondary Structure

Secondary structures are regular arrangements of the backbone of the polypeptide chain without reference to the side chain types or conformations. They are stabilized by hydrogen bonds between peptide amide and carbonyl groups.

Aggregate

↑

Globular protein

↑

Domain

↑

Supersecondary structure

↑

Secondary structure

↑

Amino acid sequence

(a)

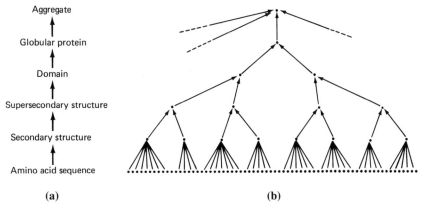

(b)

Figure 5-1. Levels of structural organization in globular proteins.

Linear Groups

The backbone of a polypeptide chain forms a linear group if its dihedral angles are repeated. Every linear group is a helix. If the regularity is such that successive peptide units assume identical relative orientations, that is, if all (ϕ,ψ)-angles (Figure 2-2) are the same, the polypeptide backbone forms a linear group. Every linear group is a helix. Helices are conveniently described by the rise per element d, the number of elements per turn n, and the distance r of a marker point on each element (here the C_α-atom) from the helix axis (Figure 5-2). Since d is taken as positive, the helix chirality can be read from the sign of n. Each helix has a polarity, because the peptide unit is polar.

The relationship between (ϕ,ψ)-angles (Figure 2-2) and helix parameters d and n is given in Figure 5-3. No helix with $\mid n \mid$ <2 is possible. There are only a few linear groups without steric hindrance that are stabilized by hydrogen bonds, either within a chain (e.g., α-helix) or between neighbor-

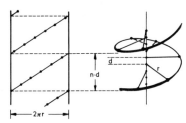

Figure 5-2. Helix with helix parameters and surface net representation: n = residues per turn, d = axial shift per residue, $n \cdot d$ = pitch of the helix, r = radius of the helix. The surface net is produced by projection of the helix onto a coaxial cylindrical sheet of paper, cutting this paper parallel to the helix axis and flattening it. The surface net is also called cylindrical plot. It allows visualization of the geometrical relationship between residues. Reversing the viewing direction onto the cut paper leads to the mirror image of the given cylindrical plot which is also in use.

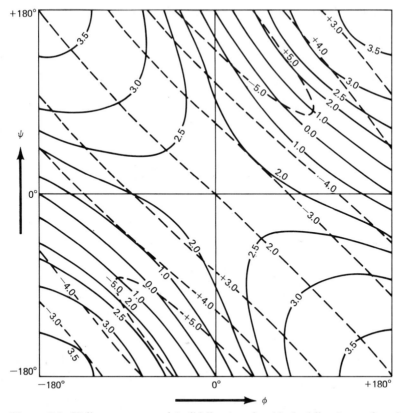

Figure 5-3. Helix parameters d (solid lines) and n (dashed lines) as a function of dihedral angles ϕ and ψ (29). Within the helix all ϕ and ψ are identical.

ing chains (e.g., β-pleated sheet, collagen). Parameters of prominent linear groups are given in Table 5-1.

Linear groups are also found in saccharides, nucleic acids, and protein aggregates. Linear groups are also found in polysaccharides, where the elements are mono- or oligosaccharides with and without side chains (178), and in RNA (DNA) where the ribose–phosphate–diester elements usually form a double helix by intertwining two antiparallel single helices (179). Tobacco mosaic virus consists of coat protein and RNA. It forms a linear group with one coat protein bound to a trinucleotide as the element (180). Cylindrical patterns of proteins as found for example in microtubules (181), T4-phage tails (182), or F-actin filaments (183) are often described as helices, or as linear groups. However, it should be kept in mind that such aggregates contain no linear chain, as, for example, the RNA connecting the protein subunits of tobacco mosaic virus, which defines a particular helix. Usually, there are several possibilities for describing such a pattern with single- and multi-start helices.

Table 5-1. Linear Groups Formed by Polypeptide Chains

Linear group	Observed	Residues per turn n and chirality[a]	Rise per residue d (Å)	Radius of helix r (Å)
Planar parallel sheet	Rare	± 2.0	3.2	1.1
Planar antiparallel sheet	Rare	± 2.0	3.4	0.9
Twisted parallel or antiparallel sheet	Abundant	− 2.3	3.3	1.0
3_{10}-Helix	Small pieces	+ 3.0	2.0	1.9
α-Helix (right-handed)	Abundant	+ 3.6	1.5	2.3
α_L-Helix (left-handed)	Hypothetical	− 3.6	1.5	2.3
π-Helix	Hypothetical	+ 4.3	1.1	2.8
Collagen-helix	In fibers	− 3.3	2.9	1.6

[a]Plus and minus correspond to right- and left-handed helices, respectively.

3_{10}-, α-, and π-Helices

Regular helices are fully described by a pair of dihedral angles. The structures of these helices are shown in Figure 5-4, and their parameters are given in Table 5-1. The 3_{10}-, α-, and π-helices are stabilized by hydrogen bonds between peptide amide and carbonyl groups of residues $(i, i+2)$, $(i, i+3)$, and $(i, i+4)$, respectively. Therefore, there are successive possibilities to form helices with hydrogen bonds between nearby chain elements. Since these helices are linear groups, they are fully described by a single point in the $\phi\psi$-map (Figure 2-3). These points represent local energy minima. Small deviations from these minima are to be expected.

The α-helix is very stable and therefore most abundant. The α-helix is the most abundant secondary structure in proteins. Therefore its conformation must be rather stable. This corresponds to its position in the center of an allowed region of the $\phi\psi$-map (Figure 2-3), and to the fact that the dipoles forming its hydrogen bonds are aligned, that is, in the minimum energy geometry (section 3.4, page 36). Moreover, the helix radius r (Table 5-1) allows for van der Waals attraction across the helical axis.

The α-helix was first postulated by Pauling et al. (25). At that time the concept of helices with nonintegral n values had to overcome a substantial intellectual barrier because it did not satisfy the expected high degree of molecular order. An immediate confirmation of this postulate came from X-ray analysis of crystalline hemoglobin (183). Later, α-helices were found in a number of fibers (184) such as α-keratin and paramyosin, and in almost all globular proteins. In fact, the notations "α" and "β" for α-helix and β-sheet originate in the classification of X-ray diagrams of fibers into α- and β-patterns, the main representatives of which are α-keratin and β-silk fibroin.

The distribution of observed α-helix lengths in globular proteins is given in Figure 5-5. The average length is about 17A, corresponding to 11

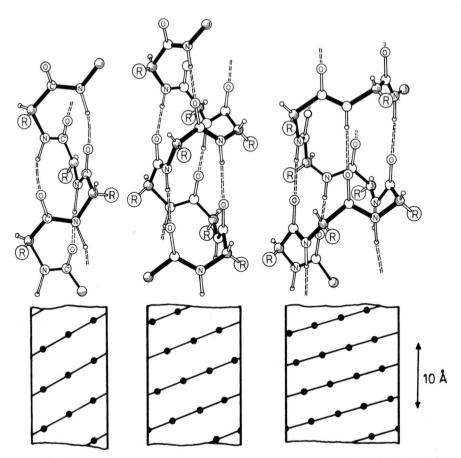

Figure 5-4. Helices of polypeptide chains with internal hydrogen bonds. From left to right: 3_{10}-, α-, and π-helix. Below: cylindrical plots of these helices; the course of the chain and the C_α-positions are marked.

residues or 3 helix turns. The large width of the distribution indicates that no length is preferred, which is in accordance with expected properties of a linear group. The relative maxima at lengths of 7, 11, and 15 residues, corresponding to 2, 3, and 4 full turns, respectively, are smaller than the error expected from a Poisson distribution ($\approx 25^{1/2} = 5$). Their significance is still questionable. α-Helices provide rather stable rods in protein structures. Kinks in these rods forming an angle of about 20° are often produced by incorporating Pro, as can be found in myoglobin (185) and adenylate kinase (186).

The 3_{10}-helix is rare; only short pieces are found. The 3_{10}-helix derives its name from the number of residues per turn $n = 3$ and from the 10 atoms in the ring closed by a hydrogen bond (187). The dipoles forming these hydrogen bonds are not aligned; they are not at minimum energy (section

3.4, page 36). The side chain packing is rather unfavorable. As shown in the cylindrical plot of Figure 5-4, side chains are on identical azimuthal positions, whereas they are well staggered in the α-helix. In the $\phi\psi$-map (Figure 5-7) the 3_{10}-helix lies at the edge of an allowed region; therefore some strain caused by steric hindrance is present. Its energetically disadvantageous geometry accounts for the rare occurrence of the 3_{10}-helix in proteins; usually pieces of about one turn are observed [sea lamprey hemoglobin has two 3_{10}-helices with two turns each (188, 189)]. These pieces tend to be at the N- and C-termini of α-helices.

The π-helix is unfavorable. The π-helix has never been observed. It is largely of systematic interest. It lies at an edge of an allowed region (Figure 2-3), giving rise to strain by steric hindrance (29,190). The dipoles forming its hydrogen bonds are reasonably aligned. With its large radius r (Table 5-1) the main chain atoms are no longer in contact across the helix axis. An axial hole is formed[1], reducing the van der Waals attraction energy appreciably. As shown in the cylindrical plot of Figure 5-4, the side chains of a π-helix are not as well staggered as in an α-helix, but they are staggered better than in a 3_{10}-helix. In conclusion, the absence of π-helices is explained by steric hindrance and by the reduction of van der Waals attraction across the helix axis.

No left-handed 3_{10}-, α-, or π-helix has been observed yet. Considering the polypeptide backbone alone, each helix has an energetically equivalent mirror image. However, side chain interactions disfavor the left-handed α_L-helix. This explains why it has never been observed in globular proteins. The same is true for left-handed 3_{10}- and π-helices.

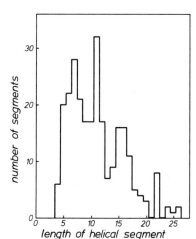

Figure 5-5. Length distribution of α-helical segments for 23 proteins [Data from Ref. (323)]. The length is given as the number of residues.

---

[1]Since the hole cannot be filled with water the binding energy is reduced by $p \cdot \Delta V$ (p = pressure, ΔV = volume held open). However, at atmospheric pressure this amounts to only 0.001 kcal/mol for a π-helix of 50 residues and is negligible.

Collagen Helix

Collagen is a superhelix formed by three parallel, very extended left-handed helices. The structure of the collagen helix is shown in Figure 5-6a. It is a right-handed helix consisting of three left-handed helical strands running in parallel. Thus it is a triple superhelix, that is a helix formed by helices. The X-ray analysis of collagen fibers has been intricate (191–194). The most important clue came from the synthetic polymers poly-Gly-II and poly-L-Pro-II (poly-L-Pro-I contains *cis* peptide bonds), the X-ray diffraction patterns of which resemble the collagen pattern. For Pro homopolymers model building is particularly restrictive because the dihedral angle ϕ is fixed around $-60°$; and no regular helix in the α-helix region ($\psi = -60°$) is possible because of steric hindrance. Thus, only the region around $(\phi,\psi) = (-60°, +140°)$ is accessible (Figure 2-3), revealing essentially the structure of the collagen single strand helix.

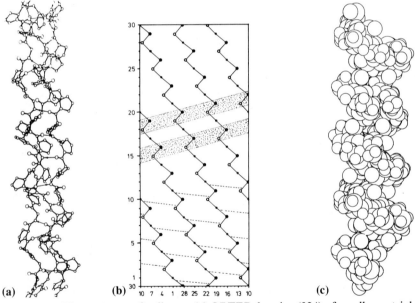

(a)   (b)   (c)

Figure 5-6. The structure of collagen. (a) ORTEP-drawing (324) of a collagen triple helix with the sequence $(Gly-Pro-Pro)_n$ and coordinates from Ref. (193). Hydrogens are only inserted if taking part in bonds (thin lines, almost horizontal). (b) Cylindrical plot of a full repeat of the collagen triple helix (325). The chain course and the C_α-atoms are indicated. The three C_α-atoms in the sequence $(Gly-X-Y)_n$ are marked by ($\cdot$ $\circ$ $\bullet$). Hydrogen bonds are given as dashed lines. (c) ORTEP-drawing of a collagen triple helix corresponding to (a), except that atoms are given with their van der Waals radii and that all hydrogen atoms are included. The helical bulge of Pro residues with a pitch of 9 A becomes visible. This bulge is also indicated by dotted stripes in (b).

The group element of the superhelix is a triplet. For a detailed analysis the knowledge of the amino acid sequence turned out to be decisive (195–197). About 96% of the chain obeys the formula $(Gly\text{-}X\text{-}Y)_m$. Position X frequently contains Pro and position Y frequently contains Hyp (hydroxyproline = Pro with an OH-group at C_γ; see Figure 1-1). The Pro and Hyp contents are each 11%. For some time three models, all with triple helices of parallel strands, were discussed (191,192). In particular, the accommodation of the frequent triplet (Gly-Pro-Hyp) finally excluded all models except for the so-called collagen-II structure (194).

The superhelix is best described using the cylindrical plot of Figure 5-6b. It is a linear group with a $(Gly\text{-}X\text{-}Y)$-triplet as group element. There are 10 triplets per strand in one superhelix turn; the superhelix pitch is 86Å. The three strands are held together by van der Waals forces and by one hydrogen bond per triplet. The dipoles forming these hydrogen bonds are aligned and approximately perpendicular to the superhelix axis. The hydrogen bond pattern is given in Figure 5-6b.

The single strand left-handed helix is roughly regular. The left-handed single strand helices are not regular in the sense of the word used above. The group element of the single strand helix is a triplet because the group element of the superhelix is a triplet. Not every but only every third (ϕ,ψ)-pair is identical. However, except for deviations of about 10°, the (ϕ,ψ)-angles of all residues are the same so that the single strand can be well approximated by a helix (linear group) with one residue as group element, and described by a single position in the $\phi\psi$-map (Figure 2-3). The parameters of the single strand helix are given in Table 5-1. With $d = 2.9$ Å this helix is so extended that no strand is stable by itself; it must aggregate with others.

Gly in every third position is required for stability. With the noninteger value $n = -3.3$ for the left-handed single strand helix and with the right-handed superhelix of 10 triplets per turn, every third C_α-atom is close to the superhelix axis. This explains the obligatory sequence $(Gly\text{-}X\text{-}Y)_m$ since a compact, hydrogen-bonded structure can be obtained only if there is no side chain at the axis that is if there is a Gly.

Collagen has a helical bulge. The external appearance of the collagen helix is dominated by the side chains at positions X and Y. This becomes particularly obvious in a space filling model of $(Gly\text{-}Pro\text{-}Hyp)_m$. As shown in Figure 5-6c, the side chains of Pro and Hyp form a helical bulge around the central cylinder with a pitch of 9 Å.

Collagen is the most abundant protein in mammals. With its fairly extended polypeptide chain collagen can withstand strong mechanical tension along its axis, because such tension is almost parallel to the covalent bonds of the backbone. This renders it very suitable for its task as a force transmitter in tendons or as a shielding texture in the skin and in other organs. Indeed, collagen is one of the most abundant proteins in nature.

Most vertebrate collagen triple helices consist of two α_1-chains and a homologous α_2-chain. Only the α_1-chain has yet been sequenced (195–197). It contains 1052 residues. Except for 16 N-terminal and 25 C-terminal residues the $(Gly\text{-}X\text{-}Y)_m$ formula is strictly obeyed. Guided by this formula collagen-like structures are easy to locate in amino acid sequences. No collagen-like structure has been found within a globular protein as yet. However, such a structure is very likely present in component C1q of the human complement system (198) which recognizes antibodies if they are bound to antigens. As determined from electron micrographs this protein contains a bundle of 18 parallel chains organized in six collagen-like fibers which feed into six globules. It is likely that more of such hybrid proteins containing collagen structures will be discovered.

Reverse Turns

The peptide chain can form a sharp reverse turn which contains a hydrogen bond. When searching for favorable conformations of three consecutive peptide units (C_α^i to C_α^{i+3}) Venkatachalam (199) found three arrangements with a hydrogen bond between O_i and N_{i+3}. They are called reverse turns I, II, and III. Reverse turn III is a piece of 3_{10}-helix which has been described above (see Figure 5-4). Reverse turn I is a deformed 3_{10}-helix; the regular (ϕ,ψ)-value of the helix has split into two different ones at C_α^{i+1} and C_α^{i+2} (Figure 5-7a). In contrast, the peptide unit between residues $i+1$ and $i+2$ of reverse turn II has flipped over (Figure 5-7b), and the (ϕ,ψ)-angles at C_α^{i+1} and C_α^{i+2} are quite different from each other (Figure 5-7a). This leads to severe steric hindrance between O_{i+1} and the side chain of residue $i+2$, so that reverse turn II requires a Gly at position $i+2$. The mirror images of reverse turns I, II, and III, denoted reverse turns I', II', and III', are disfavored because of steric hindrance.

Reverse turns are very abundant. A search for reverse turns in globular proteins showed that they are very abundant, comprising about one quarter of all residues (200,201). At closer inspection (202) the frequencies of reverse turns I, II, and III turned out to be 35, 15, and 15% of the total occurrence, respectively. As predicted, reverse turn II is found almost exclusively with Gly in position $i+2$. As expected, the total frequency of the mirror images (reverse turns I', II' and III) is low, namely 10%. Nonstandard reverse turns without hydrogen bonds occurred with a frequency of 25%. They are only defined by the more general criterion that the distance between C_α^i and C_α^{i+3} is less than 7 Å and that the chain is not in α-helical conformation. This criterion comprises all standard reverse turns but allows for a variety of other conformations. In some cases nonstandard reverse turns contain cis-Pro: Pro-93 and Pro-114 in ribonuclease (39), Pro-168 in subtilisin (40), and Pro-116 in staphylococcus nuclease (41).

Most reverse turns are at protein surfaces. When surveying the location of reverse turns in proteins, Kuntz (203) found that they are concentrated at the surface. Accordingly, they contain mostly hydrophilic residues. In the folding process reverse turns are assumed to play a passive role only. They

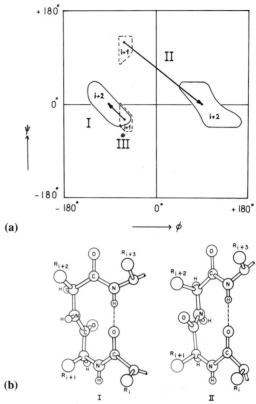

(a)

(b)

Figure 5-7. Reverse turns. (a) Regions for the dihedral backbone angles of residues $i+1$ (dashed lines) and $i+2$ (solid lines) in reverse turns of type I, II, and III as given in Ref. (199). The arrows connect corresponding areas. In the case of reverse turn III both areas are identical with each other and with the position of the 3_{10}-helix (Figure 5-4). (b) Reverse turns of type I and II (326). In type II there is steric hindrance between side chain R_{i+2} and O_{i+1} so that residue $i+2$ must be a Gly. For reverse turn III see Figure 5-4.

are points of least resistance to noncovalent forces trying to bend the chain. This assumption is supported by the large variety of observed reverse turns, demonstrating that none of them has a particularly stable conformation.

β-Pleated Sheet

Extended polypeptide chains can associate by hydrogen bonding to form sheet-like structures. Concomitantly with the α-helix, Pauling and Corey (204) postulated the (planar) parallel and antiparallel β-pleated sheets (Figure 5-8) as suitable regular hydrogen-bonded structures for polypeptide chains. In either type the chain forms a linear group with one residue as group element, the (helix) parameters of which are given in Table 5-1. In

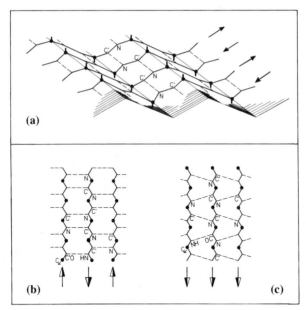

Figure 5-8. β-Pleated sheet. Hydrogen bonds are indicated by dashed lines and chain directions by arrows. C_α-atoms are marked by dots. (a) The pleated structure of a (nontwisted = planar) antiparallel β-sheet. The directions of the C_α–C_β bonds are indicated by short bars on the C_α-atoms; they are perpendicular to the sheet. (b) Hydrogen bond pattern of an antiparallel three-stranded β-sheet. (c) Hydrogen bond pattern of a parallel three-stranded β-sheet.

both cases the (ϕ,ψ)-angles are in an allowed region (Figure 2-3) and the dipoles forming the hydrogen bonds are aligned. The hydrogen bond patterns are sketched in Figure 5-8b and c. It can be seen, by viewing along a polypeptide strand, that the side chains point alternatively to either side of the sheet with the C_α–C_β bonds approximately perpendicular to the sheet plane (Figure 5-8a). Mixed parallel–antiparallel sheets are possible, although they require some deviations in the (ϕ,ψ)-angles.

Sheets are observed very frequently. Sheet structures were found in β-silk fibroin and in stretched hair (184). Globular proteins contain about 15% sheet structures (201). The lengths and widths histograms of sheets in globular proteins are given in Figure 5-9. The average length is about six residues (= 20 Å) corresponding approximately to the diameters of domains (see below). As shown in Figure 5-9a, strand lengths with uneven residue numbers are more frequent than others. This occurs because the hydrogen bond pattern (Figure 5-8b and c) leaves all strands at sheet edges with an uneven number of residues. Most sheets contain less than six strands with no preference for any particular number. The width of a six-stranded sheet is about 25 Å, corresponding to typical domain diameters. No preference of parallel or antiparallel sheets is observed, but parallel sheets with less than four strands are rare. Mixed sheets are frequent but much less frequent than is to be expected for random mixing of strand directions. Thus, sheet strands show some cooperativity; they tend to be either all parallel or all antiparallel.

Most β-pleated sheets contain a left-handed twist. With $n = \pm 2.0$ (Table 5-1) the sheet structure postulated by Pauling and Corey (204) is planar. Such a planar (antiparallel) sheet is found in glutathione reductase, for example (124). However, most observed sheets are nonplanar (43,205); they have a left-handed twist when viewed along the sheet plane perpendicular to the strands as in Figure 5-10d (that is, a right-handed twist when viewed in the direction of the sheet strands). To a good approximation a single strand of a twisted sheet forms a linear group with one residue as element. This is a very extended left-handed helix, the (ϕ,ψ)-angles and helix parameters of which are given in Figure 2-3 and Table 5-1, respectively. As sketched in Figure 5-10b, this left-handed helix corresponds to a right-handed rotation of carbonyl and amide groups of about 60° per two residues. Therefore, hydrogen bond formation between neighboring strands can be accomplished only if the strands form an angle of about 25° with each other (Figure 5-10c). This leads to the observed sheet twist. The twisted sheet length is not limited. Presumably β-silk fibroin contains very long twisted ribbons of β-pleated sheet.

The geometry of a broad (say six-stranded) twisted sheet fits approximately the large groove of a DNA double helix (206). Furthermore, a twisted antiparallel sheet of two strands fits approximately the small groove of RNA (207) and DNA (208). Thus, twisted sheets may play a major role in protein-double helical DNA (RNA) interactions (see section 10.6, page 231).

Frequently, the twist is further accentuated locally by a so-called β-bulge, where an additional residue is inserted into one of the β-strands. This feature is found only in antiparallel sheets (209).

The twist is a consequence of local optimizations. The preference for twisted sheets as compared to flat ones can be understood. In the $\phi\psi$-map (Figure 5-10) all three types of sheet are in an allowed region. As shown in

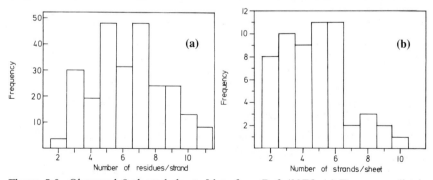

Figure 5-9. Observed β-pleated sheets [data from Ref. (327)]. (a) Frequency distribution of strand lengths. (b) Frequency distribution of the width of β-sheets.

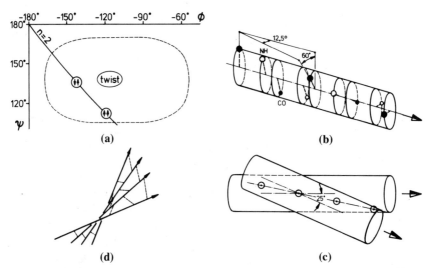

Figure 5-10. Twist of β-pleated sheets. (a) Part of $\phi\psi$-map with dihedral backbone angles for nontwisted parallel (⤩), nontwisted antiparallel (⤩), and left-handed twisted β-pleated sheet. The dashed line is the potential curve of 1 kcal/mol from Figure 2-5. (b) A single strand of a twisted sheet is twisted by itself. The backbone course is indicated by a dashed line. The directions of carbonyl groups • and amide groups ○ are indicated. (c) Two parallel chains, which are twisted as sketched in (b), can be superimposed so as to form hydrogen bonds ⊙. For this purpose they have to be tilted by 25°. (d) Schematic drawing of a parallel sheet with the usually observed left-handed twist.

Figure 5-10a, the energy E is a rather flat function of the (ϕ,ψ)-angles[2]. Hydrogen bonding between strands does not change the shape of the energy function appreciably, because hydrogen bonds retain their lengths and dipole alignment over a large (ϕ,ψ)-region. Therefore, at normal temperature the statistical weights [proportional to $\exp(-E/RT)$, see Appendix] of all chain conformations in this region are of comparable magnitude, and there is no reason to favor any particular one. With almost no main chain restrictions, now the optimization of the local side chain packing determines the exact (ϕ,ψ)-value which a residue assumes within this region. Since the effect of this optimization is almost random it does not lead to a preference and the distribution of assumed (ϕ,ψ)-angles is uniform. Consequently, the average conformation will be at the center of this region, corresponding to an overall twist of the sheet. This explanation is corroborated by the observation at the (ϕ,ψ)-angles in β-sheets show a very broad distribution, spreading over most of the region indicated in Figure 5-10a.

---

[2]The energy E is related to H_{chain} in the vacuum situation depicted in Figure 3-3 [see Eq. A-18].

5.2 Supersecondary Structures

We have seen that proteins contain regular backbone arrangements, secondary structures, which are physico-chemically favored. At the next higher level of complexity we find aggregates of secondary structures. The structures of such aggregates, so-called supersecondary structures, are difficult to predict.[3] The aggregates can be recognized if they occur frequently in proteins. Moreover, frequent occurrence indicates that they are favored for kinetic reasons during the folding process and/or for energetic reasons in the folded protein.

Coiled-Coil α-Helix

In its most regular form the coiled-coil α-helix occurs in fibrous proteins. A clear example of a supersecondary structure is the coiled-coil α-helix postulated by Crick (210). In this structure two α-helices are wound around each other, forming a left-handed superhelix with a repeat distance of about 140 Å (Figure 5-11a). Coiled-coil α-helices have been found in the fiber proteins α-keratin (211, 212), tropomyosin (213), paramyosin (214), and light meromyosin (215). Short pieces of this supersecondary structure are observed in globular proteins which contain α-helices packed in an approximately parallel or antiparallel way. Prominent examples for helices packing approximately in-line are hemerythrin (216, 217), tobacco mosaic virus coat protein (180, 218), bacteriorhodopsin (219), coat of bacteriophage fd (220), and tyrosyl transfer RNA synthetase (221).

The superhelix allows favorable side chain meshing without much α-helix distortion. The coiled-coil α-helix structure can be derived by superimposing cyclindrical plots of two adjacent α-helices. As shown in Figure 5-11b, this can be done in such a way that a contact line arises, along which side chains from both α-helices are regularly meshed. Meshing along the contact line can be prolonged indefinitely if both α-helices form a left-handed superhelix with the contact line as the straight axis. The superhelix repeat can be derived from the repeat of the contact line in one of the cylindrical plots, from that length of a cylindrical plot which allows the contact line to cross the plot from one margin to the other one. The repeat is about 140 Å. The angle between α-helices is around 10° (Figure 5-11b). The (ϕ,ψ)-angles deviate slightly from the values for a straight α-helix. The structurally repeating unit in each chain is a heptapeptide (Figure 5-11c); every seventh residue is at an equivalent position at the superhelix axis (at the contact line of Figure 5-11b). This 7.0 residue repeat is slightly smaller than the $2 \cdot 3.6 = 7.2$ residues per two turns of a straight α-helix. If the amino acid

---

[3]In a narrow sense of the definition, supersecondary structures comprise β-pleated sheets, because sheets are associated strands with regular backbone arrangements. However, usually only regularities in the strand arrangements in β-sheets are denoted as supersecondary structure.

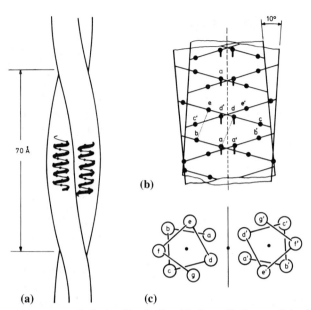

Figure 5-11. Coiled-coil α-helix. (a) Overall shape of the left-handed supercoiled helix with a repeat of 140 Å. The size of the individual α-helices is indicated. Both helices are parallel. (b) Cylindrical plots of the two parallel α-helices of (c) as projected onto the interface between them. The central axis of the superhelix is dashed. C_α-atoms are indicated as dots. Because of the 10° tilt all contacts aa', dd', etc. are aligned at the central axis of the coiled-coil structure. The side chain of d', etc. (indicated by short bars pointing opposite to the helix direction such as the C_α–C_β bond) fits in between the side chains of a, e, d and of the repeated a. Thus all side chains at the meshing lines a, d, etc. and a', d', etc. fit like knobs into holes. (c) Cross-section of a parallel coiled-coil α-helix. C_α-positions are labeled alphabetically along the polypeptide chains, with a and a' at the same height.

sequence of a chain is described by the formula $(abcdefg)_m$ all residues a (or b,c,d,e,f,g, respectively) are at structurally equivalent positions (Figure 5-11c).

The coiled-coil α-helix is energetically favorable because meshing of side chains provides for a broad and intimate contact (and appreciable van der Waals binding energy) between α-helices. If the meshed side chains are hydrophobic the free energy of this structure is particularly low, because at the superhelix axis these hydrophobic side chains are removed from the surrounding solvent.

In fact, the known amino acid sequences of tropomyosin (222) and α-keratin (212) show a pattern which allows for hydrophobic residues at the meshing positions: positions a and d (Figure 5-11c). Polar residues are usually found at the outer surface: positions b, c, and f. In tropomyosin positions e and g frequently contain charged residues forming salt bridges with their counterparts g' and e', respectively.

The model fits parallel as well as antiparallel α-helices. All arguments given above apply equally well to parallel and antiparallel α-helices. For the meshing system given in Figure 5-11b, however, only a parallel arrangement is favorable; since the C_α–C_β bond of an α-helical residue is not perpendicular to a α-helix axis but points backward forming an angle of about 45°, the side chains point backward also. Only in the parallel case shown in Figure 5-11c will the side chains mesh well. In the antiparallel case the meshing system is slightly different from Figure 5-11b. Here, a relative shift of about 1 Å along the contact line is required to release steric hindrance between side chains. If, for example, the helix with residues *abcde* in Figure 5-11c is inverted, a 1-Å shift downward releases the hindrance between side chains a and d'. Concomitantly, side chains *a'* and *d* come closer together, but no steric hindrance occurs because they are pointing in opposite directions. In conclusion, no relative helix direction should be preferred.

For tropomyosin (213, 223, 224) and light meromyosin (215) chemical data in conjunction with X-ray and electron microscope data have shown that the α-helices are parallel. Presumably this also applies for α-keratin, because the long α-keratin chain would have to be totally synthesized and stabilized before it could form a superhelix of antiparallel α-helices. In contrast, helix packing in the globular proteins hemerythrin (216, 217) and tobacco mosaic virus coat protein (180, 218) is antiparallel.

Three other preferred relative helix orientations at a contact between two α-helices have been proposed (225). The preferred orientations correspond to favorable side chain meshing as derived from cylindrical plots. However, the support from experimental data is not very strong.

βξβ-Unit

The βξβ-unit is usually right handed. Another feature widely accepted as a supersecondary structure is the βξβ-unit (also called βxβ-unit) , two parallel strands of a β-sheet with a connection ξ in between. If the connection is a nonregular chain (coil), the unit is denoted βcβ; if it is an α-helix or a third β-sheet strand on a different sheet, it is called βαβ or βββ. Such units have a chirality as shown in Figure 5-12a. Almost without exception the right-handed βξβ unit has been observed in globular proteins (226–228). This demonstrates, beyond doubt, a preferred way of chain folding and a preferred form of secondary structure aggregation, that is, a supersecondary structure.

The preferred chirality can be explained by the asymmetric steric hindrance in a random polypeptide chain. The right-handed chirality is so strongly favored that an explanation should be attempted. As discussed on (page 78) the statistical weights of a polypeptide chain favor the (ϕ,ψ)-angles found in twisted β-sheets (Figure 5-10a). Therefore, an extended chain in solution will in general assume the same (ϕ,ψ)-values; it will have

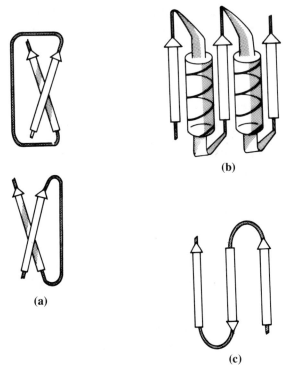

(b)

(a)

(c)

Figure 5-12. Supersecondary structures involving β-sheet strands. Arrows indicate sheet strands. (a) The βξβ unit. A left-handed unit (above) and a right-handed (below) connection ξ between two parallel strands of a sheet, which has a left-handed twist (Figure 5-10d). The right-handed connection is usually observed. (b) Rossmann-fold consisting of two consecutive right-handed βαβ-units. (c) A β-meander, defined as an antiparallel sheet of three strands.

the right-handed twist shown in Figure 5-10b. If such a chain forms a loop the overall right-handed twist will be increased for a left-handed loop and decreased for a right-handed loop. This relationship is illustrated in Figure 5-13. Increasing the twist means increasing ϕ and ψ as can be seen at the relation between planar and twisted sheets. When starting from the (ϕ,ψ)-angles for a twisted sheet it can be seen in Figure 2-5 that increasing ϕ and ψ

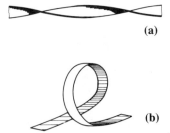

(a)

(b)

Figure 5-13. Sketch of a folding polypeptide chain. A right-handed twist along the polypeptide chain is released if a right-handed loop is formed. Try this with a belt or with a strip of paper! (a) Paper strip with right-handed twist. (b) A relaxed right-handed loop is formed if the ends of the paper strip shown in (a) are pushed together.

(going to the upper right-hand side) is much more restricted than decreasing ϕ and ψ (going to the lower left-hand side). Thus, in general, a decrease of the twist is favored and the loop becomes right-handed. After having started as a right-handed loop during the folding process, of course, it is difficult to reverse the handedness. Thus a right-handed $\beta\xi\beta$-unit is formed.

A combination of two consecutive $\beta\alpha\beta$-units is found frequently. A structural pattern which can also be denoted as a supersecondary structure is the ''Rossmann-fold'' (229) shown in Figure 5-12b. It is a special case of a $\beta\xi\beta$-unit as it consists of two adjacent $\beta\alpha\beta$-units. This $\beta\alpha\beta\alpha\beta$-unit may contain a hydrophobic core between sheet and helices. It has been found in a number of proteins (186, 230–240).

β-Meanders and Other Preferences in β-Sheets

A β-meander forms as many hydrogen bonds per residue as an α-helix of average length. In a number of proteins, as for example T4-lysozyme (241), staphylococcus nuclease (242), Ser-proteases (18, 243–246), and the catalytic domain of lactate dehydrogenase (232), β-sheets occur with three adjacent antiparallel strands and rather short connections. This feature is called a β-meander; it is sketched in Figure 5-12c. The β-meander can be classified as a supersecondary structure because it is very frequent; it is found in half of all consecutive antiparallel triple strands (247). This observation is related to the strong neighborhood correlation (section 5.3, page 86) of antiparallel sheets (Figure 5-15a). Moreover, with reverse turns at both ends, a β-meander forms about two-thirds of all possible hydrogen bonds between backbone atoms in a given piece of chain, which corresponds to the amount of hydrogen bonding in an α-helix of the average length of three turns (Figure 5-5). Therefore, with respect to free energy and also with respect to neighborhood correlation, the β-meander is comparable to an α-helix. This explains its frequent occurrence.

In β-sheets certain strand orderings are favored. The strong neighborhood correlation (Figure 5-15a) reflects some preferences in β-sheets. These preferences have been further analyzed by Richardson (247) who determined the frequencies of all patterns of two and three consecutive β-strands that occur in β-sheets with more than four strands. The frequency distribution is non-uniform, the β-meander being the most frequent pattern of three consecutive strands.

More interesting, however, is the frequency distribution of β-sheet topologies.[4] Here a meaningful empirical distribution is not yet at hand because the data base is too small. For all topologically distinct five stranded β-sheets this distribution was therefore calculated by constructing such sheets from all possible patterns of two and three consecutive β-

---

[4]The topology of a β-sheet is defined by the ordering of the strands along the chain, by the directions of these strands in the sheet, and by the handedness of $\beta\xi\beta$-units.

strands and by combining the observed frequencies of these patterns. The resulting frequency distribution is also non-uniform; certain sheet topologies are predicted to occur much more frequently than others. Such frequencies are important for detecting supersecondary structures or structural preferences, and also for evaluating the significance of a structural similarity (section 9.6, page 204).

5.3 Structural Domains

Structural domains are outlined by overall clefts in the electron density map. Inspection of electron density maps calculated in the course of X-ray analysis made it obvious that a number of proteins consist of several globular parts which are connected only loosely with each other. These parts are outlined by overall clefts in the electron density distribution. They have been denoted by the rather unspecific expression "structural domain." Clearly, the definition of a domain is ambiguous, and there are various disputable borderline cases. The immunoglobulins are globular proteins with an exceptionally obvious domain structure. They are shown schematically in Figure 4-2c. Here, the domains are ordered along the polypeptide chain like pearls on a string.

Neighborhood Correlation

Residues that are far apart along the chain tend to be far apart in the three dimensional structure. As shown in Figure 4-2c, the polypeptide chain can be divided into consecutive pieces which belong to consecutive domains (248). In this way residues that are far apart along the chain are also at a long geometric distance. This principle has been observed in all globular proteins containing clearly detectable domains (Table 5-2). Conversely expressed, the domain structure shows a high degree of "neighborhood correlation," the distance along the chain correlating with the geometric distance in a positive manner.

A neighborhood correlation plot may reveal structural domains. This has been evaluated quantitatively for chymotrypsin where a separation into two domains is not very obvious in the electron density map. As a measure we took the sum over all reciprocal geometric distances between pairs of residues that are 6 to 25 positions apart along the chain:

$$\text{neighborhood correlation } (i) = \sum_{6 \leq |i-k| \leq 25} \frac{1}{\text{distance}(i,k)}. \qquad (5\text{-}1)$$

The result is given in Figure 5-14a. Two peaks with an intermediate valley emerge, demonstrating that there are two regions in which the folding process has brought residues geometrically close together which are also

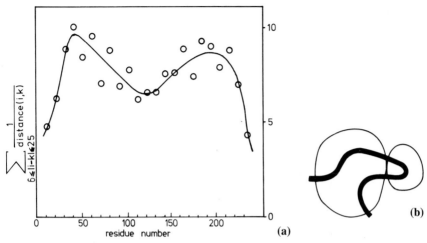

Figure 5-14. Neighborhood correlation in polypeptide chains. (a) A measure of neighborhood correlation as a function of residue number *i* in chymotrypsin. The values of 10 residues are averaged to smoothen the curve. The two-domain structure sketched in Figure 5-17d is clearly outlined. (b) Domain structure of adenylate kinase. A small and a large structural domain are linked by two polypeptide strands. The smaller domain has a stronger neighborhood correlation than the larger domain.

close along the chain. These regions should be classified as domains. A comparison with the three-dimensional structure shows that each domain contains one of the β-sheet barrels shown in Figure 5-17d.

Consecutive strands of β-sheets tend to be adjacent. Pronounced neighborhood correlation does not exist only at the domain level; it is also a generally observed phenomenon within a domain. A clear and quantitative example is the distance histogram of β-sheet strands given in Figure 5-15a. In particular, antiparallel sheet strands show a strong neighborhood correlation, with the chain often forming just two or more successive β-meanders when running through the sheet. In parallel sheets the correlation is lower but is still exceptionally significant.

Neighborhood correlation can be understood from a kinetic and from a thermodynamic standpoint. The impression is that the observed neighborhood correlation is a consequence of the chain folding process. Since there are many more conformations bringing residues together which are distant along the chain than there are conformations bringing residues together which are close along the chain, the folding process is bound to dwell on strong neighborhood correlation. Otherwise, with too many conformations to be sampled at an initial stage, a distinction between correct and false conformations would become extremely difficult, and the folding process would tend to run astray.

In the language of thermodynamics the preceding argument translates as follows. On folding the chain entropy has to be reduced (section 3.5). The

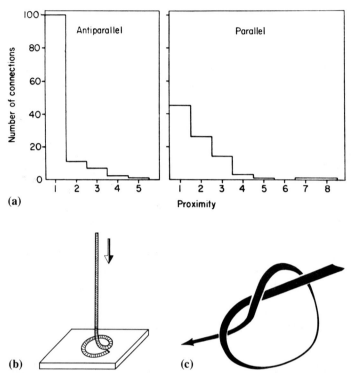

(a)

Figure 5-15. Structural preferences in globular proteins. (a) Observed neighborhood correlation in antiparallel and parallel β-pleated sheets. The number of observed connections from a given strand in a sheet (in chain direction) to its first, second . . . neighboring strand is represented. Data from Ref. (327). (b) A string is held at one end above the floor and allowed to fall down. The string will not entangle. Furthermore, it will have a high neighborhood correlation. (c) Knots in a polypeptide chain; not yet observed.

initial entropy reduction is smallest if the chain fold remains close to the conformation of an average random chain in solution, this is, if it forms a structure with a strong neighborhood correlation. With this small entropy reduction there exists no initial (activation) entropy barrier, which would be difficult to overcome within a reasonable time. Thus, the chain can fold smoothly without requiring too much binding energy and hydrophobic forces (see also section 8.3, page 156).

The observed neighborhood correlation suggests that a folding chain behaves like a string that is allowed to fall in the manner shown in Figure 5-15b. The resulting coil on the floor is not random but shows neighborhood correlation. It does not entangle and it can be easily unraveled by picking up the end. This concept is confirmed by the absence of "knots" in all structures known so far, the term knot being used in the every day sense (Figure 5-15c) and not in a mathematical sense (mathematical knots exist only in closed strings).

Structural domains are supposed to be the folding units of a protein. Because of the neighborhood correlation domains are expected to be those pieces of a long chain which fold independently of each other. If this were true, structural domains could be defined as the folding units of a chain, that is, in a more direct and less ambiguous manner than above. This notion is supported by the structure of chymotrypsin (18). This protein contains 13 water molecules between the two domains inside the molecule (Figure 5-14a). Presumably the domains fold separately and trap the water molecules during a subsequent domain-association step. The preceding argument is matched by the observation that *functional* domains (as detected by biochemical assays) indeed fold independently from each other (76, 77).

Conspicuously, most of the larger proteins split into several structural domains, containing 100–150 residues, which corresponds to a globule of about 25 Å diameter (Table 5-2). From a free energy point of view the restriction of domain size is a puzzle, because one large globule has a much smaller surface/volume ratio than several small ones. Thus, it can form a large hydrophobic core and many internal hydrogen bonds, both of which are energetically favorable. Presumably the size limitation is necessary to keep the folding process simple by keeping the length of the independently folding unit small enough.

In some cases as in adenylate kinase (186) two regions are observed which could be called domains and which are connected by two strands instead of only one. As shown in Figure 5-14b, the neighborhood correlations of such domains differ appreciably. Therefore, the domain with the stronger neighborhood correlation should fold first and it should subsequently be used as a scaffold helping to fold the other domain.

Structural Classes

Structural domains can be classified according to their secondary structures. The known domain structures can be grouped into the five classes given in Table 5-2 according to type, amount, and arrangement of their secondary structure (249, 250). Clearly, the outstanding classes are the domains containing only a single type of secondary structure. Here, the α-helical domains are expected to undergo the simplest folding process. Helix formation probably takes place at an initial stage of folding. Thus, the only remaining task is to pack α-helical rods in an energetically favorable manner.

All pure β-sheet domains (class No. 2) contain antiparallel sheets giving rise to a strong neighborhood correlation (Figure 5-15a). They show a high degree of order. Less regular structures are found in class No. 3, consisting of domains in which α- and β-structures tend to be segregated along the

Table 5-2. Domain Structure and Structural Classes in Globular Proteins[a]

Structural class	Structural domain in relation to the globular protein	Number of residues in the domain	Protein	Number of domains per globular protein	Number of connecting strands between domains
No. 1 Only α-helices	Total	104	Cytochrome c (271–273)	1	
	Total	158	TMV-coat protein (180, 218)	1	
	Total	108	Parvalbumin (59)	1	
	Total	ca. 145	Hemoglobin (188, 274–276)	1	
	Total	153	Myoglobin (185, 277)	1	
	Total	ca. 120	Hemerythrin (216, 217)	1	
	Total	51	Insulin (259)	1	
	Total	29	Glucagon (278)	1	
	C-terminal	111	Papain (279)	2	1
	C-terminal	160	Thermolysin (280)	2	1
	C-terminal	90	T4-lysozyme (241)	2	1
	N- and C-terminal	ca. 100, ca. 100	Tyr–tRNA-synthetase (221)	3	1
No. 2 Almost exclusively β-sheet	Total	237	Concanavalin A (281, 282)	1	
	Total	127	Prealbumin (206, 283)	1	
	Total	54	Rubredoxin (284, 285)	1	
	Total	151	Superoxide dismutase (286,800)	1	
	Total	62	Erabutoxin b (287)	1	
	Total	113	Subtilisin inhibitor (798)	1	

N-terminal	Papain (279)	110	2	(1)
N- and C-terminal	Chymotrypsin (18)	128, 114	2	1
N- and C-terminal	Elastase (243)	110, 119	2	1
N- and C-terminal	Trypsin (244, 245)	107, 116	2	1
N- and C-terminal	Proteinase-B (246)	80, 106	2	1
N- and C-terminal	Acid protease (288–290)	175, 145	2	1
N-terminal	Alcohol dehydrogenase (234)	175	2	1
All	Immunoglobulin (291–294)	ca. 110	2,4	1

No. 3
α-Helix and β-sheet tend to be segregated along the chain

Total	Carbonic anhydrase (295, 296)	258	1	
Total	Cytochrome b_5 hbf (297, 298)	85	1	
Total	High potential iron protein (299)	85	1	
Total	Hen egg white lysozyme (260, 300)	129	1	
Total	Pancreatic trypsin inhibitor (269)	56	1	
Total	Ribonuclease (39, 301, 302)	124	1	
Total	Staphylococcal nuclease (242)	149	1	
Total	Bacteriochlorophyll protein (303)	ca. 330	1	
N-terminal	Thermolysin (280)	156	2	(1)
N-terminal	T4-lysozyme (241)	74	2	(1)
C-terminal	Lactate dehydrogenase (232)	150	2	1

Table 5-2. Domain Structure and Structual Classes in Globular Proteins[a] (Continued)

Structural class	Structural domain in relation to the globular protein	Number of residues in the domain	Protein	Number of domains per globular protein	Number of connecting strands between domains
	C-terminal	170	s-Malate dehydrogenase (233)	2	1
	C-terminal	118	Glutathione reductase (124)	3	
No. 4 α-Helix and β-sheet tend to alternate along the chain	Total	194	Adenylate kinase (186)	1	
	Total	307	Carboxypeptidase (43)	1	
	Total	140	Flavodoxin (237, 238)	1	
	Total	275	Subtilisin (239, 240)	1	
	Total	108	Thioredoxin (304)	1	
	Total	247	Triose phosphate isomerase (305)	1	
	Total	ca. 240	Bacterial aldolase (306)	1	
	Total	ca. 230	Phosphoglycerate mutase (307)	1	
	Total	159	Dihydrofolate reductase (308)	1	
	N-terminal	181	Lactate dehydrogenase (232)	2	(1)
	N-terminal	ca. 150	s-Malate dehydrogenase (233)	2	(1)
	C-terminal	199	Alcohol dehydrogenase (234)	2	(1)

	Domain position	Number of residues	Number of connecting strands	Crossover
Glyceraldehyde-3-phosphate dehydrogenase (230, 231)	N- and C-terminal	147, 187	2	1
L-arabinose binding protein (309)	N- and C-terminal	ca. 150, ca. 150	2	(1)
Phosphoglycerate kinase (235, 310, 311)	N- and C-terminal	ca. 180, ca. 175	2	
Pyruvate kinase (80, 312)	N- and C-terminal	ca. 250, ca. 150	3	
Rhodanese (257, 799)	N- and C-terminal	ca. 130, ca. 130	2	
Glucose phosphate isomerase (313)	N- and C-terminal	ca. 360, ca. 150	2	1
Hexokinase (261)	N- and C-terminal	ca. 230, ca. 230	2	
Glutathione reductase (124)	N-terminal and central	210, 134	3	
Tyr-tRNA-synthetase (221)	Central	ca. 200	3	
No. 5				
No α-helix or β-sheet				
Ferredoxin (314)	Total	54	1	
Phospholipase (315)	Total	124	1	
Wheat germ agglutinin (316)	All	41	4	1

[a]Where the number of connecting strands is not given for multiple domain proteins, the chain tracing is not yet confirmed by amino acid sequence data. The number of connections is given as (1) if N- or C-terminal chain ends of about a dozen residues cross over to the other domain. Classes No. 3 and No. 4 are also called "α + β" and "α/β", respectively (249).

chain. Class No. 4 contains predominantly parallel β-sheets in their centers and α-helices surrounding them so that α- and β-structures tend to alternate along the chain. The Rossmann-fold (Figure 5-12b) belongs to this class. Domains with a low amount of secondary structure have been lumped into class No. 5.

Clearly, this classification is important for the explanation of the folding process. Presumably, this task is more difficult the higher the class number, that is, the lower the chain regularity.

Symmetry

Structural repeats within a polypeptide chain are frequent. Structural repeats exist in a number of proteins. The repeating unit can be a domain, a supersecondary structure, or other structural motifs. A list of examples is given in Table 5-3. In some proteins the repeating units have no symmetric relationship, as the domains of Ser-proteases (Figure 5-17d) or the cofactor-binding domains of glutathione reductase (124). Conspicuously, however, the repeating units are often approximately symmetrical to each other, suggesting that symmetry is a favorable feature.

For a large number of identical units the state of lowest free energy is a crystal. Symmetry preference in proteins can be understood. For this purpose consider crystals, in which symmetry is a consequence of favorable packing. The packing depends on the shape of the repeating unit and on its surface properties. Since there is one particular packing among all the possible packing schemes that is energetically the most favorable, the state with the lowest free energy is achieved, if this scheme can be established for all units. Invariably, this state corresponds to a crystal belonging to one of the 65 space group symmetries allowed for asymmetric elements since crystallization is the only way to aggregate an infinite number of units while forming identical contacts throughout.

For a small number of associated units point group symmetry is energetically favorable. If structures of limited size have to be built, the same principle applies. Among all contacts between identical units there is an energetically most favorable one. If this contact is used throughout, a helix, a linear group of *un*limited size is formed. Limited sizes are obtained if steric hindrance stops the helix in its first turn (Figure 5-16b). In this case, however, the last contact and therefore the whole structure is energetically disadvantageous.[5] An energetic optimum is reached if the ring is closed exactly, if symmetry is established. Consequently, symmetrical arrangements are also preferred for aggregates of limited size. The above argument has been given for point group n (Figure 5-16b), a symmetrical arrangement

---

[5] An open structure of four units forming three very similar contacts has been observed with wheat germ agglutinin (253). This shows that a particularly strong contact may be more favorable if made three times than an alternative contact which establishes symmetry and which is therefore made four times.

Table 5-3. Repeating Structures in Globular Proteins[a]

Protein	Repeating part	Number of repeating parts	Approximate symmetry
Chymotrypsin, elastase, trypsin, proteinase-B	Six-stranded antiparallel sheet barrel	2	None
	β-Meander within a barrel		2
Parvalbumin	Ca^{2+}-binding site and two α-helices	2 +1	2 None
Ferredoxin	FeS-cluster and 27 residues	2	2
Glyceraldehyde-3-phosphate dehydrogenase, lactate dehydrogenase, s-malate dehydrogenase, alcohol dehydrogenase, phosphoglycerate kinase, phosphorylase	Rossmann-fold of three parallel sheet strands and two connections	2	2
Triose phosphate isomerase, pyruvate kinase	$\beta\alpha\beta$-Unit	8	8
Wheat germ agglutinin	41 Residues with 4 S–S bridges	4	Linear group
Hemerythrin	Two antiparallel α-helices	2	2
Rhodanese	One domain	2	2
Immunoglobulin	V-domain	4 per molecule	2
	C_H1 and C_H3 domains	6 per molecule	2
Glutathione reductase	Four-stranded parallel sheet and β-meander	2	None
Acid protease	One domain	2	2
L-arabinose binding protein	One domain	2	2

[a]For references see Table 5-2

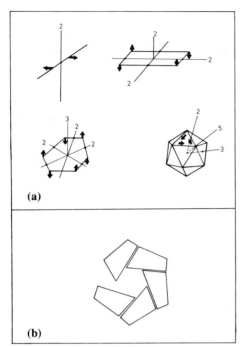

Figure 5-16. Point symmetries. (a) Representations of point groups 2 (C_2), 222 (D_2), 32 (D_3), and 532 (Y, icosahedron). The numbers denote twofold, threefold, etc. rotation axes. (b) Asymmetric elements forming a preferred contact with each other. In general such preferred contact does not lead to a point group. Also, the formation of a linear group can be inhibited by steric hindrance. In the example given the binding energy of the fifth contact is not available and thus cannot contribute to stability. Because of the additional binding energy an exact closure, giving rise to symmetric structures, is energetically favored.

around a point with an n-fold rotation axis as symmetry operator. It can be generalized for all point groups of asymmetric elements: *n, n2, 23, 432, 532* (251, 252), some of which are illustrated in Figure 5-16a.

Symmetry may reveal structural units that fold separately. In a polypeptide chain we have a structure of limited size. If in the folding process structural domains, supersecondary structures, or other structural motifs are independently formed at an initial stage and aggregated subsequently, they obey the rules for association of separate units; they will preferentially assume a symmetrical arrangement.

Since structural domains are expected to be folding units, symmetry between domains as observed in rhodanese (Figure 5-17a) or in immunoglobulins (Table 5-3) seems to be explained. In symmetrically related supersecondary structures it can be assumed that the units are essentially formed at an initial stage of the following process since they are reasonably stable by themselves. Subsequent aggregation then prefers symmetry.

Symmetrical supersecondary structures are found as the $\beta\alpha\beta\alpha\beta$-unit of dehydrogenases (Figures 5-12b and 5-17b), as the $\beta\alpha\beta$-units of triose phosphate isomerase and pyruvate kinase (Figure 5-17e), as the β-meanders in the sheet barrels of Ser-proteases (Figure 5-17d), and as antiparallel pairs of α-helices in hemerythrin.

In ferredoxin (Figure 5-17c) the polypeptide chain can be divided into two halves that are symmetrically related to each other. Each half of the chain wraps around one FeS-cluster. These clusters have also a symmetrical relationship. Here the observed symmetry reveals that the structural motif consisting of a FeS-cluster together with its surrounding polypeptide chain is rather stable by itself. A similar situation is found with parvalbumin which contains two symmetrically related structural motifs, each consisting of a Ca^{2+}-binding site and two α-helices extending from this site into opposite directions (59). At first sight such an extended structure is not expected to be stable by itself. But the observed symmetry suggests that Ca^{2+} in its site provides enough binding energy for stabilizing this structural motif in solution. In conclusion we find that symmetry reveals stable structural motifs and that it often suggests a certain order for the folding process.

5.4 Globular Proteins

Globular proteins consist of one or more structural domains. Globular protein is the term assigned to the folding result of a complete single polypeptide chain. Consequently, globular proteins consist of one or more structural domains. In some cases (Ser-proteases, immunoglobulins, rhodanese, etc.) these domains are so similar that gene duplication is very likely to have occurred. Each domain can be classified into one of the five classes given in Table 5-2. There is no preference for structural domains of the same globular protein belonging to the same class (Table 5-2). A clear correlation between structural class and the function of a domain or a globular protein has not yet emerged. However, nucleotide binding domains tend to belong to class No. 4. Moreover, it is conspicuous that all structurally known enzymes of the glycolytic pathway belong to the same class, No. 4.

Usually, active centers of enzymes contain parts of all structural domains present in the globular protein. In all known multiple domain proteins (Table 5-2) the active site is located between domains (Figure 4-1). And these domains are not only defined as globular regions of density separated by the active site cleft; they also show the other pronounced domain property of being connected by only a single chain strand (Table 5-2). Furthermore, substrates and cofactors are usually bound at different domains. In the case of NAD the cofactor binding domain always has the same significant sheet topology and NAD binds at equivalent positions

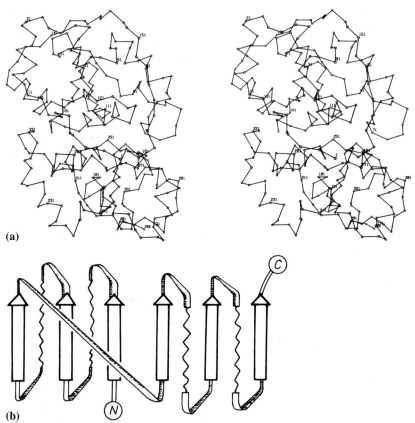

(a)

(b)

Figure 5-17. Symmetry within single polypeptide chains. (a) Stereo view of the backbone of rhodanese (799). Except for a short piece of connecting chain the upper and the lower half are related by an almost exact twofold axis perpendicular to the paper plane. (b) Sheet topology of the nucleotide binding domain of NAD-dependent dehydrogenases. α-Helices are indicated by wiggled lines. This sheet consists of two Rossmann-folds (Figure 5-12b) related to each other by a vertical twofold axis. (c) Backbone sketch of ferredoxin (314) including FeS-clusters. Part of the amino acids, the FeS-clusters and the chain fold show an internal twofold symmetry. The twofold axis is approximately perpendicular to the paper plane. (d) The topologies of the two β-sheet barrels of chymotrypsin and other trypsin-like proteases. As indicated each barrel shows approximate internal twofold symmetry. The relation between the barrels is asymmetric. (e) Sheet topology of triose phosphate isomerase (305) consisting of an eight-stranded parallel sheet barrel and surrounding helices. Neglecting the elliptic cross-section of the barrel and differences in the helices there is an approximate eightfold symmetry.

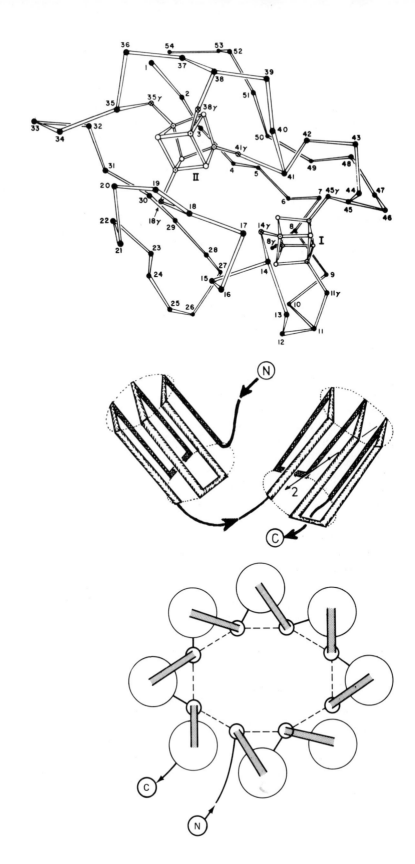

(c)

(d)

(e)

(Figure 5-17b), indicating an evolutionary relationship (254, 255). Moreover, this domain is found as the N-terminal half of three dehydrogenases and one kinase (230–233, 235), as the C-terminal half of a fourth dehydrogenase (234), and in the middle of phosphorylase (236), indicating that the corresponding gene might have been duplicated and transferred to a different place of the genome. Taken together, the partition of active sites between domains, the occurrence of cofactor-specific domains, and the possibility of domain transfer suggest that enzymes may be constructed using a modular system: to accomplish a given function the necessary cofactor- and substrate-specific domains are selected and fused to a single chain globular protein (124, 256).

5.5 Aggregates of Globular Proteins

In principle, there is no difference between aggregation of structural domains and aggregation of globular proteins. The distinction between the association of structural domains of a single chain and the association of globular proteins, that is, of separate chains, is blurred. For example, in rhodanese (Figure 5-17a) domains of one chain aggregate with almost exact symmetry (257), like subunits of a dimeric protein. Moreover, in the production of the coat of polio virus (163) and semliki forest virus (258) a large number of coat protein globules are synthesized in a single polypeptide chain, folded, and subsequently separated by a protease. After cutting, the protein aggregates symmetrically to form the virus coat. This demonstrates that single domains in a multidomain globular protein can be stable themselves and can be used like independent globular proteins. However, in general the subunits of larger proteins aggregates are synthesized and folded separately.

The reasons for aggregation are manifold. Aggregation may occur (i) to form a complex machinery which is able to fulfill complex functions, (ii) to bring enzymes of a metabolic pathway close together so that the loss of metabolic intermediates is avoided, (iii) to build structures of a given geometry, e.g., long tubes, (iv) to reduce the osmotic pressure, or (v) to enlarge the number of possible enzyme activity characteristics by introducing cooperativity between subunits and various types of regulation.

Symmetry of Aggregates

Asymmetric aggregates have complex functions. The aggregates can be subdivided into asymmetric and symmetric ones. A typical asymmetric aggregate (which also contains RNA) is the ribosome. The ribosomal function in protein synthesis is so complex that a large team of different proteins is required; and nonidentical proteins can aggregate only asymmetrically.

Aggregates can have space-, linear-, or point-group symmetry. Symmetric aggregates can be subdivided into those obeying space-, linear-, and point-group symmetry. *Space*-group symmetry is found in insulin crystals which are produced in the pancreas as a storage form of negligible osmotic pressure (259). The same kind of symmetry is observed in the striated muscle of vertebrates and insects (215). *Linear* groups are found in microtubules (181), in tobacco mosaic virus (180), and in filamentous phages (220). *Point*-group symmetry is a very common phenomenon. Because of amino acid asymmetry no point groups containing inversion centers or mirrors are allowed. Only groups n, $n2$, 23, 432, 532 with $n = 1, 2, 3 \ldots$ are possible (252, 260). Except for group 23, examples for all of them have been observed (Table 5-4).

As described above symmetric arrangements are energetically favored. Indeed, only in the case of hexokinase has asymmetric aggregation of identical units been observed, and only in the crystalline state (261). In some cases only "pseudosymmetrical" arrangements are formed. Normal human hemoglobin, for example, shows exact 2- but pseudo 222 symmetry. The pseudosymmetry would be exact if β- and α-subunits were identical. So-called "quasisymmetry" occurs if all subunits are chemically identical but have two slightly different conformations so that slightly different contacts can be made. Quasisymmetry is frequent in spherical viruses. With quasisymmetry only one type of protein is required to build an icosahedral coat (262), which is roughly spherical. In tomato bushy stunt virus the subunit contains two structural domains. The conformational difference is caused by a relative movement of these domains (263).

Intersubunit Contacts in Dehydrogenases

Only two of the three types of contacts of a 222-tetramer have to be made. After discussing symmetry, let us now consider the types of contact occurring in a symmetric oligomer. For example, take the four structurally known dehydrogenases: s-malate and alcohol dehydrogenases are dimers of identical subunits with point group 2, and lactate and glyceraldehyde-3-phosphate dehydrogenases are tetramers of identical subunits with point group 222 (see Figure 5-16a). The dimers have only one type of contact, around the twofold axis (Figure 5-18a). In the tetramers there are three types of contacts, one across each twofold axis (Figure 5-18b). Note, that only two of the three contacts have to be made in order to keep the tetramer from falling apart.

Contacts of a 222-tetramer can be represented by a triangular graph. The tetramer contacts can be conveniently described by the tetrahedron of Figure 5-18b and by the two-dimensional plot of Figure 5-18c. For distinction the subunits have been marked using a color code, and the twofold axes are denoted P, Q, and R. The "red" subunit is the reference base. It forms contacts with "blue," "yellow," and "green" across the twofold

Table 5-4. Aggregation Symmetries of Globular Proteins[a]

Protein	Number of subunits	Point symmetry	
		Crystallographic symbol	Schönflies symbol
Alcohol dehydrogenase (234)	2	2	C_2
Immunoglobulin (294)	4	2	C_2
s-Malate dehydrogenase (233)	2	2	C_2
Superoxide dismutase (286)	2	2	C_2
Triose phosphate isomerase (305)	2	2	C_2
Phosphorylase (236)	2	2	C_2
Alkaline phosphatase (317)	2	2	C_2
6-Phosphogluconate dehydrogenase (318)	2	2	C_2
Wheat germ agglutinin (316)	2	2	C_2
Glucose phosphate isomerase (313)	2	2	C_2
Tyr–tRNA-synthetase (221)	2	2	C_2
Glutathione reductase (124)	2	2	C_2
Aldolase (306)	3	3	C_3
Bacteriochlorophyll protein (303)	3	3	C_3
Glucagon (278)	(3)	3	C_3
TMV-protein disc (218)	17	17	C_{17}
Concanavalin A (281, 282)	4	222	D_2
Glyceraldehyde-3-phosphate dehydrogenase (230, 231)	4	222	D_2
Lactate dehydrogenase (232)	4	222	D_2
Prealbumin (206)	4	222	D_2
Pyruvate kinase (80)	4	222	D_2
Phosphoglycerate mutase (307)	4	222	D_2
Hemoglobin (human) (274)	2 + 2	Pseudo 222	Pseudo D_2
Insulin (259)	6	32	D_3
Aspartate transcarbamoylase (319)	6 + 6	32	D_3
Hemerythrin (217)	8	422	D_4
Apoferritin (320)	24	432	O
Coat of tomato bushy stunt virus (263)	180	532	Y

[a]In many cases a β-sheet is formed (259) or continued on aggregation (206, 281, 282).

axes *P*, *Q*, and *R*, respectively. As shown in Figure 5-18d, the two-dimensional plot can be well used for enumerating the interactions between residues of the "red" subunit with residues on the other subunits.

Contacts reveal details of an evolutionary relationship. Rossmann and co-workers (264, 265) have pointed out that (i) the four dehydrogenases have an evolutionary relationship and that (ii) a comparison between the intersubunit contacts reveals details of this relationship. The contacts can be compared because all four dehydrogenases have the same nucleotide binding domain (Figure 5-17b) which constitutes about one-half of a subunit. This domain can be used as a reference base. Lactate and s-malate dehydrogenases have almost identical chain folds, indicating that they are closely related. Glyceraldehyde-3-phosphate, lactate, and s-malate dehydrogenases have the Q-axis contact (defined in Figure 5-18d) in common. Accordingly, the dimer "red–yellow" has been well conserved during protein evolution. With respect to P- and R-axis contacts the dehydrogenases do not show any resemblance. Furthermore, there is no similarity between the contact in alcohol dehydrogenase and any other contact.[6] On the basis of these data the evolutionary relationship of Figure 5-18e can be derived.

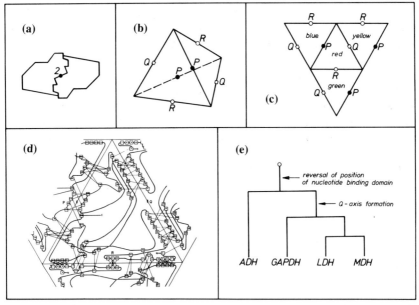

Figure 5-18. Oligomeric enzymes: dehydrogenases. (a) Sketch of a symmetric dimer, e.g., alcohol dehydrogenase. There is only one type of interface. The contacting surfaces are complementary [lock and key system (44)]. (b) Tetramer with 222 symmetry represented by a tetrahedron. The three twofold axes P, Q, R are indicated. Each subunit is represented by a triangle and each intersubunit contact by an edge. Among the three different contact types only two have to be formed to keep all subunits from falling apart. (c) Planar representation of the tetrahedron and of subunit contacts. (d) Use of the planar representation for explicit specification of the atomic contacts between subunits of lactate dehydrogenase (264). (e) Evolutionary tree of dehydrogenases as derived from subunit contacts (264). ADH = alcohol dehydrogenase, GAPDH = glyceraldehyde-3-phosphate dehydrogenase, LDH = lactate dehydrogenase, MDH = s-malate dehydrogenase.

---

[6]The special position of alcohol dehydrogenase is also reflected by the fact that its nucleotide binding domain is the C-terminal part of the polypeptide chain, whereas the other three dehydrogenases have this domain as their N-terminal parts.

Contacting Surfaces

The more contacts the stronger the interaction. A detailed analysis of protein–protein interfaces has been made for the proteins listed in Table 5-5. As can be seem from this table, there are many van der Waals contacts, some hydrogen bonds, and a few salt bridges. Generally, the more contacts there are between subunits, the stronger is the protein–protein interaction. This can be illustrated by four examples.

In hemoglobin the $\alpha_1\beta_1$ contact is by far the strongest. Table 5-5 shows that in oxy- as well as in deoxyhemoglobin the $\alpha_1\beta_1$ ($\alpha_2\beta_2$) contact is stronger than the $\alpha_1\beta_2$ ($\alpha_2\beta_1$) contact. Indeed, the cooperative change of hemoglobin on oxygenation is mainly a rotation of $13°$ of the $\alpha_1\beta_1$ dimer relative to the $\alpha_2\beta_2$ dimer. The $\alpha_1\alpha_2$ and $\beta_1\beta_2$ contacts ($\alpha_1\alpha_2$ would be equal to $\beta_1\beta_2$ if hemoglobin had exact 222 symmetry) are very weak. As mentioned above, this third type of contact in the 222-tetramer is not needed to keep the subunits together.

Frequently, contacts are continuations of β-pleated sheets. In concanavalin A, one (I–IV) of the three contact types is also much weaker than the other two. Contact I–II is particularly strong involving as many as 14 hydrogen bonds. Here, a broad (six-stranded) β-sheet is extended across

Table 5-5. Contacts between Proteins

			Number of		
	Symmetry	Name of contact	van der Waals contacts	Hydrogen bonds	Salt bridges
Horse hemoglobin					
oxy (321)	2	$\alpha_1\beta_1$	110	5	0
		$\alpha_1\beta_2$	80	1	0
		$\alpha_1\alpha_2$	0	0	0
		$\beta_1\beta_2$	0	0	0
deoxy (322)	2	$\alpha_1\beta_1$	98	5	0
		$\alpha_1\beta_2$	69	1	1
		$\alpha_1\alpha_2$	0	0	2
		$\beta_1\beta_2$	0	0	0
Concanavalin A					
(282)	222	I–II	250	14	0
		I–III	156	4	2
		I–IV	16	2	0
Insulin (259)	32	OP	111	4	0
		OQ	99	2	1
α-Chymotrypsin (18)	2	A	143	8	1
		B	57	6	0
Trypsin–trypsin inhibitor (267, 269)	None		Many	7	1

the twofold axis so that in total a regular 12-stranded β-sheet is formed. The two strands that are hydrogen bonded across the twofold axis are antiparallel, that is, the twofold axis is perpendicular to the sheet plane. Such continuations of β-sheets across protein–protein interfaces are common in aggregates of subunits and of structural domains.

Contacts in crystals of concanavalin A, insulin, and α-chymotrypsin show the structure of the dimers in solution. In solution tetrameric concanavalin A is in equilibrium with dimers and monomers. Since the dimer shows physiological activity there is interest in identifying it. Among the three types of dimers possible clearly dimer I–II is the most stable one (Table 5-5). Therefore, it is the dimer most probably found in solution.

With insulin a similar problem is encountered. In solution the insulin hexamer is in equilibrium with dimers and monomers. Since the dimer may be important for binding to the hormone receptor, it should be identified. Judging from the numbers in Table 5-5, the contact strengths of both possible dimers do not differ very much. However, the dimer OP, in addition to forming a slightly stronger contact, forms antiparallel sheet structure. Therefore, it was taken to be the dimer observed in solution (259).

Crystalline α-chymotrypsin forms two types of contacts around twofold axes A and B. Contact A is much stronger than B. At the pH value of the crystals (pH = 4), α-chymotrypsin forms dimers in solution. From the numbers in Table 5-5 it is possible to conclude that these dimers are most probably of type A.

Free Energies of Association

Free energies of association can be derived from the interface structure. The point-to-point interactions enumerated in Table 5-5 give a reasonable estimate of relative binding strengths, that is, of free energies of association. As shown above, these estimates allow the identification of the most stable oligomers. However, it is possible to analyze the interfaces in an even more detailed manner and determine the absolute values of the free energies of association $\Delta G_{assoc}^{(calc)}$ from the crystal structure. Because of the relation

$$\Delta G_{assoc}^{(obs)} = G_{oligomers}^{(obs)} - G_{monomers}^{(obs)} = -RT \ln K_{assoc}^{(obs)}$$

these calculated values $\Delta G_{assoc}^{(calc)}$ can then be compared with the experimental values $\Delta G_{assoc}^{(obs)}$ as derived from the association constant $K_{assoc}^{(obs)}$. Ways of calculating $\Delta G_{assoc}^{(calc)}$ will now be discussed.

The thermodynamics of association resemble those of folding. On association, surface areas of all partners are buried. This can be described as transfer of surface (atoms) from water to the protein interior. Furthermore, the entropy of the system decreases, because associated monomers (oligomers) show a higher order than free ones. Consequently, the free energy of association is given by

$$\Delta G_{\text{assoc}}^{(\text{calc})} = \Delta G_{\text{transfer}} - T \cdot \Delta S_{\text{assoc}}.$$

This equation can be compared with Eqs. (3-2) and (3-3): $\Delta G_{\text{assoc}}^{(\text{calc})}$ and ΔS_{assoc} correspond to ΔG_{total} and ΔS_{chain}, respectively. $\Delta G_{\text{transfer}}$ is the same. Therefore, we can refer to Figure 3-3b and c.

For the most part association is entropy driven. For nonpolar surfaces $\Delta G_{\text{transfer}}$ is proportional to the water accessible surface area (Figure 1-8). For polar surfaces $\Delta G_{\text{transfer}}$ is smaller but still of comparable magnitude (Figure 3-3). Thus, $\Delta G_{\text{transfer}}$ is roughly proportional to the total area buried. Chothia and Janin (266) determined the buried accessible surface areas of the proteins listed in Table 5-6. It turned out that more than two-thirds of the buried surfaces are nonpolar (Table 5-6). Therefore, the authors committed no large error when using the proportionality constant for nonpolar surfaces (Figure 1-8). In order to obtain ΔS_{assoc} the authors (267) used the theoretical results for free particles. The resulting values for $\Delta G_{\text{assoc}}^{(\text{calc})}$ are given in Table 5-6. They agree roughly with the experimental data. Thus, in a qualitative manner the forces governing association seem to be understood.

The change of free energy accompanying conformational changes of aggregated structure is a most interesting quantity. In many oligomers the absolute free energy of association is much less interesting than the change of this free energy on ligand-induced modifications of the aggregate. Such free energy differences may help explain cooperativity and regulation phenomena. Today, hemoglobin is the best known example of ligand-induced changes of an aggregate structure. Here, a comparison of surfaces buried in the oxy and deoxy form (Table 5-5) reveals the free energy contribution caused by the change of aggregate structure (268). However, high accuracy is required for such a comparison, because a small difference between large quantities has to be evaluated; and this accuracy is still difficult to achieve.

Table 5-6. Free Energy of Protein Association[a]

	Total accessible surface areas buried on association (Å^2)	Percentage of nonpolar area (%)	$\Delta G_{\text{transfer}}$ (kcal/mol)	$-T \cdot \Delta S_{\text{assoc}}$ (kcal/mol)	$\Delta G_{\text{assoc}}^{(\text{calc})}$ (kcal/mol)	$\Delta G_{\text{assoc}}^{(\text{obs})}$ (kcal/mol)
Horse oxyhemoglobin contact $\alpha_1\beta_1$	1720	84	−43	+27	−16	< −11
Insulin, contact OP	1130	78	−28	+23	− 5	− 7
Trypsin–trypsin inhibitor	1390	68	−35	+27	− 8	−18

[a]The surface areas were determined on the basis of crystal structures. Data are taken from Ref. (266). The accessible surface areas are defined in Figure 1-9.

Specificity of Protein–Protein Interaction

Complementarity of surfaces is necessary. What can be learned from these calculations with respect to protein–protein interactions? First of all we realize that the removal of nonpolar surfaces from water gives the largest free energy contribution (Table 5-6); entropic forces are most important. The magnitude of this contribution depends on the complementarity of the contacting surfaces; the buried surface area would be reduced appreciably if space were left for water to go in and out. Indeed, all interfaces analyzed so far show a packing density (section 3.6) that is as high as in the protein interior (267).

The interface also corresponds to the protein interior with respect to hydrogen bonding. Almost all buried polar groups form such bonds (266). Moreover, all charged groups form salt bridges. It should be remembered that efficient hydrogen bonding and salt bridge formation in the protein interior was the basis for Figure 3-3, in which $\Delta G_{transfer}$ for polar groups is roughly equal to $\Delta G_{transfer}$ for nonpolar groups.

Specificity is achieved by complementarity of surface shapes, hydrogen bond donors and acceptors, and salt bridge partners. For biological processes the strength of an interaction is usually not as important as its specificity. How is specificity achieved in protein–protein interactions? As mentioned above the contacting surfaces have to be complementary in order to get a large $\Delta G_{transfer}$. Besides, the hydrogen bond donors and acceptors have to meet. If they miss each other, the association energy is greatly reduced. This applies even more to buried salt bridges. Although their "active" contribution is small, they have a strong "passive" influence; if charges are not compensated the energy is reduced so much (section 3.5, page 41) that proteins will not associate. Thus, the specificity of a surface is determined by its shape and its pattern of hydrogen bond donors and acceptors as well as by charges. As judged from the number of different antigen-binding sites observed in immunoglobulins, a surface of 100 Å² can form on the order of a thousand patterns showing different binding specificities.

5.6 Hierarchy of Levels

A hierarchic relationship facilitates the analysis. The levels of protein structure have been introduced in Figure 5-1a. In Figure 5-1b a hierarchic scheme has been given. If this scheme were right, the evaluation of the folding process would be much simpler because it could be split into several well-separated steps (comparable to factorization in solving differential equations). The first step, for example, would be to derive the secondary structure from the amino acid sequence; as the second step the supersecondary structures would be derived from the secondary structures, and so on.

In this way, protein folding would be more amenable to analysis because it is simpler to solve the problem step by step than to solve it all at once. Unfortunately, however, this hierarchic scheme only partially applies.

The observed strictness of hierarchy of levels varies greatly. A fairly strict hierarchy is found for levels of globular proteins and aggregates, since in most cases the globular monomer is stable on its own and hardly altered during aggregation.[7] Therefore, the aggregate structure can be derived from the surface characteristics of a known monomer structure. Such a procedure has been tried for sickle cell hemoglobin (270). However, no hierarchic relation seems to apply for ribosomes because some ribosomal proteins are so extended (7) that they cannot assume a defined structure by themselves but only during aggregation.

A hierarchy is also found for the levels of amino acid sequence and secondary structure (in particular α-helices) since the formation of secondary structure is largely independent of the final chain fold. This can be inferred from the observed correlations between amino acid sequence and secondary structure described in the following chapter.

The association of domains to multidomain globular proteins is also likely to be of a hierarchic nature. In particular the predominant single strand connection between successive domains and the observed symmetry relation between domains indicate that domains are formed before association to a globular protein and that they are rather stable on their own.

Less clear are the steps from secondary to supersecondary structure and from supersecondary structures to domains. Among the secondary structures α-helices may be stable by themselves and may be formed before association to the next level. But single β-sheet strands and reverse turns are not; thus they are not amenable to separation. The symmetry between supersecondary structures found in numerous domains makes us hope that supersecondary structures may be rather stable by themselves and thus constitute a well separated level.

In conclusion, the separation into the levels of Figure 5-1a, although not strictly hierarchic, is nevertheless sufficiently hierarchic to provide an acceptable working hypothesis.

Summary

An analysis of known protein structures is necessary for an understanding of protein folding and stability. In these structures six organizational levels can be distinguished. The basic level is the amino acid sequence, which alone determines the resulting protein structure. Several types of regular backbone arrangements are very abundant in protein structures. These are

---

[7]A detailed analysis of the small alterations on association of trypsin and bovine pancreatic trypsin inhibitor was made by Huber *et al.* (269).

secondary structures and constitute the second level. Two of these arrangements (α-helix and β-sheet) were predicted on the basis of the covalent structure of the backbone, which again shows that they are simple or regular. The next two upper levels, supersecondary structures and structural domains, are very complicated and cannot as yet be predicted. At these levels there occur conspicuous regularities such as the neighborhood correlation which cannot be expressed as fixed structures but only as general rules. At the two highest levels, globular proteins and aggregates, some structure prediction has already been attempted. This is possible because the underlying structures, domains for globular proteins and globular proteins for aggregates, are expected (and in some cases known) to be stable by themselves. Thus, aggregates can be predicted by fitting the surfaces of globular proteins. The same is true for the prediction of globular proteins from domains. Moreover, surface properties as derived from protein–protein interfaces are very important for protein–protein recognition. Some rules for recognition are given. There is one general concept which appears conspicuously throughout all the levels above the amino acids: symmetry. In secondary structures and in aggregates linear-group symmetry is found. Point-group symmetry is found in all levels above secondary structures, and space-group symmetry is found in aggregates. The presence of symmetry and the presence of some hierarchical order between the levels makes us hope that one day the puzzle posed by protein structures can be solved.

Chapter 6
Prediction of Secondary Structure from the Amino Acid Sequence

To a rough approximation the secondary structure of a chain segment is a function of the constituent amino acids alone. A glimpse at Dayhoff's *Atlas of Protein Sequence and Structure* (20) shows that the number of known amino acid sequences far exceeds the number of known three-dimensional structures. Since the amino acid sequence contains the complete structural information (section 1.2 and Ref. 177), it should be possible to derive the spatial structure from the sequence alone without X-ray analysis. The first step in this direction should be to try to bridge the gap between the two lowest levels of structural organization in proteins (Figure 5-1), between sequence and secondary structure. This organization is approximately hierarchic. But the hierarchy is not strict so that the formation of secondary structure in a given segment of the polypeptide chain does not depend on the sequence in this segment alone. Other segments at a distance along the chain also exert an influence, the functional dependence between the two lowest levels being of a rather nonlocal nature.[1]

   The first indication of a relationship came from synthetic homopolymers. The first correlation between amino acid sequence and secondary structure was established by Blout *et al.* (328). They assigned α-helix-forming and -breaking properties to seven types of amino acid residues (Table 6-1) on the basis of experiments with synthetic homopolymers, polypeptides such as poly-Glu, poly-Lys, etc. Helix-forming residues were those that assumed the α-helix conformation and helix-breaking ones were those that did not.

   Davies (329) applied these results to native globular proteins and found a clear anticorrelation between helix content, as determined by optical rotary

---

[1]In this respect the functional dependence resembles a Fourier transform where *all* points of the basic space contribute to *a* given point in the transform space and vice versa. But the dependence is of course much less regular.

Table 6-1. α-Helical Propensities[a]

		Blout et al., 1960 (328)	Kotelchuck and Scheraga, 1968 (363)	Lewis et al., 1970 (368)	Robson and Pain, 1971 (346)	Chou and Fasman, 1974 (340)	Finkelstein and Ptitsyn, 1976 (371)
A	Ala	(H)	H	I	+0.09	1.45	1.08
C	Cys	C	H	I	+0.03	0.77	0.95
D	Asp	H	C	B	−0.02	0.98	0.85
E	Glu	H	H	H	+0.12	1.53	1.15
F	Phe	(H)	H	H	+0.03	1.12	1.10
G	Gly	—	Indifferent	B	−0.05	0.53	0.55
H	His	(H)	H	I	+0.08	1.24	1.00
I	Ile	(C)	H	H	+0.07	1.00	1.05
K	Lys	(H)	C	I	−0.03	1.07	1.15
L	Leu	H	H	H	+0.11	1.34	1.25
M	Met	H	H	H	+0.10	1.20	1.15
N	Asn	(C)	C	I	−0.04	0.73	0.85
P	Pro	—	Special	B	—	0.59	—
Q	Gln	(H)	H	I	+0.07	1.17	0.95
R	Arg	(H)	H	I	+0.02	0.79	1.05
S	Ser	C	C	B	−0.07	0.79	0.75
T	Thr	(C)	H	I	−0.01	0.82	0.75
V	Val	C	H	I	+0.04	1.14	0.95
W	Trp	(H)	C	H	+0.10	1.14	1.10
Y	Tyr	(H)	C	H	−0.02	0.61	1.10

[a]These propensities are derived from the observed frequencies. The amino acid residues are given in the alphabetical order of their one-letter symbols. The symbols H, I, B, and C are used for residue types that tend to be helix formers, indifferent, helix breakers, and in (random) coils, respectively.

dispersion (ORD) measurements (330) and the content of residue types (Ser+Val+Cys+Thr+Ile). As shown in Table 6-1, the first three of these are the helix-breaking residues of Blout *et al.* (328). Thr and Ile were selected because of their similarity to Ser and Val, respectively.

Correlation analyses improved and diversified with an increase of the data base. Davies' initial success encouraged searches for more specific correlations in globular proteins. However, the prospects remained poor as long as the data base, the number of known three-dimensional protein structures, remained small. Therefore the early proposals of Guzzo (331), Prothero (332), Havensteen (333), Cook (334), Periti *et al.* (335), Dunnill (336), and Low *et al.* (337) are mainly of historical interest. With an increasing data base the accuracy of the correlations improved considerably.

The proposed correlation methods, or, as they are usually called, "methods for predicting secondary structure from amino acid sequence," (in short, "prediction methods,") can be placed into two categories: probabilistic and physico-chemical. The former refers to methods which extract rules and parameters using purely statistical analyses of the data base. In contrast, physico-chemical methods apply in addition, or exclusively, structural information from outside the data base. Since the difference between the two types of methods is gradual and not absolute, there are borderline cases in which the distinction between the two is arbitrary.

We will discuss most of the current methods because (i) no method is clearly superior to all others, and (ii) even methods that are generally considered as obsolete may contain ideas useful for further improvements. Such improvement is desirable, and it still seems possible.

6.1 Probabilistic Methods

Singlet Frequencies and Propensities

The simplest approach is to consider each residue separately without attention to near or far neighbors. A simple, purely statistical approach to data analysis was adopted by Dirkx (338,339). He determined the frequency of each of the 20 residue types in the α-(helix), β-(sheet), and rt-(reverse turn) regions of the proteins in the data base. These frequencies were then defined as the "propensity of this type of residue to occur in α-, β-, and rt-conformation." As an example Dirkx' β-propensities are given in Table 6-2.

For predicting α- (β-, rt-) structure, the "α- (β-, rt-) potential"[2] at each residue position in the chain was calculated as the weighted average of the α- (β-, rt-) propensities of the m nearest residues along the chain. And whenever the α- (β-, rt-) potential exceeds a certain threshold the residue in the corresponding position is predicted to be in the α- (β-, rt-) conformation. Thus, the potential function was converted to a *yes* or *no* decision. The weighting scheme was fixed. The three thresholds as well as m are adjusted for best fit between predicted and observed secondary structure in the data base. The optimum m values were 17, 11, and 3 for α-, β-, and rt-prediction, respectively. The notions "propensity of a residue type", which is a basic value, and the "potential at a residue position", which is derived from propensities, will be used throughout this chapter in the sense defined here.

In reverse turns the relative position of the residue should be taken into account. Information exclusively from single residues (singlets) was also used for the prediction of rt. Lewis *et al.* (326) defined reverse turns as those quartets of residues i, $i+1$, $i+2$, $i+3$, in which the distance between C_α-atoms in positions i and $i+3$ is less than 7 Å and the chain is not in α-

---

[2]The word "potential" is widely used. It is not a potential in the physical sense.

Table 6-2. Relative Frequencies of Residue Types to Occur in β-Sheets and Reverse Turns $(rt)^a$

		β-sheet propensities			rt-propensities		
		Chou and Fasman, 1974 (340)	Burgess et al., 1974 (31)	Beghin and Dirkx, 1975 (339)	Lewis et al., 1971 (326)	Kuntz 1972 (203)	Chou and Fasman, 1974 (340)
A	Ala	0.97	0.29	0.37	.0.22	(T)	0.15
C	Cys	1.30	0.53	0.84	0.20	—	0.31
D	Asp	0.80	0.27	0.97	0.73	T	0.33
E	Glu	0.26	0.26	0.53	0.08	T	0.12
F	Phe	1.28	0.32	0.53	0.08	—	0.19
G	Gly	0.81	0.31	0.97	0.58	T	0.45
H	His	0.71	0.20	0.75	0.14	T	0.18
I	Ile	1.60	0.41	0.37	0.22	—	0.15
K	Lys	0.74	0.27	0.75	0.27	T	0.27
L	Leu	1.22	0.40	0.53	0.19	—	0.14
M	Met	1.67	0.38	0.64	0.38	—	0.18
N	Asn	0.65	0.23	0.97	0.42	T	0.45
P	Pro	0.62	0.34	0.97	0.46	T	0.41
Q	Gln	1.23	0.33	0.64	0.26	T	0.15
R	Arg	0.90	0.36	0.84	0.28	T	0.27
S	Ser	0.72	0.35	0.84	0.55	T	0.41
T	Thr	1.20	0.39	0.75	0.49	T	0.27
V	Val	1.65	0.50	0.37	0.08	—	0.08
W	Trp	1.19	0.23	0.97	0.43	—	0.30
Y	Tyr	1.29	0.43	0.84	0.46	T	0.33

^{a}The observed frequencies are used as propensities. In Kuntz' nomenclature T means that the respective residue type tends to occur in reverse turns.

helical conformation. In contrast to Dirkx, however, the four positions in a reverse turn were not considered as equivalent. Frequencies were derived for each type of residue at each position i, $i+1$, $i+2$, $i+3$ of a reverse turn, and defined as the "propensity of the respective residue type to occur at this particular position." The rt-potential of a given residue quartet was then defined as the product of the appropriate propensities. The threshold for the potential was adjusted for best fit in the data base (see Figure 6-2).

These frequencies were updated by Crawford et al. (200), and later by Chou and Fasman (340), who used them to predict reverse turns along the lines of the method of Lewis et al. (326). The frequencies found by Lewis et al. (326) and Chou and Fasman (340) are given in Table 6-2. However, in order to keep the table compact, only the average frequencies of occurrence in a reverse turn have been listed. For some residue types the positional differences are appreciable; Pro is almost exclusively found in position $i+1$, and Trp occurs only in position $i+3$.

The simple approach of singlet propensities was also adopted by Ptitsyn and Finkelstein (341, 342) for α- and β-prediction. These authors claimed (343) that no further information can be obtained from frequencies of pairs of residue types (doublets). This statement, however, is debatable, because the underlying data base contained only 9 proteins, and because the 20 residue types were grouped into four categories in a somewhat arbitrary manner.

Doublet Frequencies and Propensities

Doublet propensities include residue–residue interactions. A rigorous prediction based on doublet frequencies. Despite the warnings of Finkelstein and Ptitsyn (343), many prediction methods included doublet information because doublet propensities include residue–residue interactions. Periti (344) used doublets for prediction of α-helices and β-sheets with a purely probabilistic method. He considered the 27 doublets around a given residue position with relative distances of up to six residues from each other, assuming that there is no interaction beyond this distance. This gave rise to 10,800 different residue doublet types.[3] For the given residue Periti distinguished between α-, β-, and coil- ($\equiv$non-α and non-β) structure. Thus a table of $32,400 = 3 \cdot 10,800$ frequencies had to be derived from the data base. These frequencies were interpreted as propensities. In order to predict the secondary structure at a given residue in a known sequence, the propensities for α- (β-, coil-) structure of the 27 surrounding doublet types were looked up in the table and combined using a fixed statistics procedure. This yielded an α- (β-, coil-) potential. Subsequently α-, β-, and coil-potentials were compared, and the highest one was taken as the prediction. It should be noted that this method contains no adjustable parameters. Furthermore it does not handle the predictions of α- and β-structure separately, but combines them in a rigorous manner.

Rigorous potentials and adjusted thresholds. For prediction of α-, β-, and rt-conformation Robson et al. (345–352) used singlets together with the 16 doublets formed by the residue in question and residues up to a distance of 8 in both directions along the chain. Singlet and doublet propensities were derived from the corresponding frequencies in the data base. For a given residue position these propensities were combined to α-, β-, and rt-potentials using a fixed information-theoretical procedure. These potentials were compared to three different thresholds, one for each type of secondary structure. The three thresholds were separately adjusted for best fit between predicted and observed secondary structure in the data base.

---

[3]For residue position i the seven doublets $(i, i+6)$, $(i-1, i+5)$, . . . , $(i-6, i)$; the six doublets $(i, i+5)$, . . . , $(i-5, i)$; . . . and the two-doublets $(i, i+1)$, $(i-1, i)$ sum up to $7+6+5+4+3+2 = 27$ different doublets. Each doublet consists of two residues. With 20 different residue types there are $20 \cdot 20 = 400$ different residue type combinations. In total, there are $27 \cdot 20 \cdot 20 = 10,800$ different doublet types.

Predictions of α-, β-, and rt-structures were evaluated independently of each other.

Adjusted potentials and adjusted thresholds. Nagano (228, 353–356) used the doublet types described with Periti's (344) method, except for taking a maximum distance of 7 instead of 6. This gave rise to 35 doublets affecting the residue in question. Thus, the table of observed frequencies (propensities) had $20 \cdot 20 \cdot 35 \cdot 3 = 42,000$ entries. For a given residue in a given sequence the propensities of all surrounding doublet types were looked up and linearly combined to yield α-, β-, and rt-potentials.

In a similar manner to Robson *et al.*, Nagano derived his prediction by comparing the potentials separately to three different thresholds. Each threshold was adjusted for best fit in the data base. In addition, Nagano introduced a number of adjustable parameters by defining the potentials as linear combinations of propensities containing $3 \cdot 35 = 105$ coefficients, three (α, β, rt) for each of the 35 doublet types. All these parameters were adjusted for best fit in the data base. Thus, Nagano used quite a few more adjustable parameters than Periti and Robson *et al.*, who constructed their potentials from propensities using fixed probabilistic procedures.

Nagano increased the available data base appreciably by adding (with lower weight) the information contained in amino acid sequences of proteins that are homologous to structurally known ones. Thus, at a later stage (356), he was also able to incorporate triplet frequencies and propensities into this method. It should be noted that Nagano tacitly included singlets, because certain linear combinations of doublet propensities are equivalent to a singlet propensity.

Prediction of residues that break a secondary structure. Using only one type of doublet, namely, the pair of the two adjacent neighbors $(i-1, i+1)$ of the residue i in question, Kabat and Wu (357–359) developed a negative prediction method which identifies α-helix and β-sheet *breaking* residues. The propensities for α- and β-breaking were derived from frequencies in the data base. The potential was taken as the corresponding propensity. Potential thresholds were adjusted for best fit in the data base.

Triplet Information

An attempt has been made to predict the common chain fold of a protein family on the basis of triplet information. Triplets of three adjacent residues were used by Kabat and Wu to predict not only secondary structure (360), but also the complete chain fold (361,362). For chain fold prediction the (ϕ,ψ)-angles of the central residue of all triplets in the data base of proteins with known three-dimensional structures were recorded in a "(ϕ,ψ)-table." This table had $20 \cdot 20 \cdot 20 = 8000$ locations, one for each triplet type.

The method was applied only to the prediction of the common chain fold of a particular protein family (cytochromes c or, immunoglobulins, see section 9.1). In such a family numerous amino acid sequences have the same chain fold, that is, identical (ϕ,ψ)-angles. Thus, for a given residue at position i, triplets ($i-1$, $i,i+1$) could be taken from all homologous proteins and converted to (ϕ,ψ)-angles by looking up the (ϕ,ψ)-table.[4] The average of these (ϕ,ψ)-angles was taken as the predicted (ϕ,ψ)-angle for residue i. Repeating this procedure for all residues results in a predicted chain fold.

6.2 Physico-Chemical Methods

The physico-chemical methods, in contrast to probabilistic ones, do not rely exclusively on the known correlations between sequence and structure in the data base of globular proteins but incorporate other experimental and theoretical data as well. These methods are subdivided into those which depend on stereochemical considerations and those which are based on statistical mechanics. To give increased insight into statistical mechanical methods and into the physical basis of chain folding in general, some basic aspects of statistical mechanics of polypeptide chains are given in the Appendix.

Methods Based on Statistical Mechanics

Energy Calculations

Statistical weights can be calculated from geometry and van der Waals forces. The idea of basing prediction methods on statistical mechanics has been most rigorously pursued by the group around Scheraga. Thus, Kotelchuck and Scheraga (363, 364) based their prediction method on the simplifications introduced for the derivation of Eq. (A-4) as described in the Appendix. They calculated the statistical weights z^{α_R}, z^{α_L}, and z^ϵ for each residue type,[5] except for Gly and Pro which were handled separately. In these calculations they made use of the known van der Waals forces (chapter 3) between different parts of the chain. As they had to account for the whole system, chain and solvent, a term estimating the free energy contribution of the solvent was included. For each residue type the conformation with the highest z was assigned as *the* residue type propensity (Table 6-1). Since no α_L preference could be found, only "helical" ($\equiv \alpha_R$)

---

[4]Let us consider a hypothetical example. Four proteins, all members of one family, have the following sequences at triplet positions ($i-1,i,i+1$): Ala-Gly-Thr, Val-Gly-Pro, Leu-Ala-Ser, and Leu-Ala-Thr. For each triplet type the corresponding table location contains a number of (ϕ,ψ)-angles, namely the number of occurrences of the triplet type in question in the data base. The average of all (ϕ,ψ)-angles found in the four table locations is then used as the predicted (ϕ,ψ)-value at residue position i for the common chain fold of the family.

[5]The statistical weights z^{α_R}, z^{α_L}, and z^ϵ are the probabilities that this residue type assumes a backbone conformation in the α_R-, α_L-, and ϵ-(extended chain) region, respectively. These regions are defined in Figure 2-3.

and "coil" ($\equiv$nonα_R) propensities were assigned. It is noteworthy that the propensities derived with these theoretical calculations by and large fit the experimental data with synthetic polypeptides (328) and the empirical data derived from the data base of globular proteins (Table 6-1).

When calculating helix potentials from helix propensities, cooperativity can be introduced. Using these propensities Kotelchuck and Scheraga (364) devised a simple algorithm for helix prediction, according to which "helices start whenever there occur four residues with helix propensity in a row," and grow toward the C-terminus until stopped by two consecutive residues with coil propensity. Unidirectional growth was assumed, because the authors in their calculations found that a given residue can influence only its C-terminal neighbor and not its N-terminal one. With this algorithm the simple scheme of noninteracting residues of Eq. (A-4) is abandoned. Nucleation and growth are introduced. Thus neighbor interaction is implicitly assumed.

The idea of helix nucleation by four residues and unidirectional growth until breaking by two consecutive coil residues was taken up again by Leberman (365). However, the rules and the propensities were slightly modified in order to obtain a better fit in the data base.

Empirical Partition Function

Helix propensities based on frequencies are equivalent to those based on empirical energy functions. With structural knowledge of globular proteins increasing appreciably, the derivation of statistical weights z^{α_R} and z^ϵ of Eq. (A-4) from empirical data became possible. For this purpose Burgess *et al.* (31) rewrote Eq. (A-4) as

$$z^x = \int_x \left[\int \exp \frac{-E(\phi, \psi, \chi)}{RT} \, d\chi \right] d\phi \, d\psi \qquad (6\text{-}1)$$

where $x = \alpha_R$ or ϵ. Since it is very small z^{α_L} is neglected. Here, the bracketed integral is the probability that the residue type in question will assume the angles ϕ and ψ in the $\phi\psi$-map (Figure 2-3). This is a Boltzmann distribution averaged over all side chain orientations. Since the function $E(\phi,\psi,\chi)$ is not known accurately enough this probability distribution cannot be calculated. However, these probabilities should be reflected in the observed frequencies of a given residue type in the $\phi\psi$-map (Figure 2-4). Therefore, the bracketed expression is well described by a smooth function of ϕ and ψ which fits the observed frequencies. Such empirical functons can then be used to calculate the integral of Eq. (6-1) over α_R- and ϵ-regions, yielding the probability or the propensity for α_R and ϵ-conformation z^{α_R} and z^ϵ. The resulting values for z^ϵ are given in Table 6-2.[6] Although

---

[6]In Table 6-2 ϵ-propensities are listed as β-propensities. Scheraga *et al.* stress the point that only ϵ and not β-structure can be predicted because the backbone conformation gives no clue about hydrogen bonding.

this propensity evaluation is based on statistical mechanics it is closely related to a purely statistical analysis. As a matter of fact it illustrates the relationship between probabilistic methods and methods based on statistical mechanics.

It is not possible to derive reverse turn propensities in the manner described above because the (ϕ,ψ)-angles of residues in reverse turn conformation are scattered over the $\phi\psi$-map and are not confined to a definite region. Here, Burgess *et al.* (31) essentially adopted the statistical evaluation of Lewis *et al.* (326) and used it with an upgraded data base.

Propensities of nonapeptides were used to find the helix potential of the central residue. For predicting secondary structure the authors (31, 366) combined the propensities of residues $i-4$ through $i+4$ to calculate the α_R-, ϵ-, and rt-potential (for definition see page 110) of residue i. These nine residues were used because earlier energy calculations had suggested (367) that a nonapeptide is the segment length which significantly influences the conformation of the central residue. Helix $(\equiv\alpha_R)$ conformation is then predicted for all those segments longer than four residues where the α_R-potential is larger than the ϵ- and reverse turn potentials as well as larger than the threshold. This threshold was defined as a weighted average of all α_R-potentials along the chain. For ϵ- and reverse turn prediction an analogous procedure was pursued.

Zimm–Bragg Model Adapted to Heteropolymers

In its original form the Zimm–Bragg model also applies to natural proteins. A helix prediction algorithm, which rigorously follows the Zimm–Bragg model, has been proposed by Lewis *et al.* (368, 369). This method is based on Eq. (A-9) when written with individual relative statistical weights s_i for each position i of the chain; not all s_i are set to s as has been done in Eq. (A-8) for homopolymers. The probability P_i of finding the residue in position i in α-conformation is then given by

$$P_i = \frac{\partial \ln Z}{\partial \ln s_i} = \frac{s_i}{Z} \cdot \frac{\partial Z}{\partial s_i} \tag{6-2}$$

which selects from all 2^N chain conformations only those which are helical at position i [Eq. (A-6)], and divides them by the partition function Z (section A.1) for normalization. This probability P_i is the helix potential at residue position i. As an example let us derive P_1 for $N = 2$ [see Eq. (A-7)]:

$$P_1 = \frac{s_1}{Z} \cdot \frac{\partial Z}{\partial s_1} = \frac{s_1}{Z} \cdot \frac{\partial(1\cdot 1 + 1\cdot s_2 + s_1\cdot\sigma\cdot 1 + s_1\cdot s_2)}{\partial s_1}$$

$$= \frac{s_1 \cdot \sigma + s_1 \cdot s_2}{1 + s_2 + s_1 \cdot \sigma + s_1 \cdot s_2}.$$

Residues with low relative statistical weights shorten the average helix length appreciably. In order to evaluate the helix potential for a given protein the authors used a single value for the nucleation parameter $\sigma = 5 \cdot 10^{-4}$ (section A.4, page 254). Furthermore, they used three different s-values for all residue types. The s assignment is given in Table 6-1 with $s = 0.385$ for B (helix-breaker), $s = 1.00$ for I (helix-indifferent), and $s = 1.05$ for H (helix-former).[7] Using Eqs. (A-18) and (A-20), σ- and s-values are derived from slopes and transition temperatures of helix-coil transitions in synthetic polypeptides. Helical conformation is then predicted for all residue positions i with P_i larger than the average P_i-value. The resulting potential functions are smooth, since Eq. (6-2) contains the cooperativity of the Zimm–Bragg model, which forces helices to assume a certain length (Figure A-1). The prediction method yields helical segments around 10 residues long. This is much smaller than the length expected with the given σ-value for homopolymers at $s = 1$ [Eq. (A-17)], namely $1/\sqrt{5 \cdot 10^{-4}} \cong 40$, demonstrating the helix-shortening effect of inclusion of residues with low s-values.

The Zimm–Bragg model was refined in order to improve predictions. In a very similar way the Zimm–Bragg model was applied by Ptitsyn *et al.* (370). For all residue types they used a single nucleation parameter $\sigma = 5 \cdot 10^{-4}$ and six different s_i-values based on experimental data from synthetic polypeptides. Following Lewis *et al.* (368), the s_i-values of all residue types for which no experimental data were available were assigned. In subsequent papers (371–374) also stereochemical considerations were included for determining s-values (Table 6-1). The Zimm–Bragg model was modified, emphasizing structural neighborhood (for example between i and $i + 3$ in an α-helix). Furthermore, the authors introduced individual σ-values for each residue type and expanded the prediction from two (helix and coil) to half a dozen different conformations.

Zimm–Bragg Parameters from Globular Proteins and Synthetic Polypeptides

The prediction method of Chou and Fasman is probabilistic; cooperativity is included. Although the prediction method of Chou and Fasman (201, 340) refers in many instances to the Zimm–Bragg model, it is of a rather probabilistic nature. The authors derived α-, β-, and reverse turn propensities from frequencies in the data base. For a given amino acid sequence they subsequently searched for α-, and β-nucleation points by combining the α- and β-propensities in hexa- and pentapeptides, respectively. Once found, each nucleation point was then extended in both directions until residues with low propensities occur. For reverse turn prediction the

---

[7]As described in section A.2 the relative statistical weights s correspond to propensities. Following Eq. (A-8) s is the helix propensity.

authors essentially adopted the scheme of Lewis *et al.* (326). The application of Chou and Fasman's method is described in section 6.3; further details are given there.

Helix propensities derived from statistical mechanics of synthetic polypeptides correspond to those based on frequencies in globular proteins. In a further analysis Chou and Fasman (201) compared the helix propensities derived from observed frequencies in globular proteins with those derived from the helix-coil transition temperatures θ of synthetic polymers on the basis of the Zimm–Bragg model. As shown in section A.5, the transition temperatures θ can be converted to relative statistical weights s and therefore to helix propensities. Chou and Fasman showed that the s-values of the seven residue types, for which synthetic polypeptide data are known, agree within 10% to helix propensities derived from globular protein frequencies. This remarkable correspondence has been examined by Suzuki and Robson (352) in further detail.

Nucleation parameters derived from synthetic polymers correlate well with frequencies at helical ends. Chou and Fasman (201) also compared nucleation parameters σ derived from helix-coil transitions (section A.5) with residue type frequencies at helical ends in globular proteins. Here, they follow Lifson and Roig (375) in averaging over both helical ends. Moreover, they consider the frequencies of three residues at each end (and not only of one residue as in the Zimm–Bragg model) because the last *three* residues of an α-helix form only one hydrogen bond each. Comparing the resulting frequencies with the six σ-values that could be derived from helix-coil transitions, they find different magnitudes but a significant correlation (the correlation coefficient between the logarithms is $+0.75$). As coil-sheet transitions of synthetic polymers are much less well known, no analogous correlations for sheet formation are available.

Methods Based on Stereochemical Considerations

Reverse turns mostly contain polar residues. A straightforward stereochemical approach to reverse turn prediction has been adopted by Kuntz (203). He observed that reverse turns occur almost exclusively at the protein surface. Therefore, he predicts that all triplets of polar residues (designated T in Table 6-2) are in reverse turn conformation.

Helices between hydrophobic core and solvent can be recognized by the pattern of nonpolar residues. Using graphs containing five turns of helix Schiffer and Edmundson (376, 377) showed that nonpolar residues cluster on one helix side, forming nonpolar arcs. An example of such a "helical wheel" is given in Figure 6-1. For helix-prediction the authors searched for nonpolar triplets at positions i, $i+3$, $i+4$ (relative positions 1-4-5) and i, $i-3$, $i-4$ (1-2-5) in a given amino acid sequence. These positions were then taken as a helix nucleus, and a helix wheel was plotted around them. The helix is assumed to continue in both directions until the nonpolar arcs are

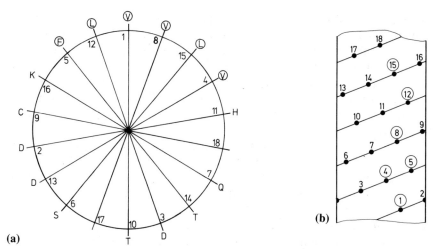

Figure 6-1. Helical wheel representation of C-terminal α-helix of adenylate kinase. (a) The wheel is the projection of all side chain positions along the helix axis onto a plane. The first residue at position 1 is Val-179, and the last residue at position 16 is Lys-194. Nonpolar residues are encircled; they all lie on one side of the helix. The three-dimensional structure (186) shows that this side points to a hydrophobic core, whereas the other side is part of the molecular surface. (b) The positions of the wheel given in a cylindrical plot of an α-helix (for definition see Figure 5-2). The positions of the nonpolar arc are encircled.

interrupted by polar residues or by Pro. This method was devised to find helices shielding a hydrophobic core from the solvent as is the case in myoglobin. Detection of helices in other locations is more difficult.

Using a much larger data base, clustering of nonpolar residues on α-helices was reexamined by Palau and Puigdoménech (378). These authors showed that the frequency of nonpolar triplets with relative positions 1-2-5 and 1-4-5 in α-helices is significantly higher than expected from the amino acid composition of these helices. Very often these triplets are enchained: 1-2-5-"1"-"4"-"5"→1-2-5-8-9, i.e., they form arcs on the helical wheel (Figure 6-1). Thus, the observations of Schiffer and Edmundson (376), which were based on only three proteins, are also valid for a large data base.

A method which recognizes residue patterns of helices and sheets. Clusters of nonpolar residues are also the basis of Lim's (379, 380) helix prediction method. As does Palau and Puigdoménech (378), the author emphasizes the intertwining of triplets in relative positions 1-2-5 and 1-4-5 and of doublets in relative positions 1-4 and 1-5. He dismisses clusters if side chains are too bulky. He also accounts for the effects of polar residues; large polar residues can stabilize adjacent nonpolar clusters by shielding them from the solvent. In β-prediction Lim distinguishes between internal, semi-surface, and surface sheet strands. Surface strands have a polar and a nonpolar side; therefore they have alternating polar and nonpolar residues.

In contrast, internal strands have nonpolar residues on both sides, revealing themselves by uninterrupted stretches of nonpolar residues. Such segments are preferentially shielded by large polar residues at their ends. Semi-surface strands are those where all side chains lie tangentially at the surface as is the case for example with the C-terminal β-strand of the central β-sheet of adenylate kinase depicted in Figure 7-6. Such β-strands should not contain large nonpolar residues as these would not be shielded from the solvent.

Lim formulated his detailed stereochemical considerations as 22 rules for α- and as 14 rules for β-prediction. He used a table of 17 antihelical doublets and 26 antihelical triplets, and he grouped amino acid residue types in 14 different ways. β-conformation is predicted only for regions where no α-helix is predicted. As his detailed rules contain a large amount of information or adjustable parameters, it seems easy to fit the data base they are derived from, even if this data base is large. Consequently, the value of this method can emerge only in tests with unknown structures.

6.3. Application of Prediction Methods

Readily applicable prediction methods are the most popular ones. Prediction methods mostly involve extensive calculations using large tables [e.g., Nagano's method (353)]; thus one has to ask the author for the computer program, the tables, and the parameters derived from the data base. Some methods can be worked out by hand but are so complicated that a long time is needed to apply them [e.g., Lim's method (380)]. Only a few prediction methods are readily applicable; probably most authors hesitated to introduce the approximations which are necessary to simplify the application. Naturally, easy availability ensures popularity. Chou and Fasman's method (340) has become the most popular. In order to provide a simple example of a prediction, its application is demonstrated for the first 24 residues of adenylate kinase.

The first step of the Chou and Fasman method is to locate helix and sheet nuclei. The prediction of helix and sheet is interdependent; therefore it is executed simultaneously. The amino acid sequence together with helix and sheet propensity values (Tables 6-1 and 6-2) and symbols (H, h, I, i, b, B) are given in Figure 6-2. As a first step nucleation centers have to be located. For helix nucleation it is necessary to search for hexapeptides, which contain four helix-forming h^α or H^α residues (I^α counts as a half h^α) and not more than one helix breaker b^α or B^α. Four of these peptides defining two nuclei are present in Figure 6-2. Similarly sheet nucleation occurs if a pentapeptide contains three h^β or H^β residues and not more than one b^β or B^β. Four such peptides defining one nucleus can be found. Both helix and sheet nuclei occur around residue 12. To resolve this

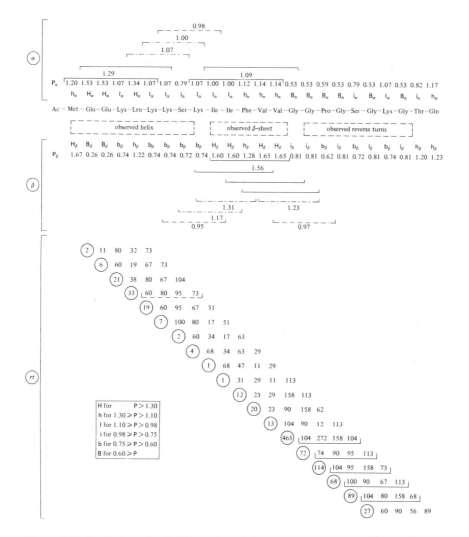

Figure 6-2. Prediction of α-helix, β-pleated sheet and reverse turns (*rt*) for the 24 N-terminal residues of adenylate kinase (389), following the procedure of Chou and Fasman (340). The α- and β-propensities P_α and P_β are taken from Tables 6-1 and 6-2. The symbols are defined in the inset. For α-predictions hexapeptides with average propensities above 1.00 are given in solid lines. Solid lines are also used for pentapeptides with average β-propensity above 1.00. The ends of helices and sheets are tested with tetrapeptides. Tetrapeptides with average propensities above 1.00 are given with dash-dot lines whereas those with averages below 1.00 are given with dashed lines. Note that in our nomenclature the average propensities could be called potentials. Reverse turn propensities depend on the position of the residue in question within the tetrapeptide to be tested. Thus, there are four propensities for each residue. All of them are multiplied by a factor of 10^{+3}. The *rt*-potential is the product of the propensities in a tetrapeptide. These potentials are marked with circles and are given on the same line as the *rt*-propensities of the tetrapeptide. All potentials are multiplied by 10^{+6}. Tetrapeptides with potentials above $50 \cdot 10^{-6}$ are marked with solid lines. A relative maximum which does not reach the threshold is indicated by a dashed line.

ambiguity the highest average propensities are compared. As the sheet propensity of 1.56 is higher than the helix propensity of 1.09 this segment is assigned as sheet nucleus.

Helix and sheet nuclei are extended to both sides. In the next step the helix nucleus at residue 4 and the sheet nucleus at residue 12 have to be extended in both directions until the average propensity of a tetrapeptide falls below 1.00. The rules do not specify exactly which of the residues of a breaking tetrapeptide should be included in helix and sheet, respectively. It seems reasonable to include only H, h, and I residues. Helix is then predicted for residues 1–7, and the sheet is confined to residues 10–14.

For reverse turn prediction the original method prevailed. For reverse turn prediction Chou and Fasman (340) adopted the original method of Lewis *et al.* (326), but used it with a larger data base. The reverse turn potential of a tetrapeptide is defined as the product of the four reverse turn propensities of residues at positions i, $i+1$, $i+2$, $i+3$ of a reverse turn (see section 6.1, page 110). All values are given in Figure 6-2. The authors apply a threshold of $50 \cdot 10^{-6}$. Therefore, reverse turns are predicted at residues 16–23.

A comparison between predicted and observed secondary structure shows good correspondence; the helix is one residue too short, and sheet and reverse turns are exact. However, the example given is not representative. Here, the α, β, and reverse turn potentials are very high and clearly separated. In general, the situation is less clear and more error prone.

6.4 Evaluation of Prediction Methods

Hazards of Judgment

Given enough parameters any data base can be fitted. If predictions are to be improved, an evaluation of the quality of a method is necessary. For this purpose many authors compared how well the method can fit the data base of structurally known proteins. But this procedure is only of limited value because fitting the data base is easier the more intrinsic information (rules or parameters) a method contains, and the amount of intrinsic information is not at all related to predictive value for proteins with unknown structure.

To circumvent this problem some authors first eliminated a given protein from the data base and then derived the parameters from this reduced data base. Subsequently these parameters were used for predicting the secondary structure of the eliminated protein. In this way the magnitudes of the adjusted parameters (the propensities given in Tables 6-1 and 6-2) became independent, but not the rules for combining these parameters. These rules, the quintessence of each method, were still based on all known proteins including the predicted one.

Some methods are not strictly defined. In addition to the difficulties in demonstrating an independent quality of fit, some methods turned out to be irreproducible (381, 382) because they had not been defined strictly enough, leaving room for individual judgments during application. Together, all these hazards kept the enthusiasm for prediction methods at a low level. However, irreproducibility and data base bias do not matter if the secondary structure of a structurally *unknown* protein is predicted. Therefore, the test cases of adenylate kinase (383) and T4-lysozyme (384), where numerous authors supplied predictions for a protein unknown to them, were decisive in overcoming widespread scepticism.

Quality Indices

There are various quality indices emphasizing different aspects. A number of quality indices have been proposed for a detailed comparison of methods. All of them sum up prediction outcomes at individual residue positions, yielding the w, x, y, z-values (for definition see Table 6-3) for a given protein. However, the arithmetic combinations of w, x, y, z are at variance, each index emphasizing different aspects. A comprehensive list is given in Table 6-4.

A very popular index yields deceptively high values. Index Q_1 severely punishes overprediction. In contrast, Q_2 largely neglects overprediction; in the extreme case that each residue is predicted as helical ($w = 1$), Q_2 even assumes its value for optimum fit ($Q_2 = 1$). Index Q_3 has become very popular, but it has the disadvantage of giving deceptively high values. Here, positive and negative correct predictions (Table 6-3) are counted with equal weight. But, since it is known that the average helix contents of proteins amounts to about 35% (201), the chance of a negative correct helix prediction is about twice as high as the chance of a positive correct prediction. Consequently, Q_3 is dominated by the negative correct prediction. This problem is illustrated by examples 1 and 2 of Table 6-4. Example

Table 6-3. Definition of the Four Cases Arising in Prediction of α-Helix (or Other Types of Secondary Structure) at a Given Residue Position

Status of residue		Notation	Percentage of this case in a given protein
Predicted	Observed		
Helical	Helical	Positive correct prediction	w
Nonhelical	Nonhelical	Negative correct prediction	x
Nonhelical	Helical	Underprediction	y
Helical	Nonhelical	Overprediction	z
			$w + x + y + z = 1$

Table 6-4. Indices in Use for Assessing the Quality of a Prediction[a]

Description	Suggesting and applying authors	Quality index	Formula	Range	Example 1 $w = 0$ $x = 0.7$ $y = 0.1$ $z = 0.2$	Example 2 $w = 0.1$ $x = 0.7$ $y = 0.1$ $z = 0.1$
Positive correct − overprediction / observed	Prothero (332)	Q_1	$\dfrac{w - z}{w + y}$	$(-\infty, +1)$	—	—
Positive correct / observed	Lewis and Scheraga (369) Nagano's [% corr. ass. 3] (353) Chou and Fasman's [% α] (340)	Q_2	$\dfrac{w}{w + y}$	$(0, 1)$	—	—
Positive and negative correct	Kotelchuck and Scheraga (364) Lewis et al. (368) Robson and Pain (346) Leberman (365) Nagano's [% res. corr.] (353) Chou and Fasman's [% α + $n\,\alpha$] (340)	Q_3	$w + x$	$(0, 1)$	0.70	0.80
Average of positive correct / observed + negative correct / nonobserved	Pitsyn and Finkelstein (342) Chou and Fasman's [Q_α] (340)	Q_4	$\dfrac{1}{2}\left(\dfrac{w}{w + y} + \dfrac{w}{x + z} \right)$	$(0, 1)$	0.39	0.69
Positive correct / 1 − negative correct	Nagano's [% corr. ass. 2] (353)	Q_5	$\dfrac{w}{1 - x}$	$(0, 1)$	0.00	0.33
all correct / all incorrect	Kabat and Wu (359)	Q_6	$\dfrac{w + x}{y + z} = \dfrac{Q_3}{1 - Q_3}$	$(0, \infty)$	—	—
Correlation coefficient between positive predicted and observed as well as negative predicted and nonobserved	Matthews (384) Argos et al. (381) Lenstra (382)	Q_7	$\dfrac{w \cdot x - y \cdot z}{\sqrt{(x + y) \cdot (x + z) \cdot (w + y) \cdot (w + z)}}$	$(-1, +1)$	−0.17	+0.38

[a]Two examples are given in order to demonstrate the differences.

1 shows a useless prediction since none of the helical residues has been found. Nevertheless, Q_3 has a value as high as 0.70. Example 2 is less extreme but is still a useless prediction, because half of the helix residues have been assigned correctly and the other half incorrectly. At $Q_3 = 0.80$ the resulting index is already close to its value for optimum fit. For sheets, the domination by the negative correct prediction is even more severe, since the average sheet contents of proteins is only about 15% (201).

Some indices scale down the negative correct prediction. Index Q_4 damps the contribution from the negative correct prediction. It weathers examples 1 and 2 better than Q_3. Compared with Q_3 the index Q_5 accounts for the negative correct prediction by multiplying with $1/(1 - x)$ instead of adding x. Thus, more appropriate values are obtained for examples 1 and 2. However, this index has never become popular. Index Q_6 is a monotonic function of Q_3 and yields no additional information. Index Q_7 is a common statistical measure. It prevents the negative prediction from dominating the result because the prediction is related to its average value (384), so that for example the negative correct α-helix prediction at a given residue position gets a low weight if the protein contains only a few helices, and a large weight if it contains many helices.

Predictions of all three types of secondary structure should be assessed simultaneously. None of these quality indices account for α-, β-, and reverse turn structures simultaneously. However, as all three predictions are interdependent (e.g., nobody will predict β-structure where a high α-potential is found) only a composite index appropriately describes the overall success of a method. A simple composite index is

$$Q_3^{\text{total}} = w_\alpha + w_\beta + w_{rt} + w_{\text{coil}}$$

where w_α, w_β, and w_{rt} are the percentages of (positive) correctly predicted α, β, and reverse turn residues, respectively, and w_{coil} is the negative correct prediction, corresponding to x of Q_3. Since proteins contain on the average about 35% helix, 15% sheet, 25% rt, and 25% coil, all four contributions are approximately balanced and no domination by negative correct prediction occurs.

Present State of the Art

In two test cases more than two-thirds of the residues were assigned to the correct secondary structures. The overall quality of prediction methods is demonstrated in Table 6-5 in which quality indices are given for the two test cases adenylate kinase (383) and T4-lysozyme (384) together with those for data base fitting. For both test cases a number of predictions are available. However, we took only the average of the two best predictions for every entry in Table 6-5. Clearly, indices $Q_{3\beta}$ and Q_{3rt} are dominated by the negative correct prediction; they are deceptively high. The more appropriate correlation coefficients Q_7 are around $+0.45$; they are best for α- and worst for reverse turn prediction. Q_3^{total} amounts to 68%, showing that the conformational status of more than two-thirds of the residues has been

Table 6-5. Quality Indices in Test Cases and in Data Base Fitting[a]

	Structural class of protein (Table 5-2)	Number of residues	α	β	rt	Total	$Q_{3\alpha}$	$Q_{3\beta}$	Q_{3rt}	$Q_{7\alpha}$	$Q_{7\beta}$	Q_{7rt}	Q_3^{total}
Test case adenylate kinase	4	194	0.54	0.12	0.24	0.90	0.77	0.92	0.86	0.56	0.58	0.60	0.67[b]
Test case T4-lysozyme	3	164	0.57	0.07	0.15	0.79	0.69	0.87	0.75	0.42	0.28	0.20	0.69[c]
			0.55	0.10	0.20		0.73	0.90	0.81	0.49	0.43	0.40	0.68
Data base	1									0.50	—	0.41	
Data base	2									0.22	0.45	0.37	
Data base	3									0.58	0.51	0.45	
Data base	4									0.51	0.50	0.43	
										0.45	0.49	0.42	

[a] In the test cases (383, 384) the two best values are averaged. The data base contains all structurally known proteins; values are taken from Nagano (356) and Argos et al. (381) as reported by Lenstra (382). Numbers below bars are the average values of the numbers above.
[b] Average of predictions of Nagano (353) and Chou and Fasman (340).
[c] Average of predictions of Ptitsyn et al. (371, 380) and Burgess et al. (31).

correctly predicted. For adenylate kinase all Q_3 and Q_7 indices are higher than for T4-lysozyme, but the Q_3^{total}-values are similar. This discrepancy occurred because in adenylate kinase the amount of overprediction is larger than for T4-lysozyme; and overprediction is more severely punished by Q_3^{total} than by Q_3 or Q_7 indices.

In general, the quality of prediction does not depend on the protein class. In both test cases the protein contains about 55% helix as compared to the average value of 35% (201). Therefore, general validity of the results obtained cannot be expected *a priori*. However, a survey of predictive quality in the data base (382) resulted in Q_7-values very similar to the test cases (Table 6-5). Hence, it can be stated that prediction methods in general reach a correlation coefficient of about +0.45 for α, β-, and reverse turn prediction; and this is far above random. In his survey Lenstra (382) divided the data base proteins into classes No. 1 through 4 as given in Table 5-2. Except for α-prediction in class No. 2 (β-proteins), which seems to be extremely difficult, the quality of prediction is similar in all classes of proteins.

If several predictions are available, they can be averaged. In evaluating the test on adenylate kinase an additional, collective prediction method was suggested (383). This consists of an unweighted average of all individual predictions. For adenylate kinase the average yielded a better correspondence with the observed data than any individual prediction. This proposal has been taken up by Argos *et al.* (381) who used it with five available prediction methods on all known protein structures.

Assessing individual success is the key to further improvements. To improve prediction in general a differential diagnosis of individual success is necessary. For this purpose the two test cases are valuable because they allow the reader to diagnose by himself. But again it should be kept in mind that the tests were performed on proteins from only two of the five classes of Table 5-2. This shortcoming has been circumvented by Lenstra (382) who compared the success of three individual and one collective prediction method in the total available data base, using Q_7 as indicator. He finds clear quality differences. In general, the collective method (383) is the best one.

Self-consistency can be used as a measure of quality. A quite different way of evaluating the quality of a prediction method is to apply it to a number of homologous proteins. It can be expected that such proteins have the same chain fold and the same secondary structures. Therefore, predictions should be invariant to the observed amino acid exchanges; the smaller the variation the better the prediction method. Such quality check has been performed for three prediction methods using 24 homologous sequences of pancreatic ribonuclease (385).

Further improvements are expected from methods based on stereochemical considerations. Can predictions be further improved? For this purpose data base extension does not seem to help much. This is particularly obvious for singlet and doublet propensities, because here the observed frequencies are already high enough to yield good averages. Therefore not much improvement is expected from probabilistic methods. Methods based on stereochemical considerations seem to stand the best chance for further development.

Moreover it is no longer worthwhile to deal with α, β, and reverse turn structures separately. The time has come for composite procedures. One should also aim at predicting the protein class. The ultimate goal, the prediction of the chain fold, still seems remote.

6.5 Emerging General Results

Local Interactions and Prediction Success

It is difficult to compare qualities of prediction of α- and β-structures if α- and β-contents are at variance. There is a widespread impression that helix prediction works better than sheet prediction. Since residue interactions via hydrogen bonds are local (along the chain) in helices and nonlocal in sheets, a reasonable explanation is available. Accordingly, quality indices should be higher for helices than for sheets. However, Table 6-5 shows no significant difference with respect to Q_7. This discrepancy between intuition and Q_7-values can be reconciled by considering that proteins contain less sheet than helix, which leads to a larger negative correct prediction (Table 6-3) for sheets. Although the negative correct prediction is somewhat damped in Q_7, it significantly increases $Q_{7\beta}$ relative to $Q_{7\alpha}$. Thus, despite a smaller positive correct prediction in sheets, which gives the impression of sheet prediction inferiority, $Q_{7\beta}$ and $Q_{7\alpha}$ are at comparable levels.

Reverse turns do not reflect local interactions. In Table 6-5 the reverse turn prediction comes out worst. This can be understood if reverse turns are determined by nonlocal interactions in spite of their geometrically local nature. By such an assumption a reverse turn should be viewed as a point of least resistance in folding, a passive kinking point, and not as an actively folding element. Energy considerations come to the same conclusion because a reverse turn contains only one hydrogen bond. Moreover, reverse turns are usually at the surface so that the hydrogen bond is exposed to water, which decreases its free energy contribution.

Prediction Success and Folding Nuclei

Folding nuclei may reveal themselves by exceptionally successful predictions. It has been suggested (383) that the best predicted secondary structures have the highest probability of being folding nuclei because good prediction indicates dominance of local interactions, and dominant local interactions allow independent folding. Such a folding nucleus consisting of a helix-tripod has been proposed for adenylate kinase (383). It is the small domain of Figure 5-14b.

This idea is corroborated by the finding that in both test cases the N-terminal half of a chain is predicted better than the C-terminal one (383, 384) because the N-terminal half is synthesized first on the ribosome and it is supposed to fold first. Using correlation coefficient Q_7, Argos *et al.* (381) put this finding on a quantitative base. They demonstrated that on the average predictions are significantly better in the N-terminal half.

Prediction as an Analytical Tool

Current methods can probably predict the structural class of a protein. Today, predictions seem to be accurate enough to allow a protein to be placed into one of the five classes of Table 5-2. The secondary structure patterns are sufficiently different in these groups so that even with a 68% correct prediction (Q_3^{total} of Table 6-4) they can be distinguished from each other.

Predictions aid in recognition of a new member of a protein family. Beyond detecting the structural class, there have been attempts to detect known chain topologies. Applying Chou and Fasman's method (340) on glutamate dehydrogenase, Wootton (386) found two domains which have secondary structure patterns similar to the nucleotide binding domain of other structurally known dehydrogenases (254). After pattern alignment he recognized amino acid sequence homology to one of these dehydrogenases (glyceraldehyde-3-phosphate dehydrogenase). This increased the significance of similarity (section 9.6) enough to indicate an evolutionary relationship (387), and consequently the existence of a nucleotide binding domain with the same chain topology. For the other domain in question no sequence homology could be found, only the secondary structure patterns resemble each other. Consequently the significance is too low to conclude topological similarity.

The same aspect was followed with Ca^{2+}-binding proteins by Argos (388) who demonstrated that predictions are a welcome adjunct for aligning amino acid sequences, especially in ambiguous cases. Moreover, he used predictions to propose Ca^{2+}-binding sites in the acyl-carrier protein of fatty acid synthetase of *E. coli* and in proteins of the blood coagulation cascade.

Summary

In experiments with peptide homopolymers it was found that some amino acid residue types tend to form α-helices and others do not. This knowledge was applied to globular protein structures as soon as these became available. It turned out that residues which form helices in homopolymers preferred to be in helices of globular proteins as well. This correspondence gave rise to various attempts to correlate the amino acid sequence of a protein with the helices of that protein. Later on, other secondary structures were also included in the correlation. These attempts are most interesting because they form the basis of any future method which tries to derive the three-dimensional structure of a protein from the amino acid sequence alone. Accordingly, we classify and describe most of the current methods for secondary structure "prediction" from the sequence. Among these methods those based on statistical mechanics need a thorough introduction. This is given in the Appendix.

The quality of the current methods is such that about two-thirds of all residues can be assigned to the correct secondary structure. There is no great quality difference between probabilistic methods, which do nothing but analyze structural data mathematically, and physico-chemical methods, which try to incorporate knowledge on the physical behavior of a polypeptide chain. This demonstrates that the correlation is not well understood as yet. For a long time individual methods were difficult to compare because various quality indices were in use. The relationship between these indices and their relative merits are discussed here. Quality comparisons are essential because they facilitate further improvements. Furthermore, it turned out that predictive success is not only useful for determining three-dimensional structures but also yields information on the folding process by suggesting those chain segments which fold first.

Chapter 7
Models, Display, and Documentation of Protein Structures

7.1 Structural Ingredients

In many cases either the covalent or the geometric structure is the only one known. Structural knowledge of a protein can be knowledge of amino acid sequence (covalent structure), knowledge of three-dimensional (geometric) structure, or knowledge of both. There are many more covalent structures analyzed than geometric ones. On the other hand, there are about a dozen geometric structures in which the amino acid sequence is not yet known (80, 124, 217, 221, 236, 261, 303, 306, 307, 310, 313, 316). All elucidated covalent structures are compiled in the *Atlas of Protein Sequence and Structure* (20). A recent survey of analyzed geometric structures is given in Table 5-2.

Covalent Structures

Since proteins consist of linear polypeptide chains the documentation of the covalent structure is simple. It can be presented by a linear string of amino acid residues, abbreviated with the one letter code (Table 1-1), which is easily printed and read. For reference purposes each residue gets a sequential number along the chain.

In general, sequence numbering of a family of homologous proteins is uniform. The documentation of homologous sequences becomes much more comprehensive if they are aligned and presented in a single printout as in Figure 7-1a. Here, all sequences refer to the same numbering scheme, which is possible only if deletions are allowed for. Difficulties arise, however, if later on homologous sequences are elucidated that have one or more additional residues. In original publications the established number-

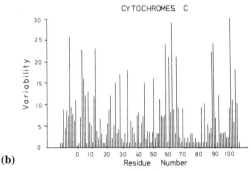

	3	4
	... 7 8 9 0 1 2 3 4 5 6 7 8 9 0 1 ...	

Horse	C H T V – – – – – – E K G G K
Neurospora crassa	C H T L – – – – – – E E G G G
Ginkgo biloba	C H T V – – – – – – Z K G A G
Euglena gracilis	C H S A – – – – – – Q K G – V
Crithidia oncopelti	C H T G – – – – – – A K G G A
Rhodospirillum rubrum	C H T F – – – – – – D Q G G A
Rhodopseudomonas palustris	C H R A – – – – – – D – – – K
Rhodopseudomonas sphaeroides	C H V I V D D S G T T I A G R
Rhodopseudomonas capsulata	C H S I I A P D G T E I V – K
Paracoccus denitrificans	C H M I Q A P D G T D I – – K

(a)

(b)

Figure 7-1. Representation of amino acid sequences. (a) Alignment of residue positions 27 to 41 in several cytochrome *c* sequences taken from Ref. (145). The dashes are deletions. (b) Variability plot of cytochromes *c* taken from Ref. (408). The variability is defined as the number of different amino acids occurring at a given position divided by the frequency of the most common amino acid at that position.

ing scheme is usually retained and the additional residues are assigned the number of the preceding residue (e.g., 27) together with letters in alphabetic order (e.g., 27A, 27B, ...). Every few years the *Atlas of Protein Sequence and Structure* (145) reshuffles all sequences so that the additional positions become regular ones. But reshuffling changes residue assignments, which becomes a nuisance for reading and writing. It would be more sensible to freeze all numbering schemes for long periods of time, say 10 years, before updating them.

Low variability at a given residue position indicates structural and functional importance of this position. Sequence alignments show that the susceptibility of a residue position to amino acid changes varies greatly. This observation is depicted in Figure 7-1b. For a given position a high variability indicates that the residue at this position is not very essential for the protein's structure or function. Since most exchanges occur at the protein surface, the variability plot also provides some information as to whether a residue position is at the surface or in the interior of a protein. This is helpful if the geometric structure is still unknown.

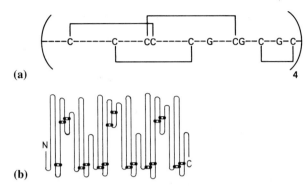

Figure 7-2. Representation of disulfide bond patterns in proteins. (a) The S–S bond pattern of wheat germ agglutinin. Exactly the same pattern occurs four times in a single polypeptide chain. Invariant Gly residues are inserted. The four repeating parts are also apparent in the three-dimensional structure (316). (b) The S–S bond pattern of human serum albumin (409). There are repeats, but they are not very exact. The amino acid sequence shows also repeats. The three-dimensional structure is not yet known.

As disulfide bridges tend to be well conserved, their patterns reveal evolutionary relationships. Many proteins, in particular extracellular ones, contain disulfide bridges as covalent cross-links between distant parts of the polypeptide chain. The documentation of these cross-links is shown in Figure 7-2. S–S-bridges tend to be well conserved during evolution so that they are characteristic for a family of homologous sequences. Moreover, they may indicate structural repeats within a single chain as in the case of wheat germ agglutinin (Figure 7-2a) where the pattern has been derived from an electron density map, or human serum albumin (Figure 7-2b) where the pattern has been determined with chemical methods.

Geometric Structures

Without knowledge of the covalent structure a polypeptide chain can be traced in the electron density map at an effective resolution of 3 Å. If an X-ray analysis of protein crystals is not supported by a covalent structure analysis, the result obtained (the electron density map) is only of limited value. At a nominal resolution of 3 Å, that is at an effective[1] resolution

---

[1]Nominal and effective resolution have to be distinguished. The nominal resolution corresponds to the resolution limit of structure factors used in the Fourier synthesis, whereas the effective resolution refers to the resulting electron density map. The effective resolution is worse than the nominal one. The difference can be appreciable if the accuracy of the structure factor phases is low or if the structure factor set is incomplete. Unfortunately, there is no unique measure for an effective resolution. In publications usually the nominal resolution is given.

somewhere above 3 Å, tracing of the polypeptide chain in the electron density map is frequently ambiguous and leads to false connections. This problem is reflected in various (often quiet) changes in the respective published geometric structures. At a higher resolution and especially after structural refinement chain tracing becomes unequivocal, and to some extent residue identities can be read off the electron density map. This considerably facilitates sequence analysis. Even at an effective resolution of 3 Å, the geometric structure can be used to overlap peptide fragments (289), which is already a major benefit for sequence analysis.

At an effective resolution of 3 Å, the electron density map allows the assignment of C_α positions with reasonable accuracy. In general the C_α-coordinates are deposited at the Protein Data Bank (390), which distributes them on request. They are further used to document the chain fold by an exact stereo drawing (section 7.2, page 142). Some chain folds are documented only by structure cartoons.

Complete Structures

A complete structure is documented as a list of atom identities and coordinates. A complete structure is a combination of covalent and geometric data. For a given atom the covalent structure reveals the identity and the geometric structure reveals the coordinates. Hydrogen atoms with their single electron are so faint in the electron density map that they are observed only in very rare cases. But most hydrogen positions can be derived to a good approximation from the coordinates of the non-hydrogen atom they are attached to.

The complete structure is usually documented as a list of all atomic coordinates. In this list the identity of a given atom is described by its internal name in the respective residue type as specified in Figure 1-1, by the residue type itself, and by the sequential number of the residue (e.g., $O_{\epsilon 2}$ of Glu-131). In some cases atom flexibilities in the protein crystal are also given. The Protein Data Bank (390) is the place where these lists are now compiled and edited. On request data are distributed via magnetic tape. A potential user has to write his own programs to evaluate and/or represent these data.

7.2. Representations of Complete Structures

The long data lists that document a complete structure can scarcely be assimilated by the human mind. Consequently their scientific usefulness remains limited until a more comprehensive representation is given. In the following commonly used representations will be discussed.

Three-Dimensional Models

Exact molecular models are unwieldy but appropriate. The only appropriate representation of the molecular structure is still a three-dimensional model. Such a model is indispensable for the design of any experiment based on the knowledge of the complete structure. In particular, it is best suited to find positions that fit substrates and effectors. However, these models are expensive and they fill valuable laboratory space. Therefore, only a few of them are built.

Wire Models

Kendrew–Watson wire models are the most accurate ones. Kendrew–Watson wire models (391) with a scale of 1 Å = 20 mm and a wire diameter of 2 mm are generally built as the most accurate three-dimensional representation of a protein. An example is depicted in Figure 7-3a. Atomic positions are at the branching points and at the ends of the wires. They are not specially marked. Bond angles, bond lengths, and the peptide unit are rigid, but they can be modified. Dihedral angles can freely rotate; they are fixed with screws on connectors. Since only thin wires and connectors are used, the model is highly transparent.

In general Kendrew–Watson models are built directly from the electron density map. For this purpose the map is plotted at equal scale (1 Å = 20 mm) on translucent plastic sheets. The model is then built "into" the map, that is, model and map are superimposed using a half-silvered mirror and appropriate illumination (392). The procedure is sketched in Figure 7-4. Model transparency helps because every part remains visible in the mirror even after completion of the whole molecule. Finally, the atomic coordinates are read from this model with telescopes running along three perpendicular axes.

Backbone wire models are easy to build. If only a chain fold representation of low accuracy is required, a linear wire model with kinks at the C_α-positions can be built. Wire bending angles (virtual bond angles and virtual dihedral angles of Figure 7-10) are calculated from the C_α-positions and applied on a special bending device (393) to a wire of 3 mm (or $\frac{1}{8}$ inch) diameter. A convenient scale is 1 Å = 5 mm. Important side chains can be added after completing the backbone. The resulting model is well suited for demonstration purposes; it can be carried around easily. An example is depicted in Figure 7-3b.

Push–Fit Models

Plastic models facilitate visualization of chemical properties. Another quite popular model consists of push–fit plastic parts designed by Nicholson (394). This model represents all atoms; non-hydrogen atoms are balls of 8–

Figure 7-3(a). Representation of protein structures. (a) Adenylate kinase as built with Kendrew–Watson wire model parts (391), (scale 20 mm = 1 Å). (b) Adenylate kinase as built as backbone model using Byron's bending device (393). Three side chain positions are indicated (scale 5 mm = 1 Å). (c) Adenylate kinase as built from Nicholson push–fit model parts (394) (scale 10 mm = 1 Å). (d) Space-filling model [CPK-model parts, scale 12.5 mm = 1 Å (395)] of ribonuclease S (courtesy of H. W. Wyckoff, New Haven, USA). (e) Space filling model (CPK) of the active site of carboxypeptidase-A in stereo, (courtesy of W. N. Lipscomb, Cambridge, USA). (f) Molecular surface display of cytochrome c [courtesy of R. J. Feldmann and T. K. Porter, (399)]. This is a space filling model in stereo as generated by a computer graphic system and displayed on a (color) television screen.

10 mm diameter, and hydrogens are indicated only by the bonds leading to them (bond diameter is 4 mm). An example is given in Figure 7-3c. At a scale of 1 Å = 10 mm Nicholson models are much more compact than the Kendrew–Watson wire models.

The plastic parts are colored so that chemical properties can be easily visualized; positively charged groups (ammonium and guanidinium) are

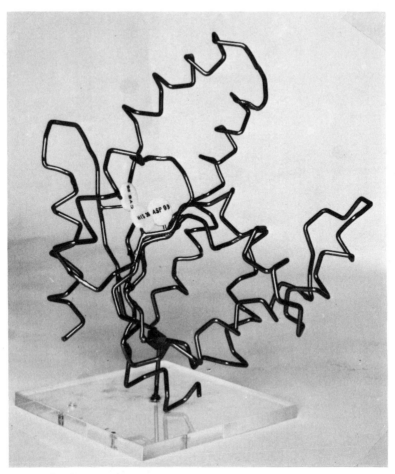

Figure 7-3(b). (Continued)

blue, negatively charged (carboxyl) and hydroxyl groups are red, sulfur is yellow, and aromatics are violet. Since the main chain is white and all side chains are grey and/or colored it is easy to follow the backbone. For these reasons Nicholson models are very suitable for anybody who wants to find his way through a molecular structure quickly.

Nicholson models lack the accuracy of the Kendrew–Watson models. Moreover, hydrogen bonds are difficult to place. Balls and sticks are large with respect to the scale so that the model lacks transparency. For all these reasons it is best to build Nicholson models from atomic coordinates and not "into" an electron density map (Figure 7-4). From the viewpoint of a user of structural results, push–fit models allow to adopt structural knowledge in considerable detail quickly and cheaply. They are also suitable for teaching purposes.

Figure 7-3(c). (Continued)

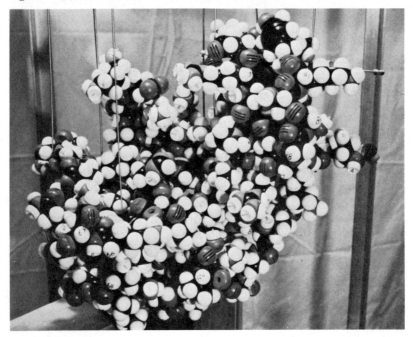

Figure 7-3(d). (Continued)

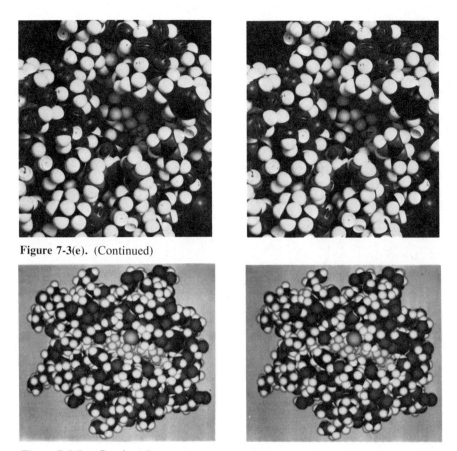

Figure 7-3(e). (Continued)

Figure 7-3(f). (Continued)

Space-Filling Models

Space-filling models show the molecular surface appropriately. Space-filling models of whole protein molecules are very difficult to build. Since the interior is not visible after completion, these models illustrate only surface properties. Therefore, they are mainly used to depict substrate and effector sites, that is, to visualize protein-substrate and protein-effector interactions. An example is shown in Figures 7-3d and 7-3e.

Unfortunately, the most popular space-filling model parts designed by Corey, Pauling, and Koltun (CPK) (395) are at a scale of 1 Å = 12.5 mm which fits neither Nicholson nor Kendrew–Watson models. Hence, these models cannot be combined, as would be desirable in many instances.

Model Maps

Model maps correspond to electron density maps. The initial representation of an X-ray structure analysis takes the form of an electron density map, a stack of transparent sheets with density contours, which has to be inter-

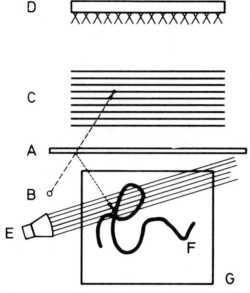

Figure 7-4. Richards optical comparator (392) viewed from the top. A, Half-silvered mirror. B, Eye of manipulator. Model and map are superimposed when looking into the mirror. C, Electron density map drawn on translucent sheets. D, Diffuse light to illuminate the map. E, Focused light shining on the part of the model that is being built. F, Model in nascent state. G, Rigid frame for model.

preted in terms of atoms. Consequently, the elucidated molecular structure can also be represented on a stack of transparent sheets. This has been done for ribonuclease (396) by printing all bonds as thick bars and all van der Waals envelopes by sets of joined circles. However, map models have not become popular, because they are not as versatile as other types of models for visualizing interactions with partners. Furthermore, they are too expensive as a means of publishing structures.

Computer Graphics

Projection of a rotating object onto a TV screen allows three- dimensional perception. For a long time computerized graphical systems were no match for hardware models. The limitations introduced by the two-dimensional screen of the cathode ray tube were too severe to appropriately represent three-dimensional structures. This has been changed with the advent of fast cathode ray tubes and of fast electronic devices dedicated to calculating coordinate changes during rotation (397). With such equipment a large set of vectors can be rotated, and the projection of this set onto a given plane (defining the view) can be calculated and displayed on the screen instantaneously. From a speed of about one rotation per ten seconds onward, the view on the screen resembles the view we have when regarding an object that is being rotated in our hands. Thus, the rotating object (or the vector set) on the TV-screen is perceived in three dimensions.

The vector set on the TV-screen can depict a covalent structure, if, for example, the vectors correspond to the wires of a Kendrew–Watson model. Alternatively, the vector set can depict an electron density map. Of course, parts of model and electron density can also be displayed together. Thus model fitting to the map can be checked. Furthermore, as all these systems are interactive, the model can be manipulated by the investigator until it best fits the map (398).During his manipulations the investigator can keep three-dimensional perception by rotating model and map on the screen.

One day computer graphics may replace molecular hardware models. With these capabilities the graphics system approaches the possibilities of an optical comparator (Figure 7-4). In addition the graphics system allows the density map to be displayed in any desired direction or at any desired level and not only in the fixed direction and levels of an optical comparator. Such directional alterations are often crucial for visualization of the electron density distribution.

Space-filling models can also be displayed on TV-screens (397, 399). But here the calculation of hidden parts is still so time consuming that one cannot rotate the model on the screen fast enough to generate three-dimensional perception. An example is given in Figure 7-3f.

For now, graphics systems are confined to large laboratories which can afford the price and also the manpower necessary for developing and adapting programs. However, the costs of electronic equipment are decreasing rapidly so that, one day, graphic systems may supersede optical comparators and possibly even molecular hardware models.

Two-dimensional Representations

Stereo Pictures

Stereo pictures are versatile two-dimensional representations of a three-dimensional object. Two-dimensional representations are necessary for publication of a structure in a journal. For this purpose stereo pictures are very convenient. In general such stereo pictures consist of two adjacent figures at eye separation (65 mm), each of which has to be viewed by the respective eye. A stereo viewer (400) is helpful. The distance of 65 mm between the pictures restricts the picture sizes to about 50 × 50 mm. Given the limited printing resolution, the desired amount of information cannot always be incorporated into this format. By printing and viewing stereo pictures in red and green colors this format can be exceeded. However, the mutual exclusion of red and green is usually not complete and color printing is expensive, so that red–green stereos are rare.

Exact Line Drawings

Stereo line drawings are appropriate structure representations. For the publication of chain folds it is convenient to prepare stereo drawings containing all C_α-atoms connected by straight lines. An example is given in Figure 4-2a. Sometimes, C_α-atoms are marked by small circles and carry their sequential number. It is astonishing how well the chain can be visualized and followed in such a stereo picture, even if it contains as many as 800 residues as in the case of phosphorylase (236).

If not only C_α-atoms and the virtual bonds between them but all bonds between any non-hydrogen atoms are drawn out, stereo pictures of whole molecules contain too many lines to be intelligible. Therefore, only molecular sections are displayed with this detail. A versatile collection of such partial stereo views of all known complete structures is available as the microfiche atlas AMSOM (401). Although it is very detailed (about 10^3 graphs per molecule) the AMSOM atlas can provide only standard views. Much more flexibility is obtained if molecules are represented on a cathode ray tube display. Stereoscopic vision can be provided by splitting the screen in two halves, or by rotating the underlying molecule in real time (397) (see section 7.2, page 140).

The graphics program ORTEP can be used for ball and stick as well as for space-filling models. A very elegant representation of molecules is achieved by the graphic program ORTEP which has been developed by Johnson (324) for the illustration of small molecules. ORTEP gives an exact view of a ball and stick model in mono or stereo. It has been applied to depict the structure of collagen in Figure 5-6a. ORTEP is very useful for presenting essential details of protein molecules as, for example, the active site of adenylate kinase in Figure 7-5. If each atom is given by a sphere with the respective van der Waals radius, ORTEP produces almost as excellent

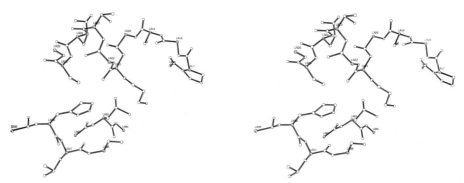

Figure 7-5. ORTEP drawing in stereo (324) of the active center of adenylate kinase. Only non-hydrogen atoms are given. C_α-atoms are labeled.

an impression of the molecular surface as a photograph of a space-filling model (Figure 5-6c). However, a small error in the intersection of the spheres of bonded atoms is committed.

Structure Cartoons

Cartoons relieve the reader of unnecessary details. Because the amount of graphic information in a protein is very large, a reader tends to lose interest if he is confronted with full detail. In general he would be happy enough to get an overview which emphasizes the essential parts. For this reason it has become fashionable to produce structure cartoons (for masterpieces see Ref. 794). Usually in these cartoons α-helices simply become cylinders, and sheet strands become arrows. An example is given in Figure 7-6. To prevent gross errors these sketches are generally drawn over backbone plots done on a computer.

Recently it has become popular (235, 247, 255) to perform a kind of molecular striptease which reduces the detailed structure to a simple β-sheet topology. Here, the whole chain fold is referred to the sheet, that is, only sheet strand directions and connections are considered. As shown in Figure 7-7, this renders the representation essentially two-dimensional. In more elaborate sketches (235, 249) α-helices in the connections between sheet strands are indicated (Figure 5-17e). These sketches and the sheet topology concept clearly helped to reveal the preponderance of the right-handed $\beta\xi\beta$-unit (section 5.2, page 81). Note that a sheet topology can easily be combined with amino acid sequence data in a single graph (186).

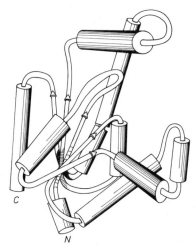

Figure 7-6. Structural cartoon of adenylate kinase. Helices are given as cylinders and β-pleated sheet strands as arrows.

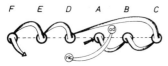

Figure 7-7. Sheet topology of lactate dehydrogenase and other NAD-dependent dehydrogenases (255). The twist of the parallel-stranded sheet (see Figure 5-10) is not shown. The six strands (circles A to F) point with their carbonyl ends to the viewer. All connections between the sheet strands are, according to the definition given in Figure 5-12a, right-handed.

Hydrogen Bond Structure

Cartoons can be efficiently applied for displaying chemical data. Apart from the information contained in the complete structure of the polypeptide chain as defined above, there are additional and essential data on hydrogen bonds and other noncovalent interactions between atomic groups on the backbone and on side chains, as well as data on any interaction (including covalent ones) of the polypeptide chain with prosthetic groups, cofactors, substrates, metals and other ligands, water molecules, etc. Usually these data are combined into a long list, from which the desired detail can be extracted. A much better documentation, however, is given in structure cartoons, as the one in Figure 7-8, which are devoted to presenting such chemical data.

7.3 Abstract Chain Fold Representations

Up to now the documentation of molecular structures as given by the cartesian coordinates of its atoms has been discussed. For several applications, in particular for theoretical structure evaluation, more abstract representations are desirable.

Plots of Backbone Angles

The backbone is fully described by all (ϕ,ψ)-angles. As described in chapter 2, the chain fold is completely described by the dihedral backbone angles ϕ and ψ at each C_α-atom (only ψ at the N-terminal C_α, and only ϕ at the C-terminal C_α). These angles can be given as points in the $\phi\psi$-map of Figure 2-4. However, for retrieving (ϕ,ψ)-angles all entries should carry the sequential number of the corresponding C_α-atom. Such a map would be crowded, difficult to read, and therefore a poor structure representation.

A more convenient representation was suggested by Balasubramanian (402). He converted the $\phi\psi$-map to a linear plot of ϕ and ψ angles versus sequential residue number, as shown in Figure 7-9. Such a plot has the advantage that the (ϕ,ψ)-angles of a given residue can be easily retrieved. In addition, it efficiently reveals regions of secondary structure.

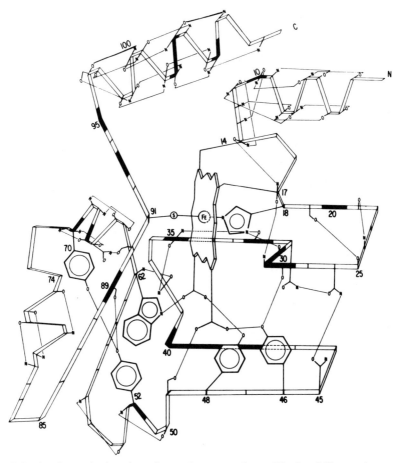

Figure 7-8. A schematic drawing of cytochrome c_2 from *Rhodospirillum rubrum* (272) showing hydrogen bond interactions. Residues lining the haem crevice are indicated by darkening the ribbon. The numbering system does not agree with that of Dayhoff (145).

Virtual C_α–C_α bonds simplify the representation of a chain fold. In another chain fold abstraction the backbone is described only by C_α-atoms and virtual bonds between them, as illustrated in Figure 7-10. The chain fold is represented by the virtual bond angle and the virtual dihedral angle (30, 403). Because of the peptide bond geometry the virtual bond angle is restricted to a range between about 80° and 160°, whereas the virtual dihedral angles can assume any value. However, steric hindrance prevents about half of the remaining angle combinations. This representation has been used for chain-folding simulations (404). It is also necessary for the production of bent wire models (Figure 7-3b). More elaborate backbone representations on the basis of virtual C_α-bonds are discussed in Refs. (403) and (405).

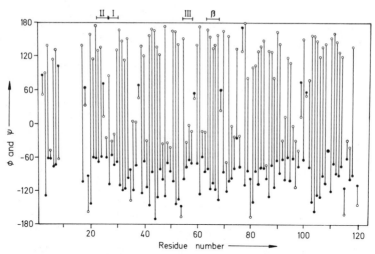

Figure 7-9. A linear representation of the dihedral backbone angles of the N-terminal part of chymotrypsin (402). Black dots show the ϕ-values whereas open circles are used for ψ. For each residue both values are connected by a bar in order to allow a better survey. Typical secondary structures are easily recognized. Examples of reverse turns I, II, III and β-sheet strands are marked. α-Helices would reveal themselves by consecutive short bars around $-60°$.

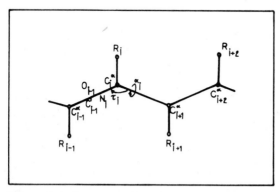

Figure 7-10. Representation of the backbone by virtual bonds between C_α-atoms. The chain is fully described by the virtual dihedral angle α_i defined by C^α_{i-1}, C^α_i, C^α_{i+1}, C^α_{i+2}, and the virtual bond angle τ_i. This description is used for building backbone wire models (Figure 7-3b). In a simplified representation τ_i can be taken as a function of α_i (30) so that the backbone is described by one free parameter per residue.

C$_\alpha$-Distance Plots

Chain folds can be represented by maps of C$_\alpha$–C$_\alpha$ distances. The C$_\alpha$-coordinates defining a chain fold are completely documented by a C$_\alpha$-distance map because all (except three) coordinates can be recovered from these distances by triangulation.[2] Since the map of mutual C$_\alpha$-distances, as shown in Figure 7-11, is sampled with a narrow grid, it can be presented with contour lines of equal distances, similar to an electron density map. In this way the chain fold is represented by a two-dimensional graph.

Secondary structures are revealed by characteristic patterns in the distance map. α-Helices give rise to small distances (many contours) along the diagonal; parallel β-sheet strands yield small distances in lines parallel to the diagonal (Figure 7-11). Antiparallel β-sheet strands as well as reverse turns correspond to lines of small distances perpendicular to the diagonal (406). Like a fingerprint, the pattern of a distance map is characteristic for a chain fold. Moreover, insertions and deletions destroy the pattern only partially by inserting or deleting strips of the matrix. Therefore, such maps are suitable for chain fold comparisons (407).

[2]There are $N\cdot(N+1)/2$ distances, i.e., equations for $3\cdot(N-1)$ unknown coordinates. The problem can be exactly solved for three C$_\alpha$-atoms and it is overdetermined for four or more C$_\alpha$-atoms.

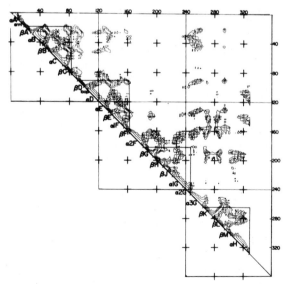

Figure 7-11. A distance plot for lactate dehydrogenase taken from Ref. (407). The distances between the C$_\alpha$-atoms of the ith and jth residues are plotted as half square matrix. However, the results are not presented as numbers but as a contour map. Contours are drawn at 16, 12, 8, and 4 Å. Different parts of the structure have been identified along the diagonal. Solid triangles mark the four half-domains which are geometrically well separated. The two N-terminal ones are $\beta\alpha\beta\alpha\beta$-units.

There are a variety of special distance plots. The neighborhood correlation plot of Figure 5-14a is also a C_α-distance plot, serving a very particular purpose. However, it cannot be inverted to the original coordinates.

The distances between C_α^i and C_α^{i+3} have been plotted versus sequential number i by Lewis *et al.* (202). Such a plot reveals all reverse turns in a polypeptide chain (section 5.1, page 74). It has the advantage of also showing those reverse turns which do not correspond to the types specified by Venkatachalam (199). However, this plot cannot be inverted to original coordinates because there are only N-3 distances for $3N$-3 coordinates.

A radial distance plot referring to a given position in the molecule, e.g., a Trp, can be versatile for spectroscopy. In such a plot the chain is spread over the 360° of a full circle and the distances of all C_α^i to the reference position are recorded radially.

All above mentioned C_α-distance plots are easily programmed using C_α-coordinates as supplied, for example, by the Protein Data Bank (390).

A very schematic distance plot, that is, an enumeration of contacts between atoms of different subunits of an oligomeric protein, is given in Figure 5-18d.

Summary

Protein structures contain a vast amount of information, namely, the identities and coordinates of several thousand atoms. So many data cannot be managed by the human mind but only by data processing machines, i.e., computers. However, the human mind has to sort, evaluate, and assimilate the available structural data. Therefore, easily intelligible representations of the atom positions are needed. Such representations should display only those structural aspects which are of interest for a given question. Unnecessary detail tends to hide essential features. Structure cartoons are the ultimate outcome of such simplifications.

Representations can be classified according to their dimensionality. Three-dimensional models of atom positions and of the covalent structure are most suitable for every investigator who tries to understand the interactions between a given protein and its ligands such as substrates and cofactors, or other macromolecules. Thus, if one is going to work with a structurally known protein for a longer period it is worthwhile to build a model on the basis of atom positions. For printing in journals or for electronic display the representation is restricted to two dimensions. Many of the two-dimensional representations imitate three-dimensional ones by introducing stereo views, or, for example, by a real time rotation on a cathode ray tube screen. Printed stereo views are cheap and provide a good overview of a structure. Besides displaying atomic positions and covalent bonds a number of representations concentrate on special structural aspects. Most of them are either two-dimensional or even one-dimensional.

Chapter 8
Thermodynamics and Kinetics of Polypeptide Chain Folding

Having thus far considered only complete structures we will now discuss how these structures form spontaneously during or after synthesis of the polypeptide chains. Although some aspects of the folding process can be described by equilibrium thermodynamics, this description is not detailed enough. If one wants to elucidate the correlation between amino acid sequence and three-dimensional structure the folding kinetics must also be understood. In the following, experimental and theoretical results applying to both aspects will be presented.

8.1 Thermodynamic Aspects

Transition between Two Thermodynamic States of the Chain

The folding process involves nothing more than conventional physical chemistry. During or after synthesis on the ribosome a polypeptide chain folds to its native globular structure. As a rule the folding process is spontaneous in the sense that no additional factor such as an enzyme or the presence of the ribosome is required. The clearest evidence for this was provided by the complete chemical synthesis and subsequent spontaneous folding of the enzyme ribonuclease (410, 411).

Originally the principle of spontaneity was formulated on the basis of renaturation experiments in which the chain refolded spontaneously after the denaturation of the native protein and the ensuing removal of the denaturing agent (94, 412). However, these experiments have always been questioned on the ground that the denaturation might not have been complete.

The behavior of synthetic ribonuclease and of denatured proteins demonstrates that completed, and not only nascent, chains fold spontaneously. Consequently the folding process can be regarded as the transition of the system "chain plus solvent" from a state of higher free energy (R, random conformation) to a state of lower free energy (N, native conformation). Only conventional physical chemistry is involved in this process.

In the vacuum situation binding energy overcomes chain entropy during structure formation. The transition from R to N is described by Eq. (3-2) in section 3.5. The interplay between chain entropy and binding energy, ΔS_{chain} and ΔH_{chain} of Eq. (3-2), will be discussed first. An example is a polypeptide chain of 100 residues that forms a single α-helix. The chain entropy is given by: $S_{chain} = R \cdot \ln W$, where W is the probability of encountering a given conformation and R is the gas constant. Accordingly, the entropy difference between states N and R is related to the probabilities of the conformations representing these states

$$\Delta S_{chain} = S_N - S_R = R \cdot \ln W_N - R \cdot \ln W_R = R \cdot \ln \frac{W_N}{W_R}.$$

For a comparison between random and α-helical state the ratio W_N/W_R can be estimated; in the R-state the dihedral angles at a given C_α-atom can assume any *allowed* value of the $\phi\psi$-map (Figure 2-3b) whereas in the N-state all (ϕ, ψ)-values have to lie in the region around $(-60°, -60°)$. Side chain conformations do not count because they are almost random in both states. If the area of the α-helical region is estimated as 10% of the total allowed area in Figure 2-3b, the probability of assuming α-helical angles is 0.1 for a given residue. For the total chain of 100 residues the probability becomes about $(0.1)^{100}$, yielding

$$\Delta S_{chain} = R \cdot \ln\left(\frac{1}{10}\right)^{100} = -0.46 \text{ kcal degree}^{-1} \text{ mol}^{-1}. \qquad (8-1)$$

If we neglect $\Delta G_{solvent}$, that is, if we consider a polypeptide chain *in vacuo*, we get

$$\Delta G_{chain} = \Delta H_{chain} + 143 \text{ kcal/mol at } T = 37°C.$$

Therefore, helix formation is induced (ΔG_{chain} becomes negative) if the binding energy per residue exceeds 1.43 kcal/mol, an energy that could be easily provided by hydrogen bonds. In this example a temperature change of 1°C corresponds to a change of a factor of 2.1 for the equilibrium constant K between native and random conformation as defined in Eq. (3-1). Thus, even small temperature variations have a strong effect on the equilibrium, in the sense that at a higher temperature it is easier for the chain to overcome binding energies and to assume a random conformation. This situation is characteristic for a system with large but well-balanced entropy and binding energy contributions.

Highly flexible chains cannot form defined structures. The preceding estimates allow us to discuss the usefulness of β-amino acids for structure formation (section 1.2). In β-amino acids a free rotation around the C_α–C_β bond exists (Figure 1-4). Therefore the probability of a given backbone structure at a certain residue is not 0.1 but only about 0.003 (taking about every 10° of the rotation around C_α–C_β as a distinct structure), and ΔS_{chain} for 100 residues becomes about -1.2 kcal degree^{-1} mol^{-1} which is 2.5 times as much as found for α-amino acids in Eq. (8-1). Such large chain entropy is difficult to overcome by noncovalent binding energy alone *(in vacuo)* and also together with solvent entropy (in the natural situation in solution).

Thus, a chain of β-amino acids would hardly be able to fold by itself to a defined structure. The same argument applies to all other types of chain molecules; a high degree of internal flexibility prevents the formation of a defined structure.

The Balance of Energy Contributions in a Globular Protein

Globular proteins are delicately balanced systems. The preceding calculations provide insight into the balancing act between the thermodynamic counterparts. But since $\Delta G_{\text{solvent}}$ is missing, they are incomplete. For the example given, however, $\Delta G_{\text{solvent}}$ is rather small because helix formation essentially removes only the polar backbone and not large hydrophobic side chains from the solvent (section 3.5, page 40).

Now consider an "average" globular protein containing 150 amino acid residues. As estimated from Eq. (8-1), in this case $T \cdot \Delta S_{\text{chain}}$ amounts to several hundred kcal/mol. This energy has to be compensated by $\Delta G_{\text{solvent}}$ and ΔH_{chain} in order to lead main chain and side chains into the native conformation [Eq. (3-2)]. As judged from the number of residues removed from the solvent and the free energy contributions of this process (Figure 1-8), $\Delta G_{\text{solvent}}$ amounts to at least 100 kcal/mol. The contribution of ΔH_{chain} is provided by hydrogen bonding in the interior of the protein [90% of all internal polar groups form hydrogen bonds (17)] and by van der Waals interactions. The resulting $\Delta G_{\text{total}} = \Delta H_{\text{chain}} - T \cdot \Delta S_{\text{chain}} + \Delta G_{\text{solvent}}$ is only a small percentage of the competing terms, or approximately 10 kcal/mol.

This order of magnitude for ΔG_{total} of the folding process has been found experimentally for a number of proteins containing around 150 residues (413). The data used for estimating ΔG_{total} include rates of hydrogen exchange in states N and R (414–416), calorimetric data of the unfolding process (417), data derived from denaturation curves [see Ref. (413) for a review], and the equilibrium constant K between native and random conformations as determined by immunologic studies (418). Because of the small value of ΔG_{total} any energy calculations trying to elucidate the correlation between covalent and geometric structure of a protein must be extremely accurate in order to yield meaningful results. Such accuracy is difficult to achieve because the noncovalent binding forces in a protein are not well enough understood (chapter 3). Furthermore, the equilibrium constant is highly sensitive to temperature changes.

Thermal Fluctuations in Protein Structures

Isolated protein molecules undergo large structural fluctuations. In any thermodynamic statement on protein structure it should be kept in mind that a single protein molecule together with its solvent environment is a very small system (419). At a given temperature the root mean square of the energy fluctuations, δH, of a system amounts to (420)

$$\sqrt{\overline{\delta H^2}} = \sqrt{kmCT^2}.$$

For macroscopic systems these fluctuations are very small indeed as compared to the internal energy. However, for a single protein molecule with a mass m of about $5 \cdot 10^{-20}$ g and a specific heat capacity of $C = 0.32$ cal degree^{-1} g^{-1} (421, 422) at 37°C the fluctuations become $7 \cdot 10^{-20}$ cal ($k =$ Boltzmann constant), corresponding to about 40 kcal/mol on a molar accounting basis. Such fluctuations are much larger than ΔG_{total}. But the relaxation times are only nanoseconds which is not long enough to induce substantial unfolding of the native structure.

Nevertheless, such fluctuations can be demonstrated experimentally. One example is the dynamic behavior of Tyr side chains in pancreatic trypsin inhibitor. As derived from X-ray analysis these side chains are rigidly encapsulated; by nuclear magnetic resonance methods, however, they were shown to flip–flop between two states differing by a rotation of 180° around the C_β–C_γ bond (423, 424). The same technique has been used to demonstrate conformational mobility of apparently rigid residues in lysozyme (425), ribonuclease (426), and myoglobin (427). It has further been found that the fluorescence of a buried Trp in ribonuclease can be quenched by O_2 (428, 429), indicating that O_2 can penetrate into the molecule. Similar results emerged from fluorescence relaxation (430), phosphorescence (431), and hydrogen exchange experiments (414). These and other observations (432) indicate that a protein molecule is extremely lively. Atoms can move from their average positions which are revealed by X-ray analysis. Note that oligomeric proteins (and protein crystals from which most data are derived) have larger masses, m, than monomers. Therefore, their fluctuations per unit mass are smaller.

The Native State: Global or Local Energy Minimum?

If proteins are only metastable, their structures must be grossly different from the most stable one. It has long been discussed whether the native structure corresponds to the absolute (global) minimum free energy [thermodynamic hypothesis of protein folding (433)], or whether it corresponds only to a local minimum [kinetic hypothesis of protein folding (248, 434)], i.e., a metastable state. Conceivably, the most stable state of a chain could, for example, be a complicated knot (Figure 5–15c), which is virtually impossible for the chain to assume spontaneously. Using the language of thermodynamics the chain cannot overcome a high barrier of $\Delta G\ddagger$, the free energy of activation. This barrier consists of a large entropic contribution because the chain must assume the small range of conformations that enable fitting of one chain end through a loop to form the knot. At this point the chain conformation is still fluffy so that there is not as much compensation by binding energies or solvent free energy [Eq. (3-2)] as in a native structure. Therefore, the $\Delta S\ddagger$ barrier emerges almost fully as a $\Delta G\ddagger$ barrier. In conclusion, the native structure may well correspond to a metastable state with a very long lifetime. And no experimental evidence could distinguish between a stable and a metastable state. Moreover, it would be of no biological relevance whatsoever. However, a metastable state must be structurally *grossly different* from the most stable one in order to withstand the large energy fluctuations discussed above.

Proving the thermodynamic hypothesis of folding is very difficult. It has been suggested that the thermodynamic hypothesis is supported by refolding experiments on pancreatic ribonuclease (435). Reduced unfolded ribonuclease was reoxidized in the presence of 8 M urea and formed disulfide bonds at random. This gave rise to some 100 different molecular species each of which is distinguished by a specific set of 4 disulfide bonds [there are $(2 \cdot 4)!/2^4 \cdot 4! = 105$ random combinations of S–S bonds (436, 437); in addition there are various theoretical possibilities to form knots]. The removal of urea and the addition of small amounts of mercaptoethanol (Figure 4-3) caused the gradual and finally the quantitative formation of the native structure. This demonstrates that from some 100 different groups of starting conformations the native structure can be reached. Although this number is not very large, it still shows that the folding result is unique; it is much too small, however, to settle the question of local or global minimum.

8.2 Speed, Precision, and Limitations of Folding *in vitro*

Chain folding may be much faster than chain synthesis. Folding *in vitro* is fast, at least for small proteins without disulfide bridges. Staphylococcal nuclease refolds within 1 sec (438) and metmyoglobin within 10 sec (439). If these values also apply under *in vivo* conditions, chain folding might proceed at least 10 times more rapidly than the biosynthesis of the amino acid sequence. Disulfide-containing proteins such as pancreatic ribonuclease refold within 1 to 10 sec if the disulfide bonds are not broken during the preceding denaturation (440). However, if such proteins are unfolded and reduced, subsequent chain folding (which includes the formation of the correct set of disulfide bonds) lasts many minutes under optimal conditions.

The precision of chain folding *in vitro* (94) is tested by comparing the properties of native and renatured proteins with respect to biologic activity and specificity (441). For instance, the O_2-carrier hemoglobin, which has been denatured in a variety of ways, can be converted back into native protein which (a) has the same solubility as the original protein, (b) is crystallizable, (c) has the characteristic absorption spectrum of native hemoglobin, (d) can combine reversibly with oxygen, (e) has the same relative affinities for carbon monoxide and oxygen, and (f) is not readily digested by trypsin which indicates that the renatured protein has a tight structure.

Refolding of enzymatically processed proteins is usually impossible. No precise refolding is obtained in many proteins that have been enzymatically processed *in vivo*. Collagen is an example of this protein category. One of the many problems in refolding collagen *in vitro* is that the three chains that are to make up the final triple-helical molecule (Figure 5-6) do not align in register. *In vivo*, the biochemist's problems with collagen are avoided by processing a soluble precursor before it is excreted. Globular extensions at both ends of each polypeptide chain guarantee that the three chains are aligned in register (Figure 4-4).

Another example is insulin which cannot be renatured once its native disulfide bonds are opened by thiols or reshuffled enzymatically (101). This observation induced the search for a precursor which was indeed found in the form of proinsulin (442). Proinsulin is stable toward the enzyme protein-disulfide isomerase (Figure 4-3), and it refolds spontaneously in denaturation–renaturation experiments (443). *In vivo*, proteolytic cleavage of proinsulin leads to insulin, and insulin requires the entropy contribution from its native set of S–S bonds for stability (section 8.3). The lability of the insulin structure may be of physiologic importance (444) since the rate of inactivation is a factor which controls the extent and duration of hormone action.

Refolding of modified proteins yields information on the folding process. The studies on refolding native proteins have been extended by refolding experiments on modified proteins. Early data (445) suggested that pancreatic ribonuclease which had been iodinated at Tyr-115, at a surface residue of the native structure, did not refold after denaturation. This indicated that the integrity of Tyr-115 is important for the folding process. In systematic experiments on staphylococcal nuclease and on pancreatic ribonuclease studies have been made on how extensively a protein can be modified (by chain cleavage and deletion of sequences or by attachment of bulky groups to side chains) without losing its folding capacity.

Furthermore, fragments of polypeptides can be tested for their contribution to the folding process by using methods of protein complementation. "Complementation" is the restoration of a biological activity by noncovalent interactions of different (poly) peptide chains (446). The classical example of *in vitro* complementation refers to pancreatic ribonuclease. When the so-called S-peptide (consisting of the 20 N-terminal amino acid residues) is cleaved off the protein and removed, the truncated protein loses its enzymic activity and its capacity of refolding from a reduced polypeptide chain. By complementing the truncated protein with the S-peptide it regains its enzymic and folding capacities although no covalent bond is reformed between the two peptide chains (94). In conclusion, renaturation experiments [cf. Refs. (94, 177) for a review] on modified and complemented forms of a given protein yield valuable information on its folding mechanism.

8.3 Structural Elements in Unfolded Chains

Until now we have referred to the unfolded chain using the term thermodynamic state. But in contrast to the native state the unfolded state corresponds to a multitude of chain conformations. It is conceivable that some order exists, i.e., some kind of structure in an unfolded chain. Such a decrease of the conformational space would be important for the folding process because it could serve as a structural nucleus to which the rest of the chain would readily associate.

Refolding experiments show a subdivision into fast and slow folding chains. A subdivision of unfolded chains into two classes, fast and slow folders, has been found with ribonuclease (440, 447, 448). In refolding experiments on ribonuclease that had been denatured without breaking the correct disulfide bonds, 20% of the chains (the fast folders) refold within 0.1 sec. This proportion is independent of the denaturing agent [guanidine hydrochloride (449), heat (450), etc.]. The remaining 80% (the slow folders) convert within 10 to 100 sec to fast folders which then refold rapidly. A similar but less quantitative observation has been reported for bovine pancreatic trypsin inhibitor (BPTI) where 15% of the chains fold significantly more slowly than the rest (451). An explanation for the difference between the two classes is not available. One hypothesis (449, 452) suggests that the fast folders correspond to chains with correct *cis–trans* isomers at all peptide bonds, whereas the slow folders contain incorrect isomers; the time is spent on converting the wrong isomers. *cis–trans* conversion happens most easily at the N-terminal side of Pro residues because here the activation energy barrier is only about 13 kcal/mol vis-à-vis 20 kcal/mol at other peptide bonds (section 2.5). Both, BPTI and ribonuclease contain four Pro; in native ribonuclease two of them are *cis*-isomers.

Antibodies can be used for identifying folding nuclei. Another way of looking for traces of structure in unfolded polypeptide chains is by splitting native polypeptide chains and checking with specific antibodies whether these fragments assume their native conformation. Most of these experiments were done on staphylococcal nuclease (418, 453); they revealed for fragments with sizes between 17 and 120 residues that the native conformation is assumed with an equilibrium constant $K = [N]/[R]$ of about 10^{-3}. According to Eq. (3-2) this corresponds to an unfavorable free energy of about 4 kcal/mol. In principle this method can be used to find folding nuclei. For this purpose fragments of the polypeptide chain must be isolated or synthesized (454). The most stable fragments, the pieces with the highest values for K, are the best candidates for folding nuclei.

Folding nuclei should have a strong neighborhood correlation (section 5.3, page 84) because neighborhood correlation enhances intermediate stability. This can be derived from Flory's formula for the chain entropy contribution of disulfide bridges (455)

$$\Delta S^{(SS)}_{chain} = 0.75 \cdot \nu \cdot R \cdot \ln (n' + 3) \tag{8-2}$$

where $\nu =$ the number of connected chains (twice the number of bridges) and

$n' =$ the number of statistical segments between bridges, which in a first approximation (456) can be taken as the number of residues between bridges.

The shorter the chain between bridges the smaller the entropy reduction on bridge formation is. This entropy reduction is not based on chemical properties of the disulfide bridge but only on the decrease of the number of possible conformations caused by the cross-link. Therefore, Flory's formula can be used to estimate the reduction of entropy necessary to bring two chain elements together.

Flory's formula shows quantitatively that a strong neighborhood correlation is favorable. As shown in Eq. (8-2), the required entropy change is smaller the shorter the chain between these elements; that is, the stronger the neighborhood correlation in the respective chain segment is. Conversely, the activation entropy $\Delta S^{\ddagger}_{chain}$, in the transition from an unfolded chain to a native format with strong neighborhood correlation is much smaller than $\Delta S^{\ddagger}_{chain}$ for the transition to a native format with weak neighborhood correlation. According to Eq. (3-2) a smaller $\Delta S^{\ddagger}_{chain}$ corresponds to a lower free energy barrier and therefore to a more stable intermediate structure. Thus, much less binding energy and solvent entropy are required to fold an extended chain to a structure with a strong neighborhood correlation than to a structure with a weaker one. In conclusion, a strong neighborhood correlation facilitates the folding process and therefore enhances the speed of folding appreciably.

Secondary structures such as α-helices or β-meanders are good candidates for folding nuclei because they have a large intrinsic binding energy and a strong neighborhood correlation. In this context it may be of significance that in refolding of reduced BPTI the first stable disulfide bridge (451, 457) links Cys-30 on a pleated sheet with Cys-51 on an α-helix. Moreover, this is the bridge with the strongest neighborhood correlation (shortest segment in between) within the native structure (Figure 8-1).

8.4 Folding Pathway

Just as a ball can roll from the mountaintop downhill to the deepest point of a valley, a polypeptide chain folds to a conformation of lowest free energy. Both, the ball and the polypeptide chain, follow a certain pathway, and both go to the lowest kinetically accessible point (local minimum) which is not necessarily the absolutely lowest point (global minimum). The question is whether the ball (chain) always takes the same pathway soon after leaving the top of the mountain (which should be viewed as a wide plateau corresponding to the large number of conformations an unfolded chain can assume), that is, whether a crevice exists that guides the ball on its way downhill.

Disulfide bonds are specific folding indicators. Elucidation of such a pathway is a difficult task because for this purpose chain conformations in solution have to be characterized. Any spectroscopic characterization of a folding chain yields only very gross results (circular dichroism, for example, gives an estimate for the amount of α-helix averaged over all conformations present) which do not permit a detailed interpretation. Only S–S bond formation gives exact though limited structural information because such bonds can be made only if two Cys residues touch each other.

At the present time the bovine pancreatic trypsin inhibitor (BPTI) seems to be best suited for such experiments (793). BPTI is a small protein of known structure which contains 58 amino acids in a single chain and three disulfide bridges. For the study of S–S bond formation and rearrangement in the course of folding it is essential that the reactions involving Cys residues are carefully controlled. This is achieved by using low molecular weight thiol-disulfide systems of defined redox potential (451), for example, the reduced and oxidized forms of highly purified dithiothreitol. Moreover, purified reagents such as iodoacetate are used for trapping free thiols. No denaturing agent is required since fully reduced BPTI is unfolded under physiological conditions.

The addition of a disulfide (e.g., oxidized dithiothreitol) to a solution of reduced BPTI marks the beginning of the folding dance. In this dance three intermediate stages, I, II, and III, can be clearly separated from each other (451); they are described in Figure 8-1. Conversion from one stage to the next shows quite different time constants, i.e., the activation energy barriers vary greatly. At stage I two abundant species, I_A and I_B, exist with a correct and an incorrect bridge, respectively. As S–S bond formation and breakage occur within microseconds the relative frequencies of these one-disulfide species reflect their relative thermodynamic stabilities. Conformation I_B is a dead end; only conformation I_A converts to the next stage, which contains three abundant species with two bridges each: II_A, II_B, and II_C. All of them keep the first correct bridge; in II_A and in II_B an incorrrect bridge, in II_C a correct one are added. Interestingly enough II_C is a dead end,

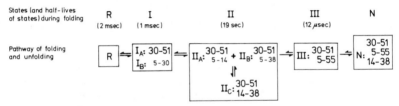

Figure 8-1. Covalent intermediates in the folding pathway of bovine pancreatic trypsin inhibitor (BPTI) (451, 793). The symbols are explained in the text. The correct pattern of disulfide bonds is shown at the outer right. In intermediates, e.g., in II_A and II_B, "correct" disulfide bonds are given in the same format as in the N-state. The folding pathway outlined here refers to the fast folders only; to 85% of unfolded molecules.

although it has two correct bridges. Only II_A and II_B convert to stage III, which contains a single species with the initial correct bridge and another correct one. This conversion is very slow, and it requires a conformational change with a high activation energy. In contrast, the continuation from III to the native protein is extremely fast.

Folding requires non-native intermediates. The most unexpected result is that folding is not a sequential addition of more and more native structure but requires non-native intermediates (II_A and II_B). This aspect is even more pronounced in the case of ribonuclease, the refolding of which was followed with a similar experimental approach (458). Like BPTI this protein can only be renatured by disulfide bridge formation. But unlike BPTI, at an early folding stage ribonuclease requires at least three bridges, and these bridges can be random which makes the existence of nonnative intermediates very likely. This indicates that the chain first has to aggregate into a small volume, that it needs the free energy ($=\Delta S_{chain}$) contributions from bridges, before further folding can take place (459). The folding process *in vitro* is very slow, lasting about 20 minutes.

For lysozyme it was found that the first two native disulfide bonds formed an order of magnitude faster than the second two which indicates that there is a preferred folding pathway (460). However, this pathway is not an obligatory one as was shown in a different study on lysozyme (162). The eight possible isomers (of reduced lysozyme) containing one permanently blocked Cys per protein molecule were allowed to renature. All isomers formed structures that were enzymatically active. This result rules out a unique role of any possible disulfide bond in the folding process. A special aspect is that no one of the four native disulfide bonds is obligatory in the formation of the other three correct S–S bonds.

Folding is cooperative. Like helix-coil transitions (Appendix) the folding and unfolding of proteins is a cooperative process. This general statement could be validated for BPTI. Here, all intermediates are unstable; they disproportionate, yielding the unfolded reduced protein and the native state with three S–S bonds. As shown in Figure 8-2, the molecular population essentially occupies only these two states; to a good approximation BPTI is a two-state system.

Folding may be quite different for proteins with and without disulfide bonds. Cystine-containing proteins fold one to two orders of magnitude more slowly than proteins of equal size without disulfide bonds (section 8.2). As disulfide bond formation itself is not a rate limiting step, these proteins seem to constitute a different folding class. Hence, the elucidation of the folding pathway of a disulfide-containing protein may be only of low relevance for the folding of proteins without these bridges.

The slow and laborious unfolding and refolding of S–S bond-containing proteins may have a physiological explanation. Such proteins are usually excreted into extraepithelial spaces where they have to withstand fluctuating environments. Making them kinetically stable enables them to survive in a quasi-native state in denaturing environments as long as these are transient with time constants smaller than that of the unfolding process. The half-time of unfolding in BPTI (step III → II of Figure 8-1) is in the order of hours (451).

In some cases slow folding and slow unfolding may control a protein's life span. After folding up in the cell such a protein would be excreted into an extracellular space. As soon as it unfolds it will become the safe victim of proteases since refolding involves loose intermediates of relatively long half-lives.

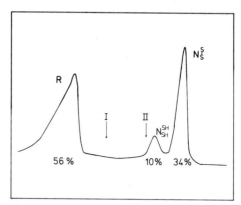

Figure 8-2. Illustration of the two-state model of protein folding [based on Ref. (451)]. Bovine pancreatic trypsin inhibitor (BPTI) was incubated in a solution containing 1 mM reduced and 30 mM oxidized dithiothreitol. The fully reduced form of BPTI which can form under these conditions is unfolded. The protein species that were present after equilibrium of the folding–unfolding transition had been attained were trapped with iodoacetate, separated by gel electrophoresis, and quantitated. The electrophoretic pattern shown in the Figure reveals that only the unfolded state R, and the native states, N_{SH}^{SH} (= III of Figure 8-1) and N_S^S (= N of Figure 8-1), are present at equilibrium; the folding intermediates I and II are absent.

In general disulfide bond-containing proteins need a fraction of these bonds for stability. Let us reconsider the question of whether disulfide bonds are necessary for the stability of the native structure. A set of cross-links which is unique and characteristic for an ordered state of a chain stabilizes this state by lowering the entropy of unfolded forms of the chain with the same set of cross-links. For a protein of 120 residues the contribution of four native S–S bonds to the stability of the native conformation is approximately $4 \times 2 = 8$ kcal/mol [Eq. (8-2)]. The findings that most extraepithelial proteins have ΔG_{total} values of about 10 kcal (413) and that they are denatured when *all* disulfides are reduced (Table 4-1) indicates that these proteins need the energy contribution from one or more disulfide bond(s) for the thermodynamic stability, the actual number of disulfides required varying with the protein in question (see also Table 4-1).

8.5 The Influence of Ligands

Usually ligands stabilize the native structure. For some proteins the attainment of the native conformation depends on the presence of specific ligands, frequently metal ions or other prosthetic groups (94, 413). The presence of ligands does not necessarily lead to a significantly different conformation of the folded protein but helps to stabilize the native form. On the basis of hydrogen exchange experiments (415) and of urea denaturation curves (461) apomyoglobin, a globular protein with a ΔG_{total} of 9 kcal/mol, was found to be 4 kcal/mol less stable than myoglobin. A contrasting example is cytochrome *c* where the presence of the (covalently bound) haem group is absolutely essential for the stability of the protein's structure. On removal of this group the protein is no longer globular (462). When only the iron ion is removed, the protein remains globular but less stable, to thermal denaturation, for example (463).

For nuclease T, a proteolytic derivative of staphylococcal nuclease (94), the influence of the ligands 3′ 5′- thymidine diphosphate (pdTp) and Ca^{2+} has been studied in some detail. The equilibrium constant for the system native conformation–unfolded conformation can be expressed as $K = k_{+1}/k_{-1}$ where k_{+1} is the rate constant for folding and k_{-1} for unfolding. For nuclease T the presence of ligands has no effect on the rate of folding (464); the rate of unfolding, however, decreases by a factor of 20 in the presence of Ca^{2+} and pdTp (465).

Two important questions have not been answered for any protein containing an essential ligand: at what stage of the folding process does it enter the game, and how does the polypeptide chain select its specific prosthetic group, such as a given metal, *in vivo,* that is, in the presence of many similar alternatives.

The folding of a pyruvate kinase monomer is primed by L-valine. The effects of the amino acid L-valine on the renaturation of pyruvate kinase (466) from yeast and of threonine deaminase from *E. coli* (468) illustrate two different roles which ligands can play in the formation of active oligomeric proteins. Pyruvate kinase (467) is a tetrameric enzyme, composed of four identical subunits each of which contains 1 molecule of L-valine noncovalently bound. The subunits dissociate and unfold in 6 *M* guanidine hydrochloride. On incubation in a renaturation medium L-valine is a specific primer of the refolding process. It induces renaturation with a pseudo-first order rate constant with respect to the monomeric species, which indicates that L-valine influences the folding of the monomeric form to its native conformation, and that the final process is a spontaneous association to the tetrameric enzyme. L-valine remains part of the integral structure of the native protein molecule (466).

Ligands induce the maturation of enzymes. In contrast, threonine deaminase, a dimeric enzyme of 360,000 daltons containing two pyridoxal phosphate groups (Figure 8-3), requires L-valine (or L-isoleucine) only as a catalyst that is not incorporated (468). Here an inactive form of the oligomer undergoes a maturation process to yield an enzymatically active species. At pH 7.5 and above the requirement of L-valine for this process, which has been termed "preconditioning" (469), is absolute.

There are other cases of oligomeric enzymes the formation of which involves obligatory, catalytically inactive intermediates. These intermediates undergo a conformational transition at a later stage to yield the thermodynamically more stable active enzyme. This maturation process represents a rate limiting and practically irreversible step in enzyme formation. It has been proposed (472) that in certain oligomeric enzymes allosteric ligands can specifically enhance maturation. This function of allosteric ligands is in contrast to their role in the modulation of enzyme activity where they stabilize either an active or inactive conformation of the protein in a freely reversible equilibrium.

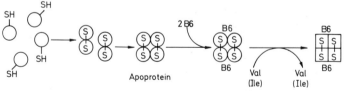

Figure 8-3. Formation of active threonine deaminase from folded monomers [based on Ref. (122)]. The circles represent the four monomers each of which contains an SH-group which is used for forming a disulfide with an identical SH-group of another monomer. Two of the resulting composite subunits form the apoprotein; the addition of the cofactor pyridoxalphosphate (vitamin B6) leads to an inactive holoprotein which matures under the catalytic influence of Val (or Ile) to become the active enzyme (468). This last step is called "preconditioning" (472).

8.6 Simulations of the Folding Process

Determination of the three-dimensional structure from the covalent structure requires understanding of the folding process. The three-dimensional structure of a protein is determined by the noncovalent forces between the amino acid residues of the chain and between these residues and the solvent (chapter 3). In principle, knowledge of these forces should allow the calculation of the native structure on the basis of the known covalent structure. However, since the native structure may not correspond to the global free energy minimum, energy calculations for all possible chain conformations may not lead to the right answer. Besides, it would be impossible to consider all chain conformations because the computer time required would exceed by far the age of the earth. Hence, such calculations are feasible only if they do not attempt to check out static structures, but concern themselves with attempting to simulate the folding process by following the folding pathway of a given chain. If this pathway is well defined (corresponding to a deep crevice in the ball analogy described in section 8.4) calculations of moderate accuracy may suffice to stay on the right track. But the calculations must be highly accurate if the pathway is less well defined.

Folding simulations were based on the hierarchy of levels of protein structure. There have been several attempts to simulate the folding process e.g. for apomyoglobin (470), for BPTI (30, 404), and for parvalbumin (471). All of them are based on strict hierarchy of levels of protein structure (section 5.6) insofar as they assume that the α-helices of the native structure are preformed in the otherwise unfolded chain. This assumption is supported by immunologic measurements on α-helix contents in unfolded chains (418, 454), and by the success of α-helix prediction methods (section 6.5, page 128). The chain folding simulations were carried out *a posteriori;* the resulting structures were known from X-ray analyses. Application of such a simulation on a protein of unknown three-dimensional structure would require information on the α-helical sections of the chain. To some extent this information can be provided by prediction methods (chapter 6).

Folding of apomyoglobin was considered as an assembly of α-helices. The folding simulations for apomyoglobin (470) were done without a computer because the underlying model was rather simple; the nine helices of the native structure were assumed to be present already (Figure 8-4a). This reduces the available conformational space appreciably. Four structural nuclei were specified as the initial stage of the folding process, each of them containing three α-helices adjacent along the chain. Within each nucleus the helices were arranged in a different way. The free energies of these helix arrangements were calculated exclusively from hydrophobic forces (section 3.5, page 38) of large nonpolar side chains; no charged side chain was allowed to be buried. From the large number of resulting helix arrangements only the 40 arrangements with the lowest free energies were included

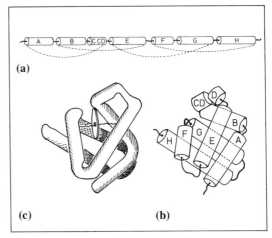

Figure 8-4. Folding simulation of apomyoglobin (470). (a) the α-helices assumed to be present at the beginning of the folding simulation. At the initial stage the four marked nuclei, consisting of three α-helices each, were taken into account. (b) The result of the chain fold simulation which showed the lowest calculated ΔG_{total}. (c) Sketch of the native myoglobin structure as derived from X-ray diffraction.

in the subsequent analysis; they spanned an energy range of 4 kcal/mol above the helix arrangement with the lowest free energy. At the following stages of the simulations further helices were added to these nuclei, or nuclei were combined, until the whole chain was folded. At any given stage only conformations with free energies not more than 4 kcal/mol above the minimal one were accepted; all others were discarded. This procedure resulted in 19 different chain folds. As expected, the helix arrangement with the lowest free energy (Figure 8-4b) corresponded to the native structure of myoglobin (Figure 8-4c). However, the difference to the structure with the next higher free energy was only 3% of the calculated free energy, which is much smaller than the errors to be expected.

The following aspects of this method are noteworthy: (a) It works only for proteins of class No. 1 (Table 5-2), proteins consisting mainly of α-helices. (b) At the starting point the existence of large nuclei representing one-third of the molecule is presumed. Experimental evidence for such nuclei is not yet at hand. (c) Helices of a nucleus are adjacent along the chain, which corresponds to a low chain entropy and a strong neighborhood correlation (section 8.3). (d) It is assumed that the native structure folds along a straight pathway; all intermediate subassemblies are immobilized and not allowed to rearrange. This assumption is in contrast to the experience with disulfide bond-containing proteins where obligatory intermediates with non-native geometry are present (Figure 8-1). However, the folding process of apomyoglobin is much faster and therefore likely to be more straightforward than those of disulfide-containing proteins (section 8.2).

In BPTI and parvalbumin a simplified chain was made to walk down the energy gradient. A quite different method is applied in the chain folding simulations on BPTI (30, 404) and parvalbumin (471). Here, the polypeptide chain is represented in much finer detail than in the simulations on apomyoglobin. Still, the approximation is crude using only the virtual C_α torsion angle (Figure 7-10) as an independent variable. The virtual C_α bond angle is taken as a function of the C_α torsion angle, and side chains are regarded as rigid spheres (Figure 7-10). This simplification reduces the number of free parameters to 1 per residue.

The simulation of the folding process was started with a chain in an extended quasi-random conformation except for the native α-helices which were considered to be preformed. Subsequently the chain was allowed to run down the free energy gradient. In this procedure the kinetic energy of the chain was not taken into account. In order to account for temperature fluctuations (section 8.1) the chain should be shaken up at short intervals. But as this would consume too much computer time, the chain was stirred up only if it had settled in an energy minimum, in order to release it from the trap if the minimum was only a local one. In contrast to the computations on apomyoglobin the method includes binding energies as well as hydrophobic forces. Moreover, subassemblies are not fixed; continuous rearrangement is allowed in all parts of the chain. As determined by this method, the chain of BPTI folded with about 50% probability to a conformation with a root mean square C_α distance of about 6 Å to the true native structure. However, even the best result had a wrong β-sheet topology and was criticized accordingly (801). It should be noted that these simulations did not account for the specific disulfide patterns assumed by BPTI during the folding process (Figure 8-1). Using this information may be helpful in future attempts. Moreover, the experimental data on renaturation of disulfide-containing proteins indicate that the folding pathway of such proteins may be more complex than the pathway of other proteins (see page 159). Therefore disulfide-containing proteins may be unfavorable for developing theoretical analyses.

In the case of parvalbumin (471), which is an α-helical protein of class No. 1 (Table 5-2), the folding simulation was less successful. This may be caused in part by neglecting the possible occupation of two Ca^{2+} binding sites. Since these sites are located in short loops between helices (59), it is expected that they influence the relative orientation of the adjacent helices and thus the folding process.

Summary

The folding process of a polypeptide chain is not yet understood in detail. We know that it occurs spontaneously, indicating that it is based on simple physico-chemical principles. Therefore, it should be possible to elucidate it and to simulate it using theoretical procedures; first attempts have already been made. Consideration of the folding process is not just a matter of satisfying curiosity. Our interest goes much deeper. To obtain insights into biological processes we must know the three-dimensional structures of various proteins which cannot be expected to crystallize and become amenable to X-ray structure analysis. The only hope for discovering these structures is to derive them from amino acid sequences. Such a derivation must follow the lines of the natural folding process; sorting out all possible final structures of a given polypeptide chain on the basis of their free energy is as hopeless as trying to answer the question of whether the native structure constitutes a local or a global energy minimum.

The most intricate problem in folding dynamics is the characterization of intermediate states. Until now the only available geometric data refer to disulfide bonds of folding intermediates. Unfortunately, it is quite possible that S–S bond containing proteins undergo a folding process that is qualitatively different from other proteins, so that the information gained is limited.

Much effort has been invested to find nuclei of the folding process which are expected to serve as scaffolds for further folding. Such nuclei should be exceptionally stable structures. Possibly, they can be detected by antibodies or in refolding experiments using modified proteins. Localization of nuclei would enormously facilitate folding simulations. In any case, many more useful experiments can be done with the hope that eventually the folding process will be elucidated.

Chapter 9
Protein Evolution

Biology is a natural science with a historic dimension. All biological species have developed continuously starting out from a single or from a very limited number of ancestral species. As chance may have played a major role in priming events of the past, we have to accept the fact that evolutionary directions may never be fully explained in terms of cause and effect. At first sight this appears to be a major drawback for scientists who love deduction from basic principles. However, quasi-random development has given rise to an awesome degree of diversity in biological species which can be considered as the outcome of numerous natural experiments. Hence one should acknowledge these experiments gratefully, analyze their results, and start interpreting them.

   Protein evolution involves changes of single residues, insertions and deletions of several residues, gene doubling, and gene fusion. In protein structure the basic step of the historical[1] process is the change of an amino acid residue in the polypeptide chain. In the course of time such changes accumulate, so that eventually all similarities between initial[2] and resultant amino acid sequence are eliminated. But as a rule even after all sequence similarities between two homologous proteins have vanished, the chain

---

[1]History means recorded history; as we regard nucleotide and amino acid sequences as records of past events we feel free to use the terms history and historical.

[2]It should be kept in mind that we know protein structures only of extant species; there are no fossils. Therefore, we can evaluate evolutionary changes of the past only from the present differences between homologous proteins.

folds still resemble each other. The tendency for replacements at a given residue position varies appreciably (Figure 7-1b). Single amino acid substitutions are not the only differences between homologous proteins. Larger steps are insertions and deletions of single residues or stretches of residues (Figure 7-1a), which affect not only side chains but also the main chain. Insertions and deletions cause great difficulties in sequence comparisons of distantly related proteins (803).

In such cases knowledge of the three-dimensional protein structures helps considerably. It reveals in general that internal residues vary slowly, whereas the differences between homologous proteins (amino acid changes or deletions and insertions of chain loops) accumulate on the surface. Consequently, the sequences of distantly related proteins can be aligned on the basis of residues that are geometrically at corresponding positions in the tertiary structures.

Very gross changes result from gene doubling, which can lead to a doubling in length of the polypeptide chain. Furthermore, cases of fusion of different structural genes are known, which implies that one or more genes have been translocated to a different position in the genome.

Homologous proteins result from speciation or differentiation. Comparisons between homologous proteins have yielded general rules for protein structures (250). Furthermore, studies on protein evolution contribute to the analysis of general biological problems. In this context it is often useful to distinguish between protein speciation and protein differentiation (473). Speciation is the evolution of homologous proteins possessing a common function in different organisms. Protein differentiation, on the other hand, is the process leading to functional diversification of homologous proteins, often within one organism.

Studies on protein speciation allow the establishment of a genealogy of organisms. In general, amino acid sequence and chain fold are well enough conserved so that even the proteins of genetically very distant organisms resemble each other. It also means that the study of protein speciation allows the establishment of the genealogy of organisms. Consequently, phylogenetic analyses based on protein speciation can be regarded as a method of taxonomy.

Protein differentiation reflects the evolution of biochemical pathways. The study of protein differentiation has other goals. Here, structural similarities usually reveal unexpected biological connections which can be used for tracing the evolution of the metabolism, for example. Furthermore, protein differentiation demonstrates convincingly that biological diversity is limited; proteins can be classified along general lines (474). A typical example of a class of differentiated proteins is the family of serine proteases (Table 9-4) which control a variety of processes either maintaining or threatening human life. In effect, the (patho)physiological and pharmacological implications of protein differentiation strongly suggests the application of structural analysis at atomic resolution to medical problems (602).

9.1 Protein Speciation

Rates of Accepted Mutations

Biological speciation works by random mutation and selection. The mutations primarily affect the DNA. Estimated mutation rates of DNA in higher organisms are given in Table 9-1. In these organisms the rates of accepted mutations at the DNA level are essentially unaffected by selection at the protein level because less than 10% of the DNA belongs to structural zones (78, 475) coding for proteins (section 4.1, page 46). Nevertheless, DNA has been found to be homogeneous within one species (476, 477), which indicates that some selection exists at the DNA level. Whether (or where) this homogeneity is due to small founder populations and "genetic drift" [the non-Darwinian hypothesis favored by "neutralists" (144, 478–480)] or whether (where) it reflects the selection of the fittest DNA for a species (the Darwinian hypothesis favored by "selectionists") is still a matter of controversy [cf. (481, 482)].

Genetic distance is measured in PAM units. At the level of macromolecules, the fixation rate of mutations (also called the acceptance rate) is generally expressed as the number of accepted point mutations per codon per 10^{10} years, which is numerically equal to the percentage of accepted point mutations in 10^8 years. The percentage of accepted point mutations (PAM units) is a common measure of evolutionary distance. An inverse measure for the acceptance rate is the "unit evolutionary period," that is,

Table 9-1. Evolutionary Rates of Macromolecules[a]

Macromolecule	Rate of evolution expressed in substitutions per codon per 10^{10} years (i.e., in PAM/10^8 years)
DNA in hypervariable nucleotide sequences (144, 483, 586)	500
Average DNA in mammals (475, 476)	45–85
mRNA of hemoglobin chains (484, 485)	100
mRNA of histone IV in different sea urchin species (486)	35
Average proteins (145)	1–50
Hemoglobin, α and β-chains, respectively (145)	14
Histone IV (145)	0.09
Cytochrome c, mitochondrial (794)	5
Hypervariable amino acid sequences in proteins (488, 586)	85

[a]All values are rough estimates and may represent special cases. DNA refers to nonrepetitive DNA (586).

the number of years for one accepted codon change per 100 residues to occur.

A selection of nucleotide sequences exists at the DNA level. Because of the degeneracy of the genetic code only 75% of all mutations in a structural zone of DNA affect the protein sequence (144, 483). The remaining 25% are so-called silent mutations which interconvert only between different codons of the same amino acid residue (Figure 1-5b). As all codons of a given amino acid are used [although with different frequencies (12, 13, 145)], there is no restriction at the protein level, and the rate of silent mutations should correspond to about 25% of the maximal DNA mutation rate. The mutation rate of messenger RNA (mRNA) coding for hemoglobin chains (484, 485) approximates one-quarter of the mutation rate in hypervariable nucleotide sequences of DNA (Table 9-1). Thus, it seems to correspond to the maximal silent mutation rate. In comparison with hemoglobin-mRNA, the silent mutation rate is smaller by a factor of three for histone IV-mRNA (Table 9-1). In sea urchins one cell contains 1200 copies of active histone IV-genes (487); the corresponding 1200 mRNA sequences appear to be homogeneous within one species (486), although they differ between species. This finding is another indication that selection of nucleotide sequences at the DNA level takes place.

The selection at the protein level is in general much more stringent than at the DNA level. This is illustrated by comparing the amino acid change rates of hemoglobin and histone IV with the silent mutation rates in the corresponding mRNAs (Table 9-1). Moreover, there are large rate variations between different proteins. Comparing the extremes, histones IV and hypervariable segments in proteins, we find a difference of three orders of magnitude (Table 9-1). This difference reflects the importance of the integrity of a given protein for its function in an organism. Histone IV is located at a central position of DNA storage and processing so that any amino acid change might have a catastrophic effect. The evolution of histone IV is strongly Darwinian; almost no mutation is accepted. In contrast, the hypervariable region of ribonuclease does not seem to have a function (488); accordingly, changes are neutral with respect to fitness and they are more easily accepted. Here, evolution may follow non-Darwinian lines (489).

Acceptance Criteria for Amino Acid Changes in Proteins

Detailed protein biographies are necessary to assess the biological importance of a given residue. As shown in the preceding example, the acceptance of amino acid changes critically depends on the biological role of the respective amino acids. However, this role is difficult to assess because it refers not only to the known function of a protein, such as the catalytic efficiency of an enzyme, but to *all* interactions of a protein with other parts of the organism over the protein's entire lifespan from activation of the corresponding gene to polypeptide degradation. And such detailed biographies of proteins are not yet available. Nevertheless, some general comments can be made[3].

Residues at the protein surface change more frequently than do internal ones. As demonstrated in Figure 7-1b, the change rates of amino acid residues in a given protein depend critically on the position along the polypeptide chain, on the location in the three-dimensional structure. As a rule residues on the surface vary faster than do internal ones. Since external residues are less crucial for protein stability than internal ones, this rule reflects the importance of stability for protein function. Exceptions are surface residues that are directly involved in the action of proteins, as, for example, residues at the catalytic site of enzymes, residues at binding sites for substrates, cofactors, prosthetic groups, allosteric effectors, or binding sites for macromolecules. Thus, cytochrome c which interacts with several other macromolecules and which accommodates a prosthetic group does not have much unspecific surface left. This explains the slow rate of accepted mutations (Table 9-1) in cytochrome c.

The effect of an amino acid change in the protein interior is often compensated for by other changes. Since globular proteins are as tightly packed as crystals of organic molecules (section 1.6), changes of internal residues affect the positions of other residues in the vicinity which usually gives rise to a lower stability. Consequently, a change in an internal position is rare. Often it is accompanied by at least one other change. Examples of such compensating changes have been demonstrated in ribonucleases (67). A very elaborate example of internal compensation is observed in two distantly related serine proteases: the internal cluster composed of Trp-29, Ser-45, Val-53, Val-200, Leu-209, Val-210, and Ile-212 in chymotrypsin becomes Ser-29, Thr-45, Met-53, His-200, Val-209, His-210, and Val-212 in elastase without any disturbance at the backbone of the polypeptide chain (490).

Changes between similar residues occur more frequently than others. Conservation of the function restricts the rate of accepted changes at a given position in the polypeptide chain. Obviously, this function is least impaired by so-called conservative replacements, changes between similar residues. In this respect, size, shape, flexibility, and charge of a side chain as well as its tendency to form hydrogen bonds are relevant. Thus, the change Lys → Arg conserves a flexible side chain with a positive charge, or Ile → Leu conserves a relatively bulky nonpolar side chain. Conversely, the principle of conservative changes has been used in section 1.6 to derive empirical similarities between amino acids from observed change frequencies (Table 1-2).

---

[3]These generalizations apply not only to proteins that act under normal conditions but also to those that are exposed to extremes of heat (491) or hydrostatic pressure (492), such as proteins from thermophilic bacteria and abyssal (deep sea) fish, respectively. Only small variations of amino acid sequences seem to be necessary to adjust to these extreme conditions.

Abnormal hemoglobins illustrate the possible impact of random mutations. Even a conservative change can have a large effect as has been found in the abnormal human hemoglobin Sydney (493), which contains Ala instead of Val in position 67 of the β-chain. The replacement of two methyl groups by hydrogen atoms loosens the haem group and strongly impairs the stability of the protein (494) and thereby the stability of the erythrocyte.

A nonconservative change, as, for example, the introduction of a negative charge in the hydrophobic interior, is bound to be incompatible with the stability of a protein's native conformation, because the cost in free energy is of the order of 10 kcal/mol, which is approximately equal to ΔG_{total} between folded and unfolded chains (section 8.1, page 151). An example is the abnormal human hemoglobin Wien (494), in which the change Tyr-130 $\rightarrow$ Asp in the α-chain has led to a large conformational change which allows the negative charge to be compensated by a positive one.

Both of the described abnormal hemoglobins cause anemic conditions. Hence, they demonstrate convincingly the possible impact of random mutations. Such mutations will probably never be accepted.

Mutation Probabilities of Amino Acid Residues

Mutation probabilities, strictly speaking probabilities of accepted mutations, vary with residue type. The mutation probabilities of a residue type are given in Table 9-2. Values on the diagonal are the probability that a given residue type will be kept. Ser has the smallest diagonal term which means it has the largest mutation probability. Ser is usually found at the protein surface. Trp has the smallest mutation probability. This is reasonable since Trp is generally an internal residue and cannot be replaced by a side chain of equal size.

A strong conservation is found for certain Gly residues. This is somewhat surprising since Gly has no side chain which could have an indispensable function. However, in each case an examination of the three-dimensional protein structure showed that at these invariant Gly positions [as, for example, at the contact between helices B and E in the globin family (145, 277) or at position 7 of Figure 7-8 between two α-helices in c-type cytochromes (495)] the *absence* of a side chain is crucial. Only without a side chain can the helices pack tightly enough. Introduction of a side chain would decrease the packing density and thus the stability (section 3.6). Invariant Gly residues were also found at the third position of six type II-reverse turns in the cytochrome family (495). As shown in Figure 5-7b, for steric reasons this position cannot accommodate a side chain.

Any evolutionary pathway starting from a random set of amino acid frequencies tends to establish the presently found frequency distribution. As shown by King and Jukes and as illustrated in Figure 9-1a, there is a strong correlation between observed amino acid frequencies and those expected

Table 9-2. Mutation Probability Matrix for the Evolutionary Distance of 2 PAM (20)[a]

	Gly	Pro	Asp	Glu	Ala	Asn	Gln	Ser	Thr	Lys	Arg	His	Val	Ile	Met	Cys	Leu	Phe	Tyr	Trp	Sum
Gly	9870	17	13	22	40	22	11	42	8	5	0	1	7	0	0	3	2	0	0	0	10063
Pro	7	9850	1	13	23	9	13	11	5	3	0	0	4	3	0	0	0	0	0	0	9942
Asp	8	1	9757	96	13	45	27	26	2	8	0	6	4	0	1	0	2	0	0	0	9996
Glu	13	17	95	9726	21	9	40	15	12	13	0	4	7	4	1	0	4	0	0	0	9981
Ala	42	54	24	37	9730	31	34	99	45	18	0	5	32	3	19	5	5	5	0	0	10188
Asn	10	10	36	7	14	9701	20	51	17	19	7	24	4	4	1	0	2	0	0	0	9927
Gln	4	11	16	24	12	15	9736	13	10	9	14	14	5	4	11	0	2	0	0	0	9900
Ser	26	15	28	16	59	67	22	9598	69	14	2	17	7	4	23	27	3	6	0	0	10003
Thr	6	8	3	14	30	25	20	76	9759	10	0	8	20	24	11	8	5	3	0	0	10030
Lys	5	6	13	21	17	37	23	22	14	9845	65	14	13	9	11	0	6	0	4	0	10125
Arg	0	0	0	0	0	5	13	1	0	23	9881	17	0	0	18	0	0	0	0	0	9960
His	0	0	4	3	2	20	15	10	5	6	19	9865	3	4	0	0	3	2	4	11	9975
Val	6	8	5	10	27	7	12	9	25	12	0	3	9783	156	82	18	22	3	0	0	10188
Ile	0	2	0	3	1	3	4	3	14	4	0	4	70	9703	22	3	22	3	0	0	9872
Met	0	0	0	0	2	0	4	5	2	2	7	0	12	7	9672	5	14	14	0	0	9737
Cys	1	0	0	0	1	0	0	12	3	0	0	0	6	2	11	9928	0	0	0	0	9964
Leu	2	0	3	7	4	3	4	5	7	6	3	6	24	52	99	0	9899	19	0	0	10140
Phe	0	0	0	0	2	0	0	5	2	0	0	4	2	18	18	0	10	9879	74	30	10047
Tyr	0	0	0	0	0	0	0	0	0	2	0	4	0	0	0	0	0	51	9909	17	9981
Trp	0	0	0	0	0	0	0	0	0	0	0	4	0	0	0	0	0	8	7	9941	9960

[a]The order of amino acids corresponds to that of Table 1-2. An element of this matrix $m(i,j)$ gives the probability that the amino acid in column j will be replaced by the amino acid in row i after an evolutionary interval of 2 PAM, i.e., 2 accepted point mutations per 100 amino acids. Thus, there is a probability of 0.0059 that Ala will be replaced by Ser, and a probability of 0.0099 that Ser will be replaced by Ala. The sum of each column is 1.0. The sum of a row represents the "growth factor per 2 PAM" of the corresponding amino acid residue; it ranges from 0.9737 (for Met) to 1.0188 (for Ala and Val).

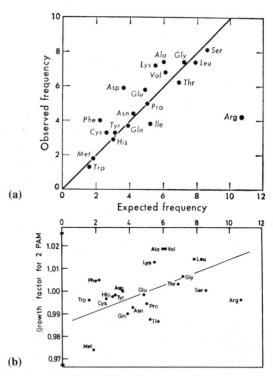

Figure 9-1. Amino acid frequencies. (a) Correlation between observed and expected amino acid frequency. The observed amino acid frequency is taken from 53 completely sequenced mammalian proteins (144). The expected frequencies are calculated from random DNA sequences on the basis of the genetic code (144). Both frequencies are given as percentages of the total amino acid content. The straight line represents an idealized equality of expectation and observation. (b) Correlation between the sum of entries in a row of Table 9-2 and the expected amino acid frequency as defined in (a). This sum represents the growth factor of a given amino acid for an evolutionary distance of 2 PAM. Thus, if in a given set of proteins Ala has a frequency of 1% it will increase to 1.019% after 2 PAM. Starting from any amino acid composition the growth factors enforce the correlation in (a) after sufficiently extensive evolutionary changes. Thus, the correlation in (a) is explained on the basis of the mutation probability matrix in Table 9-2.

from application of the genetic code (Figure 1-5b) on random nucleotide sequences in DNA (144). Apparently, this correlation indicates that amino acid changes are random, supporting the neutralists' hypothesis of evolution (496). Can these data be reconciled with the established fact that amino acid changes are conservative, that is, that protein evolution is indeed strongly affected by selection? Yes, it can.

The role of conservative changes is well demonstrated in the mutation probability matrix of Table 9-2. The entries of this matrix are very inhomogeneous, which corresponds to strong preferences in amino acid changes.

But a closer inspection of this matrix shows that the observed number of mutations leading to a given residue type (last column) also correlates with the frequency of this amino acid expected from random DNA sequences (Figure 9-1b). Thus, Figure 9-1a corresponds to Figure 9-1b, indicating that a residue type with a high observed frequency, for example, has had a high probability of appearing by mutation. Therefore, any evolutionary pathway starting from a given (random) set of amino acid frequencies changes (i.e., is modified by the mutation probability matrix) in such a way as to approach the presently observed set of amino acid frequencies.

In conclusion, the correlation of Figure 9-1a can be explained with the mutation probability matrix. Hence, the correlation holds, although in detail amino acid changes are under strong selective pressure. Therefore it cannot be deduced from the correlation between such summary values as amino acid frequencies that the evolution is neutral, i.e., non-Darwinian. Note that no attempt was made to explain the observed correlation of Figure 9-1b.

Amino Acid Changes Affecting Folding Dynamics

Protein analogues are the system of choice for studying changes affecting the folding process. The impact of amino acid changes on the function or on the *stability* of folded proteins has now been discussed. However, it is conceivable that a mutation has its worst impact on the *folding dynamics* of the polypeptide chain. For two reasons the study of this in natural mutants is difficult. If the folding pathway of a mutant protein is completely obstructed, the polypeptide chain cannot be identified and purified by conventional biochemical methods [but possibly by immunological (94, 418) or complementation techniques (446)]. Furthermore, polypeptides that do not fold at all or only fold slowly after their biosynthesis are often subject to immediate degradation *in vivo* (154). Because of these problems semisynthetic protein analogues (497–499) or proteins with modified side chains (445) are the objects of choice for studying the effect of mutations on protein folding (section 8.2, page 154).

Helix initiation residues were identified by sequence comparison. An indication of the importance of folding dynamics has been found in sequence comparisons between globin molecules (500). Here, all α-helices carry at their N-terminal portion either a Pro or a short polar side chain (Asn, Asp, His, Ser, Thr) which can hydrogen bond to the main chain. These hydrogen bonds and/or the proline that fixes the dihedral angle to the α-helix value are probably required for helix initiation. Locally, they give rise to a favorable relation between binding energy and chain entropy; hydrogen bonds increase the binding energy and Pro reduces the number of conformations and thus the chain entropy. According to Eq. (3-2), ΔG_{total} for this local piece of chain becomes highly negative and this piece of chain strongly tends to form a miniature folding nucleus.

Very strong conservation of chain folds underlines the restrictions arising from folding dynamics. The observed conservation of chain folds cannot be a consequence of keeping the delicate energy balance (or the function) of a given protein intact, because in meeting these requirements the amino acid side chains appear to be much more important than the course of the backbone. Consequently, it has to be assumed that the chain fold is strongly correlated with the folding dynamics and that these impose stringent conditions on the protein.

Amino Acid Changes as Natural Experiments

Mutant proteins can be used to elucidate structural principles. All accepted mutations in proteins can be taken as natural experiments that show us variations which affect protein stability and folding dynamics only slightly. In addition, random and probably unacceptable mutations, as in abnormal hemoglobins, exemplify variations that decrease protein stability by a measurable amount. Both types of mutations can be used to improve our knowledge of noncovalent forces in proteins. For this purpose energy minimization procedures have to be executed for the original and the mutated polypeptide chains on the basis of the known three-dimensional structures (501). The resulting energy differences and geometric deviations have to be compared to experimental data from thermodynamic measurements (413, 417) and high resolution X-ray analyses, respectively. Similar comparisons can be made using chain folding simulations (section 8.6) which may yield insight into the folding process.

9.2 Phylogeny on the Basis of Protein Structures

Each residue position is an independent trait with 20 levels of distinction. To determine phylogenetic relationships independent characters or traits have to be established, and quantitative comparisons of corresponding traits have to be carried out. The analysis of protein speciation for the deduction of phylogenetic data thus requires the definition of such independent traits in protein structures. For this purpose the sequentially ordered amino acid residues as elements of the primary structure are suitable candidates. They provide a large number of traits in any protein comparison, namely the number of residue positions. Each position can accommodate anyone of the 20 standard amino acids, each trait having 20 potential levels of distinction. Although these traits are clearly dependent on each other for single mutations (as shown in the rule of conservative changes, Table 9-2), the results shown in Figure 9-1 and numerous sequence analyses (145) indicate that, in a first approximation, such traits can be treated as independent as long as large numbers of mutations are considered.

Suitable Proteins for Phylogenetic Studies

Cytochrome c fulfills most criteria for phylogenetic studies. Although in principle many proteins can be used for phylogenetic studies, a particularly suitable one is mitochondrial cytochrome c (Figure 7-8) which was introduced by the pioneering work of Smith, Margoliash, and Fitch (502–506). The criteria that must be met for such studies and the characteristics of this protein that have made it most suitable to meet them are enumerated in (a)–(e) below.

(a) Orthodox pathway of the gene during evolution. The construction of phylogenetic trees is based on the assumption that the character (trait) in question is handed over from ancestor to descendant and that there is no other route. The transfer of structural genes from one species to an unrelated species, for example, carried out by viruses as mediators, would make protein comparisons useless as a tool for establishing phylogenetic relationships. This phenomenon of intergeneric gene transfer has not been observed as yet for structural genes in eukaryotes. But it is common in bacteria (507, 508) which means that phylogenetic trees of prokaryotes may not be deducible from structural comparisons of proteins.

(b) Singularity of the gene coding for the protein. There is only one gene for cytochrome c in the haploid genome of each eukaryotic organism (baker's yeast is the only known exception to this rule) (509). The presence of several isofunctional homologous proteins in one organism would complicate the picture for relationships between genetically distant organisms, because it is necessary to know which proteins have to be compared. Isofunctional proteins often occur at different ontogenic stages of an organism; for example, γ, ϵ, and ζ-chains of primate hemoglobin are not normally expressed during extrauterine life. Populations or species which use these chain types (510) during adult life would be grossly misplaced in a phylogenetic tree based on α- and β-chains of HbA, the hemoglobin which occurs normally in adult life (145). One should also consider Garstang's principle (511) which states that evolution often involves larval and not adult forms; only on this basis could the evolutionary pathway of guanidine kinases be sorted out (512). Incidently, Heckel's rule (ontogenesis reflects phylogenesis) has not been confirmed as yet at the protein level.

(c) Standard role of the protein in all compared organisms. Based on our present knowledge, all discovered c-type cytochromes from mitochondria have the same biological role.

(d) Ubiquity of the protein. Cytochrome c is found in all taxons for which a branch of the phylogenetic tree is under construction.

(e) Feasibility of numerous sequence analyses. The ease of protein preparation and the chain length of approximately 110 amino acid residues, which contains a sufficient number of traits but is still suitable for sequence analysis, were factors which contributed greatly to the unique role of cytochrome c.

The Phylogenetic Tree

Very distant taxons can be connected by studies on protein speciation. The point of departure for any evolutionary study based on amino acid changes is a table of differences between proteins from different species (145). These data are represented as an ordered difference matrix. The order is introduced by placing the smallest numbers (≡differences) closest to the diagonal. The basic steps leading from such a table to a family tree are described for short peptide hormones in section 9.3. In order to establish the genealogy of proteins such as mitochondrial cytochromes *c* from as many as 70 organisms (509), elaborate fitting schemes have been devised, such as the "ancestral sequence method" (513) and the "matrix approach" (504).

The resulting cytochrome *c* genealogy, which is equated to the phylogenetic tree of organisms possessing cytochrome *c*, contains genetic distances (given in PAM units) with no reference to evolutionary times. However, this tree can be calibrated by a comparison with the phylogenetic tree based on macroscopic traits and fossil records. The latter are dated using the geological time scale as derived from the radioactive decay of long-lived atomic nuclei. The calibration constant is the evolutionary rate given in Table 9-1. For the example of mitochondrial cytochromes *c* this rate is approximately constant along all branches of the tree, which gives rise to a linear calibration curve like those of Figure 9-2. If it is assumed that

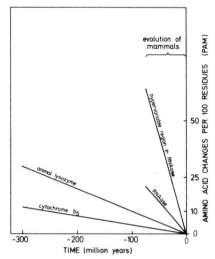

Figure 9-2. Diagrammatic representation of evolutionary rates (given in number of PAM per 10^8 years) in speciated proteins. 300 Ma (million years) ago the divergence of lines which led to reptiles and birds on the one hand and to mammals on the other occurred. Going back to 600 Ma evolutionary events as deduced from protein speciation can be checked against fossil data.

the straight lines with the lowest acceptance rates can be extended to times prior to the fossil record, phylogeny based on protein speciation is able not only to connect such distant taxons as, for example, plants and animals, but also to date the branching point. However, such extrapolations have to be taken with a grain of salt, because if there were fewer environmental constraints for earlier species, the evolutionary rate may have been much more rapid (483, 509).

Because of their small evolutionary rate (Table 9-1) cytochromes c are not suitable for clarifying relationships on twigs of the phylogenetic tree, as, for example, between cow and other artiodactyls. For this purpose faster evolving proteins such as pancreatic ribonuclease (145, 488) (Figure 9-2) have to be used.

Evolution of Behavior and Morphology in Comparison With Protein Evolution

Protein evolution is not the basis of organismal evolution. The fact that protein speciation can be used to study evolution does not mean that the evolution of organisms with respect to morphology and behavior parallels that of proteins. This is illustrated by the following examples which have been studied by Alan Wilson and his colleagues (514–517).

Placental mammals have experienced rapid organismal evolution compared to lower vertebrates such as frogs. Yet proteins seem to have speciated no faster in mammals than in frogs, and frogs are a much older group (150 Ma[4]) than placental mammals (75 Ma). As a consequence species that are similar enough in anatomy and way of life to be included within one genus of frogs (e.g., *Rana*) can differ in the sequence of a given protein as much as can a bat from a whale.

Another example concerns chimpanzee and man (514). The average human polypeptide is (more than 99%) identical to its chimpanzee counterpart; this corresponds to the genetic distance between sibling species of other mammals. However, the substantial anatomical and behavioral differences between man and chimpanzee have led to their classification in separate families.

Protein speciation provides the framework for studies at the organismal level. The current hypothesis is that the evolution of regulatory genes rather than the evolution of structural genes and the corresponding proteins is at the basis of organismal evolution (517). There are many indications that suggest that adaptive evolution results primarily from changes in the expression of structural genes relative to each other than from amino acid substitutions in the products of these genes. This hypothesis increases the significance of using protein speciation for establishing phylogenetic trees,

---

[4]Ma = mega-anni = million years.

because it implies that amino acid positions in polypeptide chains are traits which are independent of evolutionary jumps. Thus the study of protein speciation provides the framework within which evolution of behavior and morphology can be investigated more systematically; this is especially important for taxons whose fossil records are poor.

9.3 Protein Differentiation

The process by which homologous proteins with different functions evolve is called protein differentiation (473). In an enzyme the mutation of one or two amino acid positions can change the substrate specificity (508). This means that a change in availability from substrate S_x to substrate S_y can lead to the adaptation of an organism by replacing enzyme E_x by E_y. Such adaptations have been studied experimentally for bacteria (519–524). Successful examples are the evolution of an acetamidase to a phenylacetamidase (521) and of a ribitol dehydrogenase to a xylitol dehydrogenase (508, 522). In nature, proteins with new functions are constantly evolving. This is documented by the evolution of heritable resistance to toxic chemicals in insects and to antibiotics in bacteria (519).

Usually, protein differentiation starts with the duplication of the respective gene. Thereafter the amino acid sequences diverge from each other to different functions. The classical example (525) of protein differentiation is the occurrence of the multiple globin chains, e.g., myoglobin, and α, β, γ, ϵ, and ζ-chains of hemoglobin in man.

Construction of a Phylogenetic Tree for Vasotocin, Ocytocin, and Vasopressins

When comparing homologous proteins with different functions, one does not expect and indeed often does not find a constant acceptance rate for amino acid changes. This does not hamper the deduction of genealogies of proteins with different functions, but great care has to be taken when dating evolutionary events on the basis of structural comparisons between such proteins. In the following the genealogy of the peptide hormones vasotocin (VT), ocytocin (OT), bovine vasopressin (BV), and porcine vasopressin (PV) are described. Vasotocin is involved in both the reproduction and the regulation of the salt–water balance in many vertebrates; ocytocin (in mammals) serves only the former, and vasopressin only the latter function. Sequenced by du Vigneaud's group in 1953, these hormones represent the first known example of peptide differentiation (145, 526). Furthermore, they are small enough to illustrate the basic steps in the construction of a phylogenetic tree of homologous (poly)peptides.

The starting point is an ordered difference matrix. As a first step the random and the ordered difference matrix for these hormones are set up as shown in Figure 9-3. As a further refinement the matrix is then converted to Minimum Base Changes (MBC) between sequences and again ordered. The introduction of MBC increases the numbers, because more than 33% of the most frequently accepted amino acid changes are inconsistent with single nucleotide changes [see Figure 9-11 of Ref. (20)]. To some extent this also increases the accuracy because double and triple nucleotide changes may point to more than one accepted amino acid change at this position. Moreover, multiple nucleotide changes may occur with strongly conserved residues because such residues can probably wait for a multiple nucleotide change if it happens to be necessary for an acceptable replacement. Hence, the MBC-scale puts larger weights on changes at strongly conserved positions which is appropriate.[5]

The topological relationship between molecular species has to be established. Before constructing a phylogenetic tree in our example, the topological relationship between the four peptide species has to be established. This can be done graphically as shown in Figure 9-3d. Here, crosses indicate single base changes, so that in going from any molecular species (e.g., PV) to any other one (e.g., OT) the crosses add up to the number in the MBC difference matrix (e.g., 3). If the same approach is tried using the amino acid difference matrix, no solution can be found for our example, even if molecular species on the topological graph are rearranged (Figure 9-3d).

---

[5]Objections to the use of the MBC scale in comparisons of distantly related proteins have been put forward in Ref. (509).

Figure 9-3. Construction of a phylogenetic tree. (a) Sequences of four peptide hormones. Changes occur only in two positions. (b) Occurring amino acid changes, the corresponding codons, and the minimum base changes (MBC). Codons related with MBC = 1 are encircled. (c) Processing of an amino acid difference matrix: without any order → ordered → conversion from amino acid differences to MBC → reordered. (d) Topological relation between the four peptide hormones. Other topologies are given in parentheses. The nodes α and β correspond to intermediate, not necessarily extant species. The crosses denote amino acid changes (or alternatively MBCs) along the corresponding branches. The encircled cross is a special case in which an amino acid change is assumed to be at the same residue position as the change in the branch PV-α. Consequently this change does not show up in the entry PV–OT of the amino acid difference matrix. (e) The system of six linear equations corresponding to the topology given in (d); pv stands for the number of crosses in branch PV-α, $conn$ denotes the number of crosses in branch α–β, etc. The system has a solution if the first equation contains a three and not a two, i.e., if the amino acid change at the same residue position is counted. No solution is possible for the other topologies. (f) Phylogenetic tree derived from the topology of (d).

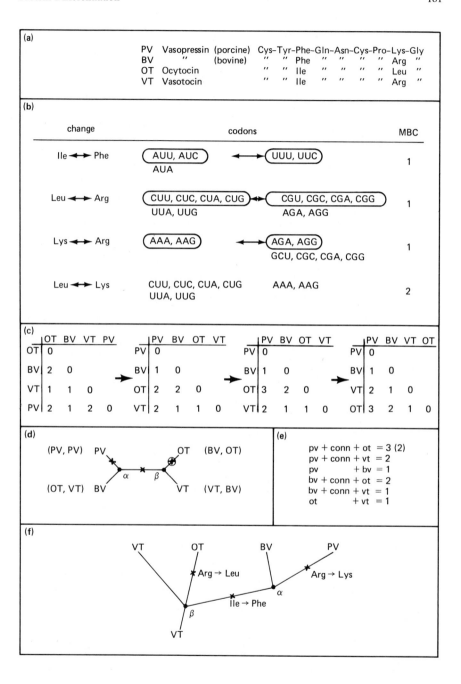

(a)

PV	Vasopressin	(porcine)	Cys–Tyr–Phe–Gln–Asn–Cys–Pro–Lys–Gly
BV	"	(bovine)	" " Phe " " " " Arg "
OT	Ocytocin		" " Ile " " " " Leu "
VT	Vasotocin		" " Ile " " " " Arg "

(b)

change	codons		MBC
Ile ⟷ Phe	(AUU, AUC) ⟷ (UUU, UUC) AUA		1
Leu ⟷ Arg	(CUU, CUC, CUA, CUG) ⟷ (CGU, CGC, CGA, CGG) UUA, UUG AGA, AGG		1
Lys ⟷ Arg	(AAA, AAG) ⟷ (AGA, AGG) GCU, CGC, CGA, CGG		1
Leu ⟷ Lys	CUU, CUC, CUA, CUG AAA, AAG UUA, UUG		2

(c)

	OT	BV	VT	PV
OT	0			
BV	2	0		
VT	1	1	0	
PV	2	1	2	0

→

	PV	BV	OT	VT
PV	0			
BV	1	0		
OT	2	2	0	
VT	2	1	1	0

→

	PV	BV	OT	VT
PV	0			
BV	1	0		
OT	3	2	0	
VT	2	1	1	0

→

	PV	BV	VT	OT
PV	0			
BV	1	0		
VT	2	1	0	
OT	3	2	1	0

(d)

(PV, PV)   PV   OT   (BV, OT)

(OT, VT)   BV   VT   (VT, BV)

α   β

(e)

pv + conn + ot = 3 (2)
pv + conn + vt = 2
pv   + bv = 1
bv + conn + ot = 2
bv + conn + vt = 1
ot   + vt = 1

(f)

VT   OT   BV   PV

★ Arg → Leu   ★ Arg → Lys
α
β   Ile → Phe

VT

A closer inspection shows that the problem of fitting a difference matrix to a topological graph corresponds to a system of six linear equations (six matrix elements of Figure 9-3c) and only five unknowns (number of crosses in each branch), which has solutions only in exceptional cases. In our example we get a straightforward solution for the MBC matrix for one of the three possible topologies. The amino acid difference matrix can also be solved if it is assumed that the mutation in branch OT occurs at the same residue position as the one in branch PV, so that there are three changes when going from PV to OT although only two occur in the difference matrix.

The topological relationship has to be converted to a phylogenetic tree. Since all molecular species are extant, the ancestral species have to be placed at one of the branching points. A branching point is chosen so that the number of changes along all branches to the extant species is as similar as possible. In our example both branching points α and β are equivalent in this respect (Figure 9-3) leading to BV and VT as respective ancestors. A glance at the organisms, however, decides the problem in favor of vasotocin (VT) as ancestor since it occurs in more ancient taxons (145). The tree shows further that the sequence of bovine vasopressin (BV) is older than that of porcine vasopressin (PV).

Note that this method provides us with ancestral sequences at the branching points. It represents the basic step of the "ancestral sequence method" mentioned above. For a comparison of more than four (poly)peptides there is, in general, no exact solution. Here, and in particular in cases such as cytochrome c (70 speciated proteins, see above) and serine proteases (20 differentiated proteins, see below), the ancestral sequence method (513) is used to accomplish the best possible fit by establishing the best phylogenetic tree on the basis of the present data. In this tree the ancestral sequences tend to become ambiguous in more and more positions when going back to successively earlier branching points.

α-Lactalbumin and Lysozyme

The comparison between lactalbumin (527, 528) and lysozyme (518, 529–531) is the textbook example of two proteins with similar sequences but with different functions and different rates of mutation acceptance. The structural similarity of the two proteins was suspected in 1958 and proved 10 years later (532, 533) by sequence comparisons. Some relevant properties of the two proteins are shown in Table 9-3. The three-dimensional structure of bovine lactalbumin was deduced from the structure of hen egg white lysozyme by model building (534) and subsequent energy minimization (501, 535). This procedure assumes that both chain folds are identical, which is a rather safe assumption if the sequence correlation is that strong (Table 9-3). Furthermore, this example demonstrates how data on one

Table 9-3. Properties of α-Lactalbumin (528) and Lysozymes (518, 531)

	α-Lactalbumin	Lysozyme
Probable time of divergence from a common ancestor (145)	$3 \cdot 10^8$ years ago	
Fate after divergence	Involved in the development of a new physiological function	Preservation of the ancestor's function
Rate of mutation acceptance (145)	22 PAM/10^8 years	10 PAM/10^8 years
Amino acid differences of the two proteins in man	$82 \equiv 62\% \equiv 126$ PAM	
Function	Serves as an on/off switch for lactose synthesis by interaction with a (Golgi) membrane-bound protein	Hydrolyzes polysaccharides of bacterial walls
Occurrence	In milk (15% of the total milk proteins)	In many tissues and secretions of all major groups of organisms; it is not clear (241) whether all enzymes with lysozyme activity are homologous

protein can be used for the structural analysis of a distantly related homologous protein.

Trypsin-Like Serine Proteases

Trypsin-like serine proteases (138, 536) are a family of protein-cutting enzymes which control many important physiological processes (Table 9-4). The digestive enzyme trypsin, for which the term "enzyme" was coined (1), is the most prominent member of the family, because it has been known for more than hundred years and because its specificity for cleaving lysyl and arginyl bonds resembles that of most other members. Most of the relatives of trypsin, however, are much more specific than trypsin itself;

Table 9-4. Serine Proteases Essential for Maintaining, and Others Threatening Life in Man (602)

Protease	Location where active	(Patho)physiological process governed
Trypsin	Small intestine	Protein digestion
Thrombin	(a) Blood	Blood clotting
	(b) Inflamed tissues	Walling off the area
Plasmin	Wounds and inflamed tissues	Dissolution of the provisional fibrin clot
Acrosomal protease of sperm	Ovum	Fertilization
Components of the complement system	Cytoplasmatic membranes of bacteria; infected tissues	Destruction of alien cells, initiation of inflammatory processes, etc.
Kallikrein	Blood and tissues	Circulatory regulation and pain sensation
White cell protease	Inflamed tissues	Destruction of pathogens and of elastic connective tissue
M-protease	Sarcoma cells	Tumor growth
D-protease	Myocardial and other cells	Liberation of the deadly diphtheria enzyme from the toxin

each of them cleaves only one (or very few) peptide bond(s) in one specific protein. The structural homologies in serine proteases were reviewed and previewed by Hartley in 1970 (490). Pairwise comparisons of trypsin, elastase, chymotrypsin, and thrombin were shown to exhibit about 40% sequence identity (58 PAM). Today the atomic structures of the first three of these enzymes are known. As predicted, all of them have the same chain fold (18, 243–245).

The sequences of protease B and chymotrypsin could be aligned in the light of the three-dimensional structures. Trypsin-like enzymes are also found in microorganisms. *Streptomyces griseus,* for example, possesses the so-called prote(in)ase B which has only 186 residues (versus 234 in chymotrypsin). X-ray diffraction analysis showed that the chain fold of protease B closely resembles the chain fold of chymotrypsin (246). As inferred from the three-dimensional structures, deletions and insertions occur only at the surface of the molecule. This is shown in Figure 9-4. As indicated before, conservation of the protein interior can be considered as a general rule. After clarifying the linear alignment of the sequences on the basis of the geometrical alignment, only 13% of the residues are identical, which corresponds to a genetic distance between chymotrypsin and protease B of 432 PAM. At such a distance homology could not have been deduced from the amino acid sequences alone (page 201, and Ref. (803)).

The arrangement of the residues at the catalytic site, the chemical knife for severing peptide bonds, is precisely conserved in all serine proteases, whereas the substrate specificity has diverged widely. This may indicate that the catalytic mechanism represents a physico-chemical optimum (537,

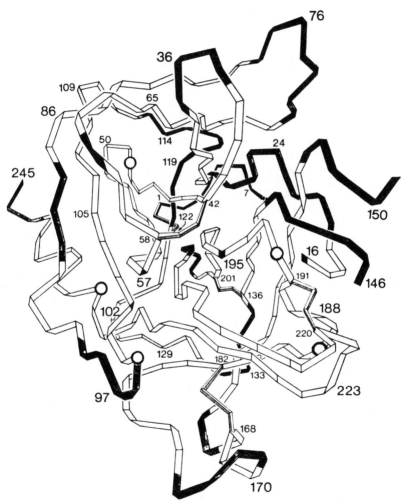

Figure 9-4. The polypeptide chains of α-chymotrypsin and *Streptomyces griseus* protease B [Ref. (18)]. The chain of chymotrypsin is represented by a ribbon kinked at each C_α-atom. The first β-sheet barrel (Figure 5-17d) appears at the upper left of the figure, and the second one appears at the lower right. Protease B has the same chain fold. However, the parts of the ribbon marked in black are deleted. Moreover, there are insertions at the circles in the ribbon. Note that all insertions and deletions are at the molecular surface.

538). However, other solutions also exist for severing peptide bonds; different families of proteinases (145,539) use different catalytic mechanisms.

Immunoglobulins, Transplantation Antigens, and a Superoxide Dismutase

A chain fold of about 100 residues which is present in quite different proteins is the "Ig-fold" (540–543). It was named after the immunoglobulins in which it was first discovered (75, 292, 544, 545). The Ig-fold is the exclusively used building block in immunoglobulins, in cytosolic superoxide dismutase (546), and probably also in transplantation antigens (HL-A proteins) (86–88, 547). The properties of these proteins are shown in Figure 4-2 and in Table 9-5. The sequence homology between an HL-A light chain and a C domain of immunoglobulins is so significant [the genetic distance being 217 PAM (541)] that a common evolutionary origin and chain fold similarity are clearly indicated. Between V and C domains of immunoglobulins there is only a marginal sequence homology but a very close chain fold similarity. The rms distance between superimposed C_α-atoms of V_L and C_H1 domain is 1.2 Å using 56 equivalenced residues (548) which indicates another case of protein differentiation (see section 9.6). There is no sequence homology between cytosolic superoxide dismutase and the other two types of proteins. However, the chain folds resemble each other so closely that a distant evolutionary relationship is very likely (548). The significance of homology in this case will be discussed in section 9.6.

A glance at the functions of these three types of proteins shows that immunoglobulins and HL-A proteins defend the integrity of an individual organism against intruders. Superoxide dismutase also traps troublemakers, but in this case they are small O_2^- radicals. It fulfills a function which had to be developed with the advent of oxygen in the Earth's atmosphere. Since all three protein types are protective, they constitute a case of protein differentiation in which the function has changed considerably but not totally.

The "Globin Fold"

Globins are monomeric or oligomeric haem-containing proteins. They occur in a wide variety of organisms including bean, midge, and man (145). Members of the globin family are involved in the transport of O_2, in nitrogen fixation by root nodules of leguminous plants, in filling the swim bladder of certain fish species, and in other physiological functions (549, 550). Three-dimensional structures of a number of species have been elucidated (185, 188, 274–277) and all of them possess the same chain fold.

Table 9-5. Proteins Possessing the Ig-Fold

	Immunoglobulins	Transplantation antigens	Superoxide dismutase
Properties	Involved in the recognition and elimination of harmful agents such as toxins, microorganisms, or cancer cells. An immunoglobulin G contains 12 Ig-domains, two times V_L, C_L, V_H, C_H1, C_H2, and C_H3 (Figure 4-2). Some other classes of immunoglobulins contain multiples of 12 domains (145, 542, 544)	Cell surface proteins playing various roles in cell–cell interactions. Represent main barrier against organ grafting. Also called HL-A proteins (human leukocyte antigens) or H2-proteins in mice. Each of the two identical heavy chains is likely to contain three Ig-domains (86, 547). All light chains are identical; they consist of a single Ig-domain, identified as "β_2-microglobulin" in blood and urine before its origin was recognized (88)	The Cu^{2+} plus Zn^{2+}-containing dimeric enzyme (546) is located in the cytosol of eukaryotes; mitochondria contain an unrelated isozyme. Constitutes the primary defense against O_2^- radicals, an important agent of oxygen toxicity, by converting it to O_2 and H_2O_2. Each subunit of the enzyme contains one Ig-domain
Sequence data	Numerous sequences known; see Ref. (541) for a review	Only sequence of light chain is known (541, 603)	Sequence is known (604–606)
Three-dimensional structure data	About half a dozen structures are known (42, 292, 294, 543, 545, 607, 608)		
Established sequence homologies	Among C-domains (541), among V-domains (541)	Between light chain and C-domains of immunoglobulins (541)	None
Intradomain S–S bridge across β-cylinder, encompassing 60–70 amino acid residues	Present in all domains (541, 543)	Present in all domains (86)	Absent (286)
Position of active site	Between V_L and V_H	Not yet identified (547)	Two out of the three loops between β-strands which define the antigen binding site in V-domains of immunoglobulins correspond to loops forming the Cu^{2+} and Zn^{2+} binding sites in the active center (548)

Historically, globins were the pacemakers for the ideas of protein specia-tion and protein differentiation. As early as 1865 a hypothesis was put forward that the globins from muscle (551) and from red blood cells (552) were identical or related proteins. Almost 100 years later X-ray diffraction analysis of hemoglobin (553) and myoglobin (185) revealed that the two proteins possess a common chain fold, the typical "globin-fold" (Figure 8-4). The sequence analyses of the polypeptide chains of myoglobin (554) and of hemoglobin (555) were finished somewhat later showing a difference of about 73% (192 PAM).

Thus, the first objects of three-dimensional structure analyses already showed that chain folds are extremely well conserved. Moreover, they indicated that proteins of different function, of different tissue distribution, and of different organisms may resemble each other and therefore may have a common origin. These findings constituted an important stimulus to the use of protein structure analysis, and in particular sequence analysis, for tracing evolution at the level of macromolecules (145).

Morphological Similarities between Globins and Cytochromes *b*

Homology between cytochromes in different electron transport chains may show homology between transport chains. There are some indications that globins are related to certain *b*-type cytochromes[6]; cytochrome b_2, b_5, and b_{562} are of particular interest in this respect. Data on these cytochromes are given in Table 9-6. As predicted on the basis of proton magnetic resonance spectra (556) and as established by sequence comparisons (557), cyto-chrome b_5-hbf (haem binding fragment) and cytochrome b_2-hbf are homol-ogous proteins with a sequence difference of about 72% (185 PAM). The similarity becomes even more striking when the sequence of b_2-hbf is fitted to the known crystal structure of b_5-hbf (297, 557). The observed homology between b_2-hbf and b_5-hbf suggests that not only these two proteins but the whole corresponding electron transport chains (Table 9-6) may share a common origin.

There are structural similarities between *b*-type cytochromes and glo-bins. Cytochrome b_{562} shows no sequence similarity with b_2-hbf or b_5-hbf. However, its sequence appears to be homologous with myoglobin (558). This and other findings (559) (which indicated that there may be a contin-uum of homologous structures ranging from *b*-type cytochromes to glo-bins) induced comparative studies (406, 560) on the well-known[7] tertiary structures of b_5-hbf and hemoglobin β-chain.

---

[6]Cytochromes are classified on the basis of their absorption spectra in the region between wavelengths of 400 and 600 nm. These spectra depend on the type of haem (cytochromes *b*, for example, contain protohaem IX) and on the mode of attachment of the haem to the polypeptide chain. According to this nomenclature the designation of proteins as *b*-type cytochromes does not imply that they possess a common chain fold (509, 561).

[7]X-ray analyses of cytochrome b_2 (562) and b_{562} (563) are in progress.

Table 9-6. Properties of Cytochromes b_2, b_5, and b_{562}

	Cytochrome b_2 from yeast mitochondria (561), identical with L-lactate dehydrogenase (EC 1.1.2.3) from yeast	Cytochrome b_5 from the endoplasmic reticulum (microsomes) of calf liver (561)	Cytochrome b_{562} from *Escherichia coli* (561, 609)
Whole protein	A soluble tetrameric protein; each subunit (molecular weight 58,000) consists of one FMN-containing domain and one haem-containing domain	A membrane-bound monomeric protein (molecular weight 16,700) which may form aggregates *in vivo*. The 40 to 50 C-terminal residues are located in the membrane	A monomeric protein (molecular weight 12,000) from the soluble fraction of *E. coli* extracts
Possible function of the whole protein	b_2 mediates electrons from L-lactate to the respiratory chain; the pathway of the electron(s) being: L-lactate $\rightarrow$ b_2 (FMN-containing domain) $\rightarrow$ haem-containing domain of b_2 $\rightarrow$ cytochrome c	b_5 mediates electrons from NADH to certain functional units of the metabolism [e.g., to the "fatty acid desaturase" (610)], the pathway of the electrons being: NADH $\rightarrow$ FAD-containing protein $\rightarrow$ haem-containing domain of b_5 $\rightarrow$ "cyanide sensitive factor"	Unknown
Haem-binding portion	b_2-hbf (cytochrome b_2 core), the haem-binding fragment, is obtained by specific proteolysis of b_2 and contains 95 amino acids	b_5-hbf, the haem-binding fragment, is obtained by specific proteolysis of b_5 and contains 93 amino acids	The hemoprotein contains 110 amino acids
Known structural properties of the haem-binding portion	The sequence of b_2-hbf (557) is homologous with that of the haem-binding fragment of b_5 (611, 612) and could be fitted to the known tertiary structure of b_5-hbf	The sequence of b_5-hbf (611, 612) is homologous with that of the haem-binding fragment of b_2 (557). The tertiary structure of b_5-hbf is known (297)	The sequence of b_{562} (613) may be homologous (558) with that of sperm whale myoglobin (554). The tertiary structure analysis of b_{562} has been started (563)

This comparison yielded the following results (560). Among the 85 three-dimensionally characterized residues of cytochrome b_5-hbf, 51 have their counterparts in the globin. When the two structures are superimposed the haem irons are found to be separated by less than 1.4 Å and the angle between the haem normals is about 10°. This corresponds to the values for the comparison of erythrocruorin (an insect hemoglobin) versus myoglobin (277), and for the comparison of reduced cytochrome c versus oxidized cytochrome c (273, 564). The iron-liganded histidine and two adjacent helices in hemoglobin correspond to a histidine and adjacent helices of cytochrome b_5-hbf.

Whereas the orientations of the haem normals are similar, a rotation of the haem ring by 53° around this normal is found in the comparison between cytochrome b_5-hbf and hemoglobin (560). The authors regard this difference as energetically minimal and conservative as the haem side chains are not directly involved in the functions of either protein. The latter point must be reinvestigated (cf. section 10.3, page 212). In fact, the structural comparisons of haem-containing proteins could help to clarify the role of the haem side chains (565–568).

In conclusion, there are similarities between globin and cytochrome b_5-hbf extending over parts of helices A, B, E, F, and G (Figure 8-4) and the haem position. However, the question remains as to whether these similarities are homologous, that is, based on a divergent evolution from a common ancestor or analogous, that is, caused by a convergent evolution to a physically favored supersecondary structure for noncovalent haem binding (569). Presumably this problem will be settled by solving the tertiary structure of cytochrome b_{562} (563), which could turn out to be the desired missing link in the genealogy (Table 9-6). The question of recognizing very distant relationships between proteins is discussed further in section 9.6.

c-Type Cytochromes

c-Type cytochromes have undergone protein differentiation. As described in section 9.2, mitochondrial cytochromes c constitute a well-analyzed example for protein speciation. Other c-type cytochromes have different functions, most of them being involved somehow in electron storage or electron transport (509). All c-type cytochromes together may represent a case of protein differentiation; however, as the term cytochrome c is a spectral and not a structural classification, no generalization is possible. Not counting the structure of mitochondrial cytochromes c, X-ray diffraction analyses have been completed for three bacterial cytochromes (Table 9-7) and have been initiated for three more prokaryotic cytochromes, namely, cc_3, c_{555}, and c' [for a review see Ref. (509)].

c-Type cytochromes are subdivided into Small and Large variants. As indicated by their three-dimensional structures and by sequence alignments, all known cytochromes c share a common chain fold. However, there are large insertions and deletions at the surface similar to those found

Table 9-7. Structurally Known Small and Large Variants of c-Type Cytochromes[a]

Representative(s) Source	Small variant c_{551} Pseudomonas aeruginosa	Large Variants		
		c_2 Rhodospirillum rubrum	c_{550} Micrococcus denitrificans	c Mitochondria of tuna
Molecular weight	8,100	12,500	14,890	11,500
Available structural information	Sequence and tertiary structure	Sequence and tertiary structure	Sequence and tertiary structure	Sequence and tertiary structure
Structural information on close relatives	> 6 sequences	> 5 sequences	> 1 sequence	> 67 sequences, 2 tertiary structures
Important amino acids for the haem attachment	Cys–X–Y–Cys–His near the N-terminus and Met closer to the C-terminus of the polypeptide chain. Both Cys bind covalently to vinyl side chains of the haem group. His and Met are axial ligands of the haem iron			
Pattern of aromatic side chains	Only two or three positions in the structure are directly comparable with Large variants	Identical in eight positions		
Insertion at the bottom of the molecule (corresponding to positions 40 to 55 in mitochondrial c) giving rise to the designation Small or Large variant	Absent	Present		
Metabolic role	O_2-respiration	Photosynthesis	O_2-respiration	O_2-respiration
Metabolic role of a rather close relative	Photosynthesis (c_6)	Photosynthesis and O_2-respiration (c_2 from Rhodcpseudomonas spheroides)	O_2-respiration	O_2-respiration (c from all eukaryotes)

[a]Data are from Ref. (509).

for protease B and chymotrypsin (Figure 9-4). Cytochromes c are classified as "Large" and "Small variants" according to the presence or the absence of loop 40–55 (Figure 7-8), respectively.

Photosynthetic and respiratory cytochromes c are structurally as similar as chymotrypsin and trypsin. The classification of cytochromes c with known three-dimensional structures is given in Table 9-7. Among the Large variants, the primary and tertiary structures of c from tuna mitochondria, c_2 from *Rhodospirillum rubrum,* and c_{550} from *Paracoccus denitrificans* (Table 9-7) do not show greater differences than are found between trypsin and chymotrypsin; hence c_2, c_{550}, and mitochondrial c can be regarded as relatives. Structural comparisons between these proteins support the hypothesis (570) that mitochondria originated from a bacterium closely resembling *Paracoccus*.

Attempts to align the sequence of c_{551}, a Small variant, with the sequences of the Large variants were unsuccessful. Only X-ray analysis of c_{551} showed how the two proteins are related (571). With these X-ray data, however, the sequence alignment was no problem and could even be extended to other cytochromes of known sequence, for example, the photosynthetic cytochromes c_6 (also called f) of prokaryotic and eukaryotic algae.

Among the Large variants there are cytochromes which are involved in a respiratory electron transport chain, and others which are parts of a photosynthetic electron transport chain; among the Small variants the same functional diversity is found. This means that the two electron transport chains may have a close evolutionary connection. This hypothesis is supported by the finding that respiration and photosynthesis are coupled at a molecular level in *Rhodopseudomonas spheroides;* only one (Large variant) cytochrome c_2 is used for both functions (572). An organism in which a Small variant serves both functions has not as yet been found.

c-Type cytochromes may reveal the evolution of metabolic pathways. c-Type cytochromes have introduced us into the realm of prokaryotes. In principle, these protein structures could be used to establish some order among bacteria, in the same way as mitochondrial cytochromes c were employed as taxonomic tools for eukaryotic organisms. Initial attempts at classifying bacteria on this basis have already been undertaken (509, 571). However, as intergeneric gene transfer may be common in bacteria (507, 508) the establishment of a phylogenetic tree is hindered by genes that jump like squirrels from branch to branch.

Nevertheless, in spite of gene transfer, c-type cytochromes are expected to aid in tracing the development of modern photosynthesis and respiration from more ancient precursors (e.g., photosynthesis based on H_2S, respiration based on sulfate, etc.); they may help (571) to elucidate the evolution of metabolic pathways which is one [early postulated (573, 574)] final goal of protein differentiation studies.

9.4 Gene Fusion

Whereas amino acid changes, insertions, and deletions are comparatively small structural modifications in evolution, the process of gene fusion gives rise to very substantial changes; either it combines different polypeptide chains with each other or it doubles a given chain.

Gene Fusion in Action

Fusion of structural zones in DNA are not just rare events of the past. Substantial evidence exists to indicate that gene fusions occur regularly during ontogenesis of modern organisms, for example, as a prelude of immunoglobulin synthesis.

The fusion of V and C genes of immunoglobulins takes place at the DNA level. The light (L) chain of an immunoglobulin molecule consists of a variable (V) and a constant (C) portion (Figure 4-2c). The structural zones coding for V and C are separate along the DNA but are on the same chromosome. In spite of this separation at the DNA level, the mRNA corresponding to an L-chain was found to be a single continuous molecule (576). The same observation was made for heavy (H) chains, which consist of one V and three C regions (477). Moreover, pulse-labeling experiments showed only a single growing point in L-chain synthesis indicating that there is no splicing of V and C-chains at the protein level.[8] Furthermore, using L-chain mRNA as a probe in hybridization experiments, rearrangement of DNA was observed in precursors of immunoglobulin-producing cells (577). The last finding suggests that the fusion of V and C occurs exclusively at the DNA level and not at the RNA level. This question is not, however, finally settled (78).

Fusion of different DNA chains is also common with lysogenic bacteriophages which incorporate their DNA into a bacterial genome and keep it dormant there for some time. Furthermore, DNA-splicing is a basic step in all genetic engineering experiments (542, 575).

Gene Fusion in Evolution

As documented in a number of cases gene fusion was an important process in protein evolution. This implies that the composite gene has to establish itself in germ line cells unlike the case for a V–C gene. Representative examples are found in the pathways of amino acid synthesis and fatty acid synthesis. The textbook example of one polypeptide chain containing two enzyme functions is aspartokinase I—homoserine dehydrogenase of *E. coli* (578). Comparative studies on the enzymes involved in Trp synthesis

[8]Fusion at the protein level occurs during the biogenesis of bacteriophage λ (142).

revealed a great variety of arrangements of several enzyme activities on polypeptide chains (579) which suggests that gene fission may occur as well as gene fusion.

Fatty acid synthase from baker's yeast is one of the most spectacular cases (76, 580). There are two gene loci, fas 1 and fas 2, in which two polypeptide chains, but eight catalytic functions, are encoded. There is no apparent correlation between the sequence of biosynthetic reactions and the order of the corresponding functional domains along the two polypeptide chains. For example, the coding order of domains in fas 1 corresponds to the biosynthetic steps 5, 6, 2, 8.

Gene fusion does not drastically change protein properties. The first example of an artificial fusion of two enzymes at the genetic level was the splicing of two enzymes which are involved in the biosynthesis of histidine in *Salmonella typhimurium* (581). This example demonstrates that gene fusion alone does not drastically change the properties of proteins. Each of the two enzymes consists of two identical subunits and this structural feature is conserved when the genes are fused. The enzymic and spectroscopic properties of the individual proteins also remain unchanged.

Fused proteins have several advantages. One is that the functional domains are synthesized in stoichiometric amounts. Another is the possible development of desirable domain–domain interactions which (in contrast to such interactions in oligomeric proteins) are independent of protein concentrations (76). A third is that in specialized biosynthetic pathways it may be desirable to have a labile intermediate channeled directly to the next enzyme.

Gene fusion might have played an important role during the evolution of the major metabolic pathways. Each of the following enzymes of the energy metabolism probably originated by splicing a copy of a primordial (di)nucleotide-binding domain with one or more different other domains: phosphoglycerate kinase (235, 310, 311), the dehydrogenases specific for glyceraldehyde-3-phosphate, lactate, malate, and alcohol, respectively (91), and glycogen phosphorylase (236). As emphasized before (section 5.4), the (di)nucleotide-binding domain represents the N-terminal part of the first four enzymes, whereas this domain is located in the C-terminal part of alcohol dehydrogenase and in the middle of the phosphorylase chain. This indicates that the constraints for arranging domains as modules of sophisticated proteins posed no problems in the evolution of proteins.

Gene Multiplication

Up to now we have been concerned with fusion between different genes. In many cases, however, gene fusion is found between identical genes, which indicates that the gene was multiplied before fusion occurred.

Contiguous multiple genes code for gene products such as ribosomal RNA and histones. Multiplication (without fusion) of a structural gene resulting in distinct new genes is a known evolutionary process. Contiguous multiple genes (582, 583) code for gene products such as ribosomal RNA or for histones (487). In the case of ribosomal RNA the advantage vis-à-vis a single gene is obvious; the transcription product is (apart from modification reactions) the final product; no amplification of the final gene product is possible at the ribosomal machinery. In the case of histones it has been argued that transcription of multiple genes into histone-mRNA helps to provide the large amounts of these proteins which are required at the early stages of embryogenesis; apparently the numerous copies of histone-mRNA exclude other mRNA-species from the ribosomes (542).

Presumably, gene multiplication is important for protein differentiation. It is assumed that gene multiplication played an important part in the evolution of protein structure and function (523, 525, 582, 584, 585, 592). After multiplication of a gene one copy served the original function whereas the other one(s) could evolve to a related or to a new function. The best-known examples are human genes coding for the α, β, γ, ϵ, and ζ-chains of hemoglobins and for myoglobin. Controversy exists concerning the questions [(586) vs (523)] of how many generations a redundant gene will be tolerated in the genome and whether nonfunctioning (silent) forms of this extra copy are possible intermediates during the development of a protein with a new function.

A structural repeat in a protein can also be caused by unequal crossing over. Gene multiplication followed by gene fusion gives rise to gene products with two or more identical substructures (587). However, as shown in the following example, other events may lead to the same result. A (partial) structural duplication is present in the rare α_2-chain of human haptoglobin (145, 588). Since the amino sequence of both parts are identical and are identical to a large portion of the common α_1-chain, this structural duplication is a very recent event. Most likely it was caused by a chromosomal aberration (unequal crossing over) in a (human) founder population. If this event had occurred much earlier, so that sequence homology was blurred by amino acid changes, insertions, and deletions, it would no longer be possible to distinguish between gene doubling and subsequent fusion on the one hand and chromosomal aberration on the other. Consequently, ancient cases of structural repeats are generally referred to as "gene duplications" without trying to distinguish between different mechanisms.

Some structural repeats reveal themselves by repeating disulfide bond patterns. Structural repeats are often obvious in the pattern of S–S bonds as shown for wheat germ agglutinin in Figure 7-2a. The fourfold repeat was confirmed by a corresponding repeat in the three-dimensional structure (316). A complex genealogy has been postulated for serum albumin on the basis of internal sequence homologies (82, 589), which are reflected in the disulfide pattern of Figure 7-2b. The following evolutionary stages are likely to have occurred: primordial gene coding for a protein with around 80 amino acids and an intertwining S–S bond → triplication of this structure → deletion of approximately 30 residues → duplication of the resulting structure which gave rise to a protein of 400 amino acids, and another duplication of one-half of this structure.

The abundance of structural repeats indicates that new chain folds are rare, precious innovations. A duplication of amino acid sequence and of the chain fold has been found for ferredoxin (section 5.3, page 96). In parvalbumin the sequence repeat is more or less restricted to the Ca^{2+}-binding residues; the overall chain fold repeat is clear, however, in the tertiary structure (59, 590). Simple chain fold repeat without traces of sequence homology occurs in rhodanese (Figure 5-17a) and, as represented by the two barrels (Figure 5-17d), in trypsin-like proteins. All these examples indicate that chain folds were rare innovations; a "working" chain fold was multiplied and afterwards conserved in each copy.

A gene duplication was postulated for the NAD- binding domains of four dehydrogenases (91), where a Rossmann-fold (Figure 5-12b) is repeated. But the evolutionary significance of this observation (section 9.6) is not very high because the Rossmann-fold is a physically preferred supersecondary structure (section 5.2). Similar cases of low evolutionary significance are the structural repeat within each of the barrels of serine-proteases (Figure 5-17d) and the structural repeat within triose-phosphate isomerase (Figure 5-17e), respectively.

The hypothesis that all proteins evolved from oligopeptides has not as yet been substantiated. In view of the observed structural repeats it was postulated (591) that all proteins have evolved by repeated gene duplications and fusions from shorter segments of around 15 residues. However, this hypothesis could not be substantiated when the sequences of 50 globular proteins were tested by statistical methods (587). Three-dimensional structures are also of no help in settling this question. Here, short repeats cannot be detected, because the number of possible chain folds of such a short piece of chain is small. Therefore, any structural coincidence is only of low significance (see section 9.6). The number of possibilities would become large when comparing exact dihedral backbone angles and not only general chain courses. However, these angles are not exactly conserved during evolution.

Some proteins contain multiple repeats of short sequences. Repeats of short sequences have been found in so-called periodic proteins (145, 593), which include collagen, wool keratin, histones, tropomyosin, human lipoprotein A1, and a freezing point depressing glycoprotein of an antarctic fish. In the latter case the repeating unit is Ala–Thr–Thr throughout the sequence. In some cases periodicity may reflect peculiar features of the corresponding DNA (593), and in other cases structural requirements (formation of the collagen triple helix shown in Figure 5-6, appropriate binding of tropomyosin to filamentous actin and of histones to a DNA double helix) may have exerted selective pressure.

9.5 Convergent Evolution in Proteins

The number of possible amino acid sequences and chain folds is so large that similarity is not expected to occur by chance but to reflect the historical development, i.e., protein speciation or differentiation. However, the vast number of structural options does not exclude the possibility that certain structures are strongly favored and that several evolutionary processes may converge on such structures. Clear examples of convergence with respect to protein structure are α-helices and β-sheets; supersecondary structures are also likely candidates.

Evolutionary convergence to a common function on the basis of different structures is rather frequent. Since functions can be fulfilled in various ways, a number of different proteins acquired the ability to serve similar functions; their evolutionary pathways converged with respect to protein function. An example is the reversible binding of O_2 to hemoglobin, hemerythrin,[9] and hemocyanin[9]. Dioxygen is always bound to iron or copper, but these ions are incorporated in completely different protein structures (594). Similarly, Mn^{2+}-containing superoxide dismutase and $Cu^{2+}-Zn^{2+}$-containing superoxide dismutase are examples of convergence to common function on the basis of different structures (546, 595).

The textbook examples of convergent evolution are the catalytic sites of the serine-proteases chymotrypsin and subtilisin (18, 239, 240). In both enzymes an incoming substrate would encounter the same arrangement of hydrogen bonds for fixing the backbone of the substrate chain, the same position of the side chain binding pocket, the same charge relay mechanism of bond cleavage, and the same set of hydrogen bond donors (backbone NH-groups) for fixing the carbonyl-O^- of a catalytic intermediate (537). Nevertheless, both enzymes are completely uncorrelated with respect to amino acid sequence, chain fold, and, for example, the order of the charge relay residues Asp, His, Ser along the chain (18, 239, 240). Thus, all data indicate that the catalytic surface structure of these enzymes is so far superior to others that functional requirements have forced it onto both chymotrypsin and subtilisin. As described in detail in Figure 11-1, the catalytic mechanisms of the protease papain and of glyceraldehyde-3-phosphate dehydrogenase, an enzyme of the glycolytic pathway, may be analogous to that of serine proteases. On the other hand it should be kept in mind that there are also other ways of cutting polypeptide chains as demonstrated by thermolysin, cathepsins, and acid proteases (539, 596).

---

[9]These names are derived from haima = Greek for blood; they do not imply that the proteins contain haem groups.

Even in homologous proteins details can be a consequence of convergent evolution. Proteins diverging from a common ancestor in their overall architecture can also exhibit aspects of convergent evolution (273, 597). For example, the innermost haem propionate is hydrogen bonded to Trp-56 in the Small variant cytochrome c_{551}, but to Trp-59 in the Large variant mitochondrial cytochrome c (509)[10]. Here, a function is maintained with Trp residues located at nonequivalent positions of the homologous polypeptide chain. This demonstrates that a sequence alignment which fixes the positions of "essential" residues can be misled.

9.6 Recognition of Distant Relationships

Amino Acid Sequence Comparisons

The _a priori_ significance is an upper limit for the significance of a relationship. Similarities between distantly related proteins are not always obvious; efficient criteria are needed to establish them. As an example take the α-chain of human hemoglobin and sperm whale myoglobin as aligned by Dayhoff (20), neglecting insertions and deletions. Here, the overlapping chain length is 142 residues, 37 of which are identical. The "_a priori_ probability" of finding common residues at any 37 positions of these chains is very small:

$$\left(\frac{1}{20}\right)^{37} \cdot \left(\frac{19}{20}\right)^{105} \cdot \frac{142!}{37!\,105!} = 5 \cdot 10^{-17}.$$

Conversely, the observation of such a similarity tells us that it cannot have occurred by chance; it is highly significant. Accordingly an "_a priori_ significance" is defined as the inverse of the _a priori_ probability. For biological material a highly significant similarity points to an evolutionary relationship.

Biological constraints reduce the mathematically derived _a priori_ significance appreciably. The "_a priori_ probability" has been derived in a mathematical manner. To derive more relevant numbers, biological constraints should be taken into account. For example, not all amino acids occur with the same frequency (Table 1-1), consequently at a given residue position they are not accepted with a probability of 1/20 each. Moreover, all residues are integral parts of three-dimensional structures. For example, no charged residue is acceptable at an internal position. This increases the probability for other residues at this position from 1/20 to a higher value.

---

[10]These two Trp are located at nonhomologous positions of the chain; the numbering schemes in both cytochromes are not adjusted to each other as is the case for the schemes of Figure 7-1a. The Trp in question corresponds to Trp-62 of Figure 7-8.

Biological constraints are reflected in relative substitution frequencies. These constraints are best taken into account by basing the analysis on experimental findings in natural proteins. For this purpose, the matrix of relative substitution frequencies (Table 1-2) can be used. The entries in this matrix are the ratios between observed amino acid change frequencies and those expected by random changes for a given amino acid frequency distribution (Table 1-1). Therefore, these entries on the average reflect the natural constraints in amino acid changes.

When two sequences are compared in order to find similarities, according to this matrix, the answer at any sequence position is not yes (fit) or no, but is more gradual: Ile $\leftrightarrow$ Ile is a strong contribution to the significance, and Ile $\leftrightarrow$ Val (a frequent change), Ile $\leftrightarrow$ Thr, and Ile $\leftrightarrow$ Arg are successively lower ones corresponding to the magnitudes of the matrix elements (Table 1-2). The contributions at all residue positions have to be added and related to a comparison between random sequences. Here, care has to be taken that the random sequences have the same amino acid frequencies as the proteins in question. Using this scheme, two totally different sequences which contain many frequent changes would still be recognized as related. In contrast, sequences with 15% identity and a large number of unlikely changes would be designated as unrelated.

Such a scheme has been devised by McLachlan (598). He converted the matrix elements $m(i, j)$ of Table 1-2 to integers between 0 and 7 by reducing all nondiagonal elements linearly to levels 0–5. The diagonal elements are set to 6 and 7, which renders a residue identity not much more significant than a frequent change. To obtain a measure for similarity the matrix elements $m_r(i, j)$ for every position r of the chain containing amino acid i on one chain and amino acid j on the other are summed up. In our example (myoglobin versus α-chain of hemoglobin) this measure is the total score

$$M = \sum_{r=1}^{142} m_r(i, j).$$

Subsequently the distribution of all scores M expected from random sequences with the same amino acid frequencies is calculated in a combinatorial manner [see Ref. (598)]. This calculation is appreciably facilitated by the introduction of small integers for the $m_r(i, j)$.

Since primary interest is not in the probability of getting exactly a score M but in the probability that the obtained score M indicates a relationship, it is necessary to compare the cumulative probability (sum of all probabilities) of all scores greater than or equal to M with the cumulative probability for all scores less than M. The latter cumulative probability can be set to 1.0, because interest is only in high M values, i.e., low cumulative probabilities of all scores greater than or equal to M, and because the sum of both types of cumulative probabilities is 1.0. A distribution of cumulative probabilities of all scores greater than or equal M is given in Figure 9-5a. For the comparison of sperm whale myoglobin with the α-chain of human hemoglo-

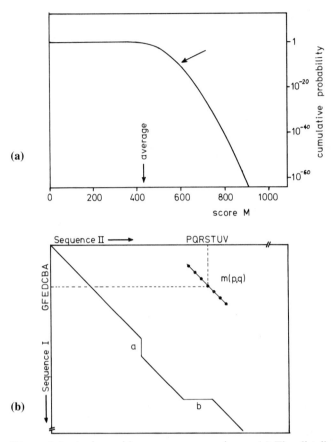

Figure 9-5. Amino acid sequence comparisons. (a) The distribution of cumulative probabilities of all scores greater or equal to M for random sequences in chains of 142 residues (598). For the calculation of M see text. The score and the corresponding cumulative probability for a comparison of sperm whale myoglobin and the α-chain of human hemoglobin are indicated by an arrow. (b) Sketch of a comparison matrix for two sequences I and II (598). The two spans ABCDEFG and PQRSTUV of length 7 are compared to give a single entry in the matrix. The diagonal line shows a correlation between the sequences which reveals insertions in sequence I at position a and in sequence II at position b.

bin, this cumulative probability amounts to $2 \cdot 10^{-9}$ (598), which is eight orders of magnitude larger than the *a priori* probability calculated above.

McLachlan's method yields the standard significance for the relationship between two aligned sequences. Since this cumulative probability takes account of the biological constraints in amino acid changes (although in a rather wholesale manner) it is much more reliable than the *a priori* probability. Such biologically adjusted probability has been called "standard probability" and its inverse "standard significance" (387). The level above which a standard significance indicates an evolutionary relationship can be disputed. Clearly for values above 100 such a relationship should be accepted as a working hypothesis. In principle, a high standard significance should exclude a case of convergent evolution because convergence is forced by biological constraints and these should have been already considered. In practice, however, it is never certain as to whether all constraints are known.

The comparison matrix may help to align sequences. The standard significance calculated above assumes perfect alignment without insertions and deletions, which can no longer be expected in a case with such a large evolutionary distance as between myoglobin and hemoglobin. In order to cope with insertions and deletions, stretches of the sequence have to be shifted for adjustment. However, allowing such shifts increases the probability of high M scores, because there is an automatic search for a good fit. Therefore, the original probability distribution (Figure 9-5a) can no longer be applied.

In such a case it is necessary first to locate insertions and deletions. For this purpose a certain sequence length (for example, seven residues) is defined, and it is assumed that no shift (i.e., insertion or deletion) within this length occurs. Subsequently all possible pairs (p,q) of stretches centered at position p of the first chain and at position q of the second chain, respectively, are superimposed. Each superposition gives a score $M(p, q)$ as shown in Figure 9-5b. This score (or corresponding symbols) are inserted into a comparison matrix (Figure 9-5b). In this matrix lines of high scores indicate the best alignment; dislocations of these lines indicate insertions and deletions.

After establishing the alignment with the comparison matrix of Figure 9-5b, the cumulative probabilities and thus the standard significance of the relationship can be calculated by following the procedure described above for the comparison of myoglobin with hemoglobin.

A qualitative measure of relationship can be deduced from the observed distribution of scores M. However, very often no clear alignment can be deduced. For nonaligned sequences a qualitative measure of significance can still be derived. In such cases the observed distribution of *all* $M(p,q)$ values (20,000 for a comparison of a chain of 100 residues with a chain of 200 residues) can be compared with the distribution expected from a comparison of random sequences. If high $M(p,q)$ values occur much more frequently than expected there are many fitting pieces of sequence and the proteins are related to each other. However, it is difficult to quantitate the strength of this relationship because no standard significance can be calculated.

The scoring system overcomes spurious markers. It may reveal structural repeats. Since this method places a very high value on amino acid similarity and since it does not put too much weight on amino acid identity, it is not sensitive to wrong alignments based on spurious marker residues as, for example, Trp-59 and Trp-56 in cytochromes c and c_{551}, respectively (section 9.5).

This method can also be applied to internal comparisons of a chain in order to find repeats which may indicate gene duplications. Such repeats would show themselves as lines of high significance parallel to the diagonal.

In special cases in which strictly regular repeats are to be expected, as in collagen or in tropomyosin, such regularities can also be detected by a one-dimensional Fourier analysis (599).In such an analysis each residue type is assigned a certain value, for instance, values 1 to 20 for the amino acids in the ordering of Table 1-2. This ordering is a one-dimensional representation of amino acid exchange probabilities. With such an assignment the given sequence is converted to a continuous function, value *versus* position, in which the height of the function is characteristic for amino acid property (very high values, for example, correspond to aromatic side chains). A Fourier transform of this function then reveals periodicities, that means repeats of amino acids with similar properties.

Chain Fold Comparisons

Chain folds reveal distant evolutionary relationships. It has been shown that during evolution the three-dimensional fold of a polypeptide chain is much better conserved than the constituting amino acids. Hence the chain fold should allow detection of those distant relationships that have been obliterated in the amino acid sequences. For this purpose it is necessary to quantitate chain folds and chain fold comparisons.

Mean C_α-distances are a common measure of chain fold similarity. A chain fold is well represented by the coordinates of the C_α-atoms. Thus chain folds can be compared by superimposing two chains in such a way as to minimize the distance between corresponding C_α-atoms (600). The root mean square (rms), or, occasionally the normal average of the residual distances between corresponding C_α-atoms can be taken as a similarity index for the comparison. This index, the rms C_α-distance, is now in common use. It was applied for example to checking the results of chain fold simulations (section 8.6, page 164). On the basis of C_α-distances also more complicated similarity indices were defined and used in chain fold comparisons (601). These indices, however, have not become very popular.

A standard significance for chain fold comparisons is not yet at hand. In a comparison between chain folds the rms C_α-distance can be used in a similar manner as the score M introduced above for amino acid sequence comparisons. But whereas the cumulative probability for M could be calculated by comparing random amino acid sequences (Figure 9-5a), the situation with chain folds is more difficult because random chain folds are not available. Consequently, it is not yet possible to assign a standard significance to such a comparison as was done in the sequence comparison between myoglobin and the α-chain of hemoglobin given in Figure 9-5a.

The generation and comparison of random chain folds for the determination of cumulative probabilities of rms C_α-distances is a tedious task (387). Moreover in calculating such distributions structural similarities caused by physico-chemical constraints (i. e., the similarity between secondary and supersecondary structures) and similarities between freely variable structures must be distinguished. But the insight into structural principles, which is necessary for this distinction, is not available as yet. The chain folds of structurally known proteins could be used instead of randomly generated ones. However, the number of known chain folds is not large enough for this purpose.

A comparison matrix of C_α-distances may reveal chain alignment. Protein structures which are only distantly related tend to deviate from each other by large insertions and deletions as shown in Figure 9-4 for the structures of chymotrypsin and protease B. Like amino acid sequence comparisons also chain fold comparisons require the alignment of corresponding residues although the rms C_α-distance is not as badly affected by a misalignment as the score M. Such an alignment can be accomplished using the same scheme as the one given for the sequence comparison in Figure 9-5b; but $M(p,q)$ has to be replaced by the rms C_α-distance between chain segments of a given length centered at p and q (802). After establishing the alignment the overall rms C_α-distance between corresponding residues can be determined. But again, it cannot be converted to a standard significance of the similarity between the two proteins in question, because the cumulative probabilities for rms C_α-distances are not yet known.

A comparison matrix of rms C_α-distances can also be used for revealing structural similarities between parts of proteins. Such a similarity was found, for example, between parts of bacteriophage T4 lysozyme and hen egg white lysozyme (802). The applied segment lengths were 40 as well as 80 residues.

Chain fold comparisons should account for insertions and deletions. When calculating the rms C_α-distance between those parts of the chain which have been aligned on the basis of the comparison matrix, insertions and deletions are completely neglected. This is overly permissive because insertions and deletions have to be accounted for when deriving a measure of similarity. After all, if too many of them are present in a comparison, the respective proteins are dissimilar. In a pairwise comparison of chain folds insertions and deletions can be included with appropriate weights by deriving the mean C_α-distance from the minimal area subtended between the compared chains as depicted in Figure 9-6. This method does not require alignment. However, minimizing the subtended area between two chains is computationally more intricate than minimizing distances between two equal-sized sets of atoms, which is the basis of determining the rms C_α-distance.

Sheet Topology Comparisons

Comparisons of sheet topologies emphasize the importance of β-sheet moieties in protein structures. Given the obstacles in converting mean C_α-distances to standard significances, a less rigorous approach to chain fold comparisons based on β-sheet topologies and restricted to proteins containing β-sheets is worth mentioning. Such a topology is defined as the pathway

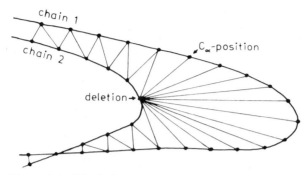

Figure 9-6. Calculation of a mean C_α-distance between two chains which includes systematic insertions and deletions (387). The mean C_α-distance is defined as the minimal area stretched between both chains (corresponding to the area of an elastic film stretched between two wires) divided by the average chain length. The sum of the triangles between C_α-atoms, which are indicated in the figure, is a good approximation to this minimal area.

of the chain as referred to a β-sheet (to a substructure taken as rigid) without reference to exact coordinates (255) (Figure 7-7). For example, a different ordering of the strands along the chain or connections between parallel sheet strands over different sides of the sheet would give a different sheet topology.

Sheet topologies can be quantitated in a combinatorial manner. As an example the number of different sheet topologies for six-stranded parallel sheets is $6! \cdot 2^{6-2} = 11,520$ (255). Thus, the *a priori* probability of finding identical sheet topologies in two different proteins with six-stranded parallel sheets is 1/11,520, and the *a priori* significance is 11,520. This comparison was restricted to proteins with six-stranded sheets only. Taking account of all other proteins as well increases this significance by, say, a factor of 10. An example for identical sheet topologies is the NAD-binding domain of dehydrogenases (91). In this case the NAD-binding sites as referred to the sheet (above the carbonyl ends of the sheet strands, see Figure 7-7) are also identical, giving rise to a further increase (255) by about a factor of 10, that is, in total to an *a priori* significance of 1,152,000.

In two cases approximate standard significances of relationship have been derived from sheet topologies. The *a priori* significance can be reduced to an approximate standard significance of 3500 by allowing for the observed preference for the hand of $\beta\alpha\beta$-units (section 5.2, page 81) and for the observed neighborhood correlation in β-sheets (Figure 5-15a), both of which reflect physico-chemical constraints which reduce the number of realizable sheet topologies. Without doubt this significance of 3500 indicates an evolutionary relationship among dehydrogenases.

A standard significance has also been determined for the similarity of immunoglobulin and superoxide dismutase sheet topologies (548). With a value of 30,000 (3000 calculated by the authors and a factor of 10 for the

restriction of the group of compared proteins as in the case of the dehydrogenases) it clearly points to an evolutionary relationship. Although sheet
topology comparisons will remain special cases, they nevertheless demonstrate how to proceed in recognizing and quantitating very distant relationships from chain folds.

Summary

Protein diversity is a consequence of protein evolution. This evolution is
the result of numerous natural experiments, (mutations and subsequent
selections), which can be used to study principles of protein structures. The
basic mutational step of protein evolution is the change of an amino acid
residue; larger steps are insertions and deletions of one or more residues;
very gross changes result from gene multiplication and gene fusion.

In comparing extant proteins, selection criteria can be recognized;
changes at protein surfaces are much more tolerable than changes in the
interior, important positions such as active sites are almost invariant, etc.
Moreover, it becomes obvious that very similar three-dimensional structures can result from quite different amino acid sequences. Conspicuously,
the chain fold is extremely well conserved indicating that not only the final
structure but also the folding process is subject to strong selective pressure.
Additional structural information can be obtained from mutations upon
which selection has not worked yet, as, for example, from mutant human
hemoglobins. Here, structural principles are revealed by deviations from
normal properties.

Protein evolution can be subdivided into protein speciation and protein
differentiation. Speciated proteins fulfill the same function in different
organisms and can be used in establishing the genealogy of organisms. It
should be noted, however, that protein speciation does not *drive* organismal evolution. Protein differentiation is the process leading to functional
diversification of homologous proteins. Thus, the study of protein evolution
not only gives structural insight but also connects proteins of quite different
parts of the metabolism. This introduces more order into the large list of
existing proteins. Furthermore, it may become possible to reveal the
evolution of metabolic pathways. Important mechanisms of protein differentiation seem to be gene multiplication and gene fusion.

Apart from divergent evolution (of homologous proteins) convergent
evolution (of analogous proteins) is omnipresent. For example, convergence with respect to function is a rather common phenomenon. Convergence to similar structures may also occur but is difficult to evaluate on the
basis of our present structural knowledge. In particular, in searching for
very distant evolutionary relationships, it is not always possible to distinguish analogous from homologous similarities.

Chapter 10
Protein–Ligand Interactions

Proteins are selective in their interactions with cell constituents. In contrast to naturally occurring proteins, chemically synthesized polypeptides of random sequence behave like small children; they touch, bind, and break many low molecular weight metabolites (614). Natural proteins were educated by evolution to touch only a small selection of molecules (586). This was only possible because they learned to form defined compact structures in contrast to synthetic polypeptides. Specific binding is an individual property of individual proteins. Nonbinding, rather than binding, is the result of organized protein structures.

Relatively few types of molecules play a role in living systems as cofactors or metabolites. Apart from ions (615, 616) typical examples are the derivatives of adenosine, of glucose, of porphyrin, and of α-amino acids. Consequently, most proteins are specialized in interacting with one or more of such standard building blocks (19).

10.1 Ligand-Binding Sites of Immunoglobulins

The only category of proteins which has specialized in creating binding sites for practically all kinds of molecules are the immunoglobulins ($\equiv$Ig). Well-studied immunoglobulins are the antibodies of the blood plasma and myeloma proteins. The latter are produced by tumor cells. Otherwise they are normal immunoglobulins; many myeloma proteins from men and mice have been shown to bind ligands[1] in a reaction which closely resembles that of a specific ligand–antibody system [see Refs. (540) and (617)].

---

[1]In biochemistry a ligand is a small molecule bound to a larger one; in coordination chemistry ligands are molecules or ions grouped around a center such as an iron ion. Both meanings of the term occur in this chapter.

The three-dimensional structures of the binding sites have been elucidated for two different myeloma proteins, for the human immunoglobulin
New which binds derivatives of vitamin K (618) and the murine immunoglobulin McPC 603 (607) which binds phosphoryl choline. The following
conclusions have resulted.

**Random pairing of n_1 H-chains and n_2 L-chains gives rise to $n_1 \cdot n_2$
different antigen-binding sites.** In both proteins the ligand-binding site is
located between the V_L domain of a light chain and the V_H domain of a
heavy chain (Figure 4-2c). This means that the binding site is formed by two
different polypeptide chains. The use of two chains might reflect the
solution to the problem of forming a practically unlimited number of
different binding sites with a limited amount of genetic material. If the
number of different L- and H-chains is n_1 and n_2, respectively, then there
are $n_1 \cdot n_2$ combinations of V_L- and V_H-domains and therefore $n_1 \cdot n_2$ different binding sites (540). The number of different binding sites, $n_1 \cdot n_2$, was
estimated to be 10^7 (542); accordingly n_1 and n_2 are in the order of 10^3 to
10^4.

**There are six hypervariable segments each containing 5 to 10 residues per
binding site.** The chain fold is very similar in all variable domains of
immunoglobulin New and immunoglobulin McPC 603 and of other immunoglobulins (540, 543, 544). Structural differences occur predominantly in
the so-called hypervariable loops (619), three segments of 5 to 10 residues
each, in both V_L and V_H. The amino acid residues forming the site which is
complementary to a given ligand are contributed by these six hypervariable
segments.

The pattern of insertions and deletions characteristic of hypervariable
regions determines the dimensions of a binding site. The combining site of
Ig New consists of a shallow groove of about 15Å × 6 Å with a depth of
about 6 Å. The phosphoryl choline binding center of Ig McPC 603 is
approximately 12 Å deep.

Model Building of Immunoglobulin–Ligand System

**Structurally known immunoglobulins are used to study immunoglobulins
with other specificities.** The V_L/V_H region of immunoglobulins can be
regarded as consisting of a rigid framework to which the hypervariable
loops are attached. This provides a basis for the comparative analysis of
immunoglobulin–ligand systems by model building, using the coordinates
of the framework (from structurally known V_L/V_H regions) and the amino
acid sequences of the hypervariable regions of the immunoglobulin which is
to be analyzed. The first protein studied by this approach (291, 620) was the
mouse myeloma protein IgA MOPC 315 which is known to bind 2,4-
dinitrophenyl derivatives. These studies were refined by Dwek et al. (621)
who applied model building in conjunction with nuclear magnetic resonance, electron spin resonance, and chemical modification reactions.

A systematic analysis of any immunoglobulin–ligand system may become possible. The model building studies can be pursued in a systematic manner since the synthesis of antibodies against any specifically designed organic molecule (of low or high molecular weight) can be induced in animals such as rabbits or goats. Using affinity chromatography the antibodies produced can then be purified. In general, this procedure gives rise to a degenerate immune response, i.e., to the production of several antibody species by several clones of so-called plasma cells (542). These antibodies differ in affinity constants for the ligand which is also reflected in different chemical, spectroscopic, and immunologic properties of the binding sites. Sequence analyses further revealed that the differences correspond to amino acid changes in the hypervariable regions on both V_L and V_H domains (622). At first glance the degenerate response seems to complicate the structural elucidation of binding sites. In effect, however, data on several similar binding sites can be combined and checked against each other, which increases appreciably the probability of finding correct models.

Immunoglobulin–Ligand Systems as Models for Other Protein–Ligand Interactions

Our knowledge of protein–ligand interactions is still so limited that an X-ray structure analysis is necessary to elucidate them in every given case (623). This is not a systematic approach since most proteins of interest, such as membrane proteins, have resisted crystallizing. Therefore, it is important to extract general rules for ligand binding from the available systems. Immunoglobulin–ligand complexes are well suited for this purpose, because, as described above, they allow a systematic approach to the problem.

Immunoglobulin–ligand complexes resemble enzyme–ligand complexes. There are, for example conspicuous parallels between the binding properties of immunoglobulins and enzymes (85, 624, 625):

(a) As indicated by the values of rate constants (up to $10^8 \ M^{-1} \ sec^{-1}$) the binding step is often a diffusion controlled reaction.

(b) The bound ligand has little rotational freedom; it is firmly and specifically held by noncovalent interactions. The standard free energies of binding range from -6 to -15 kcal/mol which corresponds to dissociation constants of $K = 10^{-4} \ M$ to $K = 10^{-10} \ M$ for the ligand–enzyme complexes.

(c) Ligand binding of immunoglobulins seems to obey the lock and key model (44). This is also true for many enzyme–ligand interactions however, it does not apply to those enzymes which undergo major conformational changes (induced-fit) on ligand binding (626).

The immunoglobulin architecture may serve as a basis for *in vitro* syntheses of peptides with desired binding properties. For theoretical and practical reasons it would be desirable to synthesize a polypeptide chain *in vitro* with defined specificity for and affinity to a given compound. One possible approach could start out with a natural or synthetic V_L/V_H region without hypervariable loops as a framework. By inserting appropriate sequences at the positions of the hypervariable segments a specific binding site for the ligand in question would be molded without disturbing the folding process and stability of the frame (498). The example of $Cu^{2+}-Zn^{2+}$-containing superoxide dismutase (286) might be regarded as a natural precedent for this peptide engineering approach. The coordination geometries around the metals at the active site possess very close analogues in crystal structures of copper-imidazole and zinc-imidazole complexes (661). Thus the two basic features of this enzyme are the Ig-architecture and a metal complex which could have been designed by an organic chemist.

10.2 Substrate-Binding Sites of Serine Proteases

The mechanism of substrate binding is similar for all known serine proteases (537) including subtilisin (627). Since the substrates of serine proteases are proteins, the complex formation between these enzymes and their substrates also provide information on protein–protein interaction.

Complex Formation between Chymotrypsin and Its Substrate

Substrate-binding can be divided into several steps. Interactions based on backbone atoms of chymotrypsin play a role in each step. Consider the prelude to the cleavage of a peptide bond at the carbonyl-end of amino acid residue *i* by chymotrypsin (537, 538). This residue can be either an aromatic or a bulky aliphatic residue. On complex formation the scissile bond of the substrate is oriented precisely relative to the attacking groups (537). Starting from the fixed side chains this process can be divided into several steps. The importance of interactions involving immobile backbone atoms in these steps should be noted.

(a) The side chain of *i* including the C_β-atom (Figure 10-1) is trapped in a deep pocket of the enzyme, sandwiched between three peptide bonds, and secured by hydrophobic interactions. The shape of the pocket entrance leaves little positional freedom for the C_α-atom of this residue.

(b) With the side chain fixed, the dihedral angle χ_i^1 (Figures 2-2 and 10-1) defines the position of the substrate's backbone. The available range of χ_i^1 is restricted by the enzyme surface. χ_i^1 is fixed by the formation of an H-bond between the NH-group of substrate residue i and of the backbone carbonyl group of residue 214. As a consequence, the angle ϕ_i is also fixed.

(c) Once these interactions are established the dihedral angle ψ_i has to be set. Again the coarse control is based on the profile of the enzyme surface which permits only two ranges of this angle. One of the two possibilities enables the carbonyl group of i to form H-bonds with the backbone NH-groups of residues 193 and 195. This brings the amide group of the scissile bond (i.e., the NH-group of substrate residue $i+1$) close to the chemical team of the enzyme, i.e., to the linear "charge relay system" (628) consisting of side chains of an Asp, a His, and a Ser. The enzyme substrate complex (Michaelis complex) is now formed, and the peptide bond between i and $i+1$ is ready for enzymatic attack.

Nonproductive binding prevents the hydrolysis of peptides consisting of unwanted D-amino acids. A peptide consisting of D-amino acids would also bind tightly to chymotrypsin. However, a rather unreactive enzyme–substrate complex or an unreactive intermediate result since the scissile bond is not properly located relative to the catalytic apparatus (629); in this way the free energy of binding is utilized to inhibit the reaction of a substrate analogue which would lead to undesired products. Nonproductive binding may be a general mechanism for ensuring enzyme specificity (630, 631).

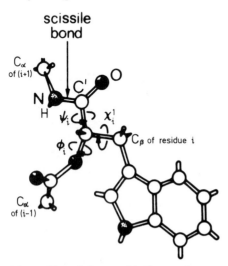

Figure 10-1. Substrate-binding to chymotrypsin [Ref. (537)]. The dihedral angles of the substrate which have to be adjusted before cleavage of the peptide bond between residues i and $i+1$ can take place are shown.

The Substrate-Binding Site of Trypsin

Electrostatic interaction contributes to trypsin's specificity for Lys- and Arg-residues. Trypsin (244, 245, 536) binds its substrates in essentially the same mode as does chymotrypsin. Trypsin, however, is specific for positively charged residues of the substrate; the side chain of a Lys or an Arg is bound electrostatically to an aspartate residue at the bottom of the enzyme's binding pocket. Crystallographic studies on complexes of trypsin and proteinaceous trypsin inhibitor(s) (269, 632) showed that the binding mode resembles closely the formation of a serine protease–substrate complex. Evidently the inhibitor mimics a substrate perfectly. The mechanisms leading to the cleavage of the substrate by trypsin and to the stabilization of the trypsin–inhibitor complex (269, 536) are dealt with in section 11.2.

10.3 Haem-Binding Sites

Protohaem IX occurs in hemoglobin, in b-type and c-type cytochromes, and in catalases and peroxidases. We shall concentrate on members of the following haem-protein families since three-dimensional structures of these proteins are known: hemoglobins (85, 550, 634) which include monomeric species such as myoglobin from mammals (185, 635, 795) and erythrocruorin from insect larvae (277), c-type cytochromes (509) and b-type cytochromes (562, 297). Other examples of haem proteins are catalases and peroxidases. The best known of the many naturally occurring haems (637) is protohaem IX[2] which is encountered exclusively in the examples below. The polypeptide chain forming the haem-binding domain is about 100 residues long for the cytochromes, and about 150 residues for the globins. However, there is an example of unusual haem-binding capacity. In cytochromes c_3 of *Desulfovibrio* species four haem groups are bound by a single polypeptide chain of about 120 amino acids. This means that only 20 to 30 amino acid residues are used for binding one haem group [for a review see Ref. (509)]. The four haems act more or less independently of each other (638).

Haem proteins are carriers of electrons or of small molecules such as O_2. In hemoglobins the function of the haem and the polypeptide chain around it is to give the iron its special dioxygen[3]-binding properties and to protect the liganded ferrous ion from oxidation (639). The function of the haem iron in cytochrome c is not to coordinate a small molecule but to carry elec-

---

[2]Protohaem IX is also called ferroprotoporphyrin IX (633). The tetrapyrrol ring of this compound can be regarded as planar for our purposes; eight side chains—four methyl groups, two vinyl groups, and two propionic groups—are attached to this ring system (see Figure 10-2). "Proto" refers to the nature and "IX" to the sequential order of these side chains.

[3]Dioxygen = molecular oxygen = O_2.

trons; it is reduced ($Fe^{3+} \rightarrow Fe^{2+}$) and oxidized enzymatically by the appropriate protein partners of cytochrome c in the energy metabolism (509). Cytochrome b_5 is a component of another electron-transporting ensemble of proteins which is involved in the desaturation of fatty acids and in other chemical reactions (297). It has even been suggested (640) that cytochrome b_5 might interact with cytochrome c *in vivo*. However, so far the reduction of cytochrome c by cytochrome b_5 has been shown only *in vitro*. A structural model of this interaction has been recently proposed (640).

Bonds between Haem and Apoprotein

Sixty atoms of the apoprotein are in van der Waals contact with the porphyrin ring. The major similarity between hemoglobins, cytochromes c, and cytochrome b_5 is that the haem group sits in a crevice which is lined with side chains of apolar amino acids. In a subunit of hemoglobin, there are some 60 atoms making van der Waals contact with the atoms of the porphyrin ring (60).

The interaction of haem side chains and apoprotein can involve all known bonding types. The haem side chains play different roles in different haem proteins. In cytochromes c the two vinyl groups [in exceptional cases (509) only one] are bound covalently to Cys-14 and Cys-17 of the apoprotein; one charged propionic group is at the surface of the molecule but the other one is hydrogen bonded to the interior of the molecule which appears to be of functional significance. One or both propionate groups also seem to be essential for structure and/or function of horseradish peroxidase (641) and cytochrome b from Mung bean (642).

For the cases of cytochrome b_5 and hemoglobin earlier findings had indicated that the side chains play only a minor and a nonspecific role (560, 568, 637). In the light of recent X-ray diffraction studies, however, the importance of the haem's individual side chains is under reinvestigation (565, 566). In cytochrome b_5-hbf, for example, one propionate group may be directly associated with the oxidation–reduction cycle by neutralizing the extra positive charge of the iron in ferricytochrome b_5 and by binding to another cationic group in ferrocytochrome b_5 (566).

Side chain atoms of His and Met are coordinated to the haem iron. In cytochrome b_5 the coordination of the iron is symmetrical, consisting of six octahedral ligand sites occupied by four pyrrol nitrogen atoms of the haem and by the two ϵ-nitrogen atoms of His-39 and His-63 in the axial positions (297). The δ-nitrogens of His-39 and His-63 are hydrogen-bonded to main chain carbonyl oxygens so that both imidazole rings are firmly held in place by the rigidity of the backbone structure.

Hemoglobins and cytochromes c have the same ligands except that one of the axial coordination sites is occupied by dioxygen and methionine sulfur (634), respectively. When dioxygen is absent the sixth coordination site (the "ligand pocket") of hemoglobins is empty (60, 636, 643). In a bacterial cytochrome c' both axial ligand sites are assumed to be unoccupied; apparently the polypeptide chain encapsulates the haem in such a manner as to prohibit potential ligands from binding at the faces of the haem (644).

The Influence of the Haem Group on Protein Structures

The major difference between the structures of cytochrome c on the one hand and hemoglobin and cytochrome b_5 on the other is in the folding of the polypeptide chain and the importance of the haem for the structural integrity of the protein.

Apocytochrome c is completely unfolded. Cytochrome c can be thought of as consisting largely of an extended polypeptide chain wrapped around the haem and forming a shell one polypeptide chain thick (Figure 7-8). Apocytochrome c with the haem group cleaved off is completely unfolded. The central role of the haem as an integral part of the structure is emphasized by the fact that the most highly conserved amino acid side chains, when comparing cytochromes of known sequence from 70 sources (509), are those which attach to the iron and the vinyl group, as well as the hydrophobic residues packed around the haem.

Apomyoglobin possesses a helix-rich globular structure similar to that of myoglobin. In contradistinction to the flimsy cocoon-like envelope of the haem in cytochrome $c,$ the haem-crevice in hemoglobin is formed by a framework of solid self-supported α-helices. The haem group does not serve as an essential folding template; apomyoglobin, for example, assumes a tertiary structure in the absence of the prosthetic group (see chapter 8) but the haem group adds extra stability (415, 461, 645, 646). The relaxed structural requirements for the interaction of haem and protein in hemoglobins are also reflected in the fact that only the general pattern of apolar haem contacts is invariant among the members of the globin family (277, 634). Structurally equivalent positions are occupied by different amino acid residues.

In apocytochrome b_5 the scaffolding for haem-binding is as solid as in hemoglobin. In the case of the b-type cytochromes the data base is too small; therefore it cannot be determined whether there are invariant apolar residues at specific locations for interaction with the haem group. As in the globins, the scaffolding for binding the haem is solid; the walls of the haem crevice are formed by two pairs of roughly antiparallel helical segments and the floor is formed by a pleated sheet (297). This solid structure of the haem-binding site might indicate a certain freedom in how to line the crevice with apolar residues. Preliminary results which are based on fitting the sequence of the haem-binding fragment of b_2 to the tertiary structure of b_5 (557) indicate that only the overall pattern of apolar contacts but no specific side chain is conserved within the cytochrome b family, as is the case for the globins.

Chemical Events at the Haem Iron in a Protein-Made Microenvironment

Fe^{2+} in hemoglobin is protected from oxidation by its nonpolar environment. In a haem protein the polypeptide chain creates a special microenvironment for its prosthetic group, so that one distinct chemical event can take place and other reactions are prevented from occurring. The example of the hemoglobins illustrates the importance of the microenvironment. In water, ferrous haem binds O_2, but it is immediately oxidized to ferric haem which does not bind dioxygen (637). However, the haem can undergo reversible oxygenation without oxidation if it is embedded in an apolar environment such as benzene where it is much more difficult to strip an electron away from the ferrous iron than it is in water (639, 647). It may be concluded that the major function of the nonpolar haem-binding pocket is to protect the ferrous state of the haem from oxidation by excluding water.

Chemical model compounds possess hemoglobin activity. Models of the active site of hemoglobins have been synthesized with the idea of replacing the protein by another polymer (639, 727) (Figure 10-2) or by hydrophobic side chains attached to the haem in the form of a picket fence (797). Biomimetic systems of this type might prove valuable models for the structure–function relationship of natural macromolecules. Furthermore they extend the principles of protein structure to conditions (extremes of temperature, pH, and solvent composition) which are incompatible with the functional integrity of proteins themselves.

O_2 and CO are bound as bent ligands to hemoglobin. This binding mode is possibly a mechanism for discriminating against CO. Hemoglobins bind O_2 with a twofold lower affinity than simple iron–porphyrin models such as haem dissolved in benzene. With respect to binding of CO, the two systems differ by a factor of 100 (648). Hence, the protein-made microenvironment discriminates against CO without affecting O_2 binding appreciably. This selection is based partly on different consequences for CO and O_2 of the electronic trans-effect originating from the other axial ligand of the iron (649), i.e., from the so-called proximal His (Figure 10-2).

However, another aspect seems to be more important. Both in simple iron–porphyrin systems (650, 651) and in hemoglobins (690) dioxygen is bound as a bent ligand (Figure 10-3a) and may be best described in this form as superoxide, O_2^- [(652, 653), but cf. (650)]. In contrast CO binds as a linear ligand in simple systems (648, 654) but as a bent ligand in all hemoglobins tested so far (655–657). Thus, the protein environment enforces a mode of binding, which is favorable for O_2 but unfavorable for CO. It should be mentioned that in spite of this discrimination the affinity of hemoglobins for CO is still about 500 times higher than for O_2 (634).

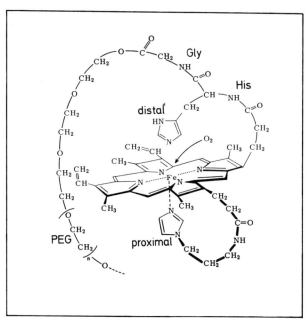

Figure 10-2. A chemical analogue of the active site in hemoglobin [from Ref. (727)]. Protohaem IX is at the centre of the sketch. One propionic side chain (upper right) forms a peptide bond with the analogue of the distal His in hemoglobin. This His is linked via Gly to polyethylene glycol (PEG). The other propionyl group (lower right) is linked via a spacer to an imidazole which represents the analogue of the proximal His. The chemical O_2-carrier qualitatively resembles hemoglobin in the mode of reversible dioxygen-binding and in certain spectral properties.

The evolution of a ligand site discriminating against CO was probably directed against endogenous CO, which is produced during the breakdown of porphyrin to bile pigments (85). Without the precaution described, this CO would occupy up to one-third of the hemoglobin-binding sites (649). A modern aspect of the discrimination against CO is that hemoglobins were given more than 10^8 years for adaptation to an otherwise deadly hazard of tobacco smoking and air pollution.

In cytochromes the haem iron is protected against unwanted ligands. In cytochromes the protein-made microenvironment has two functions: to protect the haem from unwanted ligands and to adjust the redox potential of the iron to the required level. A protective role has been suggested for Met-80 which ligands the haem iron in cytochrome c (Figure 7-8) (509). Without this contact the iron is easily and uncontrollably reduced by agents such as ascorbate (658). Positively speaking, the polypeptide chain guarantees that the transported electron goes along its biological pathway (509). In cytochrome b_5 the inertness of the iron ion toward ligands is extraordinary (297). This is caused by the rigidity of the axial ligands (section 10.3, page 212).

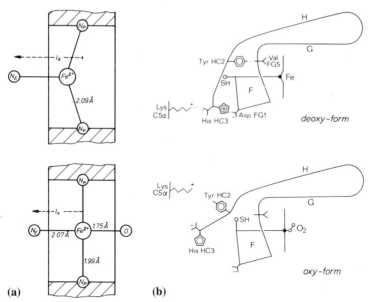

Figure 10-3. Changes on oxygenation of hemoglobin. (a) Probable positions of the iron-ion relative to the porphyrin plane (hatched) in deoxyhemoglobin (804, 654) and oxyhemoglobin (550, 651). The data (center-to-center distances) were taken from references quoted by Phillips (690). It should be noted that the haem geometry of oxyhemoglobin was deduced from so-called picket-fence complexes (651), i.e., from model compounds. The positions of two porphyrin nitrogen atoms (N_P) and the N_ϵ of the proximal His are indicated. On oxygenation N_ϵ is displaced by $l_d - l_o = 0.6 \pm 0.1$ Å. (b) Conformational change in a subunit of hemoglobin on oxygenation (courtesy of M. F. Perutz). The movement of helix F toward helix H leads to the expulsion of Tyr HC2, the penultimate residue of a β-subunit, from a pocket between these helices. The position of the terminal carboxylate group (of His HC3) is directly affected. Val FG5 is Val-98.

The redox potential of the haem iron in cytochrome *c* is determined by the axial ligands. The influence of the apoprotein on the level of the redox potential was studied by Moore and Williams (659). By comparing the redox potentials of haem model-compounds (660), by using structural information on various cytochromes, and by using nuclear magnetic resonance studies on distantly related *c*-type cytochromes, the authors concluded that the axial ligands of the haem-iron dominate the redox potential. Thus, two histidine side chains as axial ligands, found, for example, in b_5, mean a lower redox potential than a histidine/methionine pair, commonly found in *c*-type cytochromes. When different *c*-type cytochromes are compared, variations in the length of the Fe–S bond are found (659). A shortening of this bond by 0.1 Å is associated with a decrease in redox potential of 400 mV, which reflects an increase in electron donor power of the methionine sulfur.

Haem Ligands as Triggers for Structural Changes of Mammalian Hemoglobin

Deoxyhemoglobin contains high spin Fe^{2+} whereas oxyhemoglobin contains low spin Fe^{2+}. There is still controversy as to what conformational changes take place in the course of the reduction–oxidation cycle of cytochrome c (495, 509, 564). Therefore our discussion is restricted to the well-established case of mammalian tetrameric hemoglobin (60, 662, 666).

Crystallographic studies have revealed that the quaternary structures of oxygenated and deoxygenated hemoglobin differ appreciably. A list of differences is given in Table 10-1. For instance, on oxygenation the four subunits move relative to each other so that the distance between the iron atoms of the β-chains decreases by 6.5 Å. This dramatic change in quaternary structure is probably based on a phenomenon of pure iron–porphyrin

Table 10-1. Differences between Deoxyhemoglobin and Oxyhemoglobin (550, 666)

	Deoxyhemoglobin	Oxyhemoglobin
Sixth coordination position of an Fe^{2+}	Empty	Occupied by O_2
Configuration of the iron's d-electrons	High spin	Low spin
Length of the bond between Fe^{2+} and a pyrrol nitrogen	2.09 Å	2.01 Å
Displacement of the iron from the plane of the porphyrin ring	0.75 Å	0.05 Å
Distance between the ϵ-nitrogen of the liganded His and the porphyrin plane	2.67 ± 0.1 Å	2.07 Å
Position of the penultimate tyrosine's side chain	Between helix F and helix H	Expelled from this site (Figure 10-3b)
Position of the terminal carboxylate groups	Fixed by electrostatic bonds to other subunits	Have freedom of rotation
Distance between Fe^{2+} of β_1-subunit and Fe^{2+} of β_2-subunit	39.8 Å	33.4 Å
Position of diphosphoglycerate	Fixed between β_1- and β_2-subunit	Not bound
Ligand pocket in a β-chain	Blocked by the side chain of Val-67; can be oxygenated only when opened by thermal energy	Contains O_2

chemistry (637, 663, 664). In an iron-porphyrin complex two types of d-electron arrangements are possible. The high spin configuration found in deoxyhemoglobin is characterized by a greater number of electron orbitals with unpaired electrons than the low spin configuration of oxygenated hemoglobin. One consequence is that the high spin iron is too fat to be accommodated in the plane of the porphyrin ring, and that is displaced by 0.5 Å from this plane (550) (Figure 10-3a).

The mechanism of O_2-uptake by hemoglobin is known in atomic detail. Let us now follow the oxygenation process of deoxyhemoglobin (60, 550). Dioxygen approaches the high spin iron of one α-subunit (in the β-subunits of deoxyhemoglobin the sixth ligand position is blocked by Val-67 and is unavailable for O_2) and induces the low spin configuration. This brings the iron back by 0.5 Å to the plane of the porphyrin ring. The imidazole of the (proximal) histidine which is liganded to the iron moves the same distance or perhaps even 0.7 Å since the iron-axial ligand bond may shorten by 0.1 to 0.2 Å during the transition from high spin to low spin ferrous iron.

Helix F to which the proximal His is attached is pushed so close to helix H that the side chain of the penultimate tyrosine can no longer be accommodated between the two helices and is expelled, pulling along the C-terminal residue which had been clamped by a salt bridge to the other α-chain. Thus a major tie fixing the deoxyform is broken (Figure 10-3b). Binding of O_2 to the next subunit has the same effect on links between subunits and precipitates the change of quaternary structure. One consequence of this change is the withdrawal of Val-67 from the ligand pocket of the β-subunits so that these iron atoms also become available for O_2. Energetic aspects of this transition have been discussed by several authors (60, 268, 634).

The example of deoxyhemoglobin indicates why it is difficult to define the chemical event which triggers gross changes in the quaternary or the tertiary structure of any protein; the triggering event must be expected to be structurally inconspicuous. In deoxyhemoglobin it took many years to detect the trigger, the haem iron, which makes the subunits of hemoglobin move (666).

In tetrameric hemoglobin the O_2-affinity of the haem is controlled by the concentrations of O_2, CO_2, H^+, and 2,3-diphosphoglycerate. What was gained in the evolution of a monomeric hemoglobin of the myoglobin type to the sophisticated mammalian hemoglobin? The main advantage is the enhanced physiological utility of the tetrameric protein which has been achieved by bringing the intrinsic oxygen affinity of the binding site progressively under the control of external influences (276, 549, 667) (Figure 10-4).

One of the achievements is an improved Haldane-effect (668, 669), that is the release of protons on binding of dioxygen to hemoglobin and vice versa. This effect is important for the dioxygen transport from the lungs to dioxygen-consuming tissues and for proton transport from these tissues to

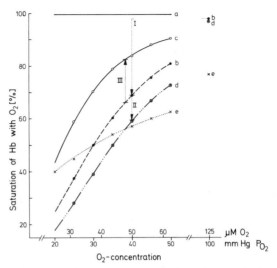

Figure 10-4. O₂-binding curves of hemoglobins (85). The saturation of hemoglobin (Hb) with O₂ at pH 7.2 is plotted against the concentration of free oxygen. The O₂-concentrations in the capillaries of the lung (125 μM) and in the capillaries of O₂-consuming tissues (50 μM) are fixed within narrow limits. Curve a: in the absence of diphosphoglycerate (DPG) hemoglobin is saturated with O₂ in the lungs but cannot supply the tissues. Curve b: at physiological DPG levels (4.5 mM), approximately 30% of the O₂ taken up in the lung is released in the tissues (arrow I). Curves b and c: as fetal hemoglobin (curve c) has a lower affinity to DPG than maternal hemoglobin, O₂ released from the maternal blood can be taken up by fetal hemoglobin (arrow III). Curve d: a high DPG-concentration (8 mM) leads to an increased supply of the tissues with O₂ (arrows I & II). Curve e: in the absence of cooperativity between subunits of hemoglobin less O₂ would be transported from the lungs to the tissues. For determining the hypothetical binding curve e, a dissociation constant of 38 μM was assumed for the HbO₂-complex.

the lung. Furthermore, by employing different types of hemoglobin, which are influenced by the hydrogen ion concentration differently, certain fish species can use O₂ for two separately controllable functions: to fill their swim bladder with O₂ and to make O₂ available to the metabolism (136). Another achievement is the use of specific organic phosphoryl compounds such as 2,3-diphosphoglycerate (DPG) or inositol hexaphosphate (670, 671) as allosteric effectors which favor the deoxy-conformation. These effectors play a decisive role in O₂-transfer from maternal to fetal blood (274, 671) in adaptation to high altitude (672), and in disease processes where the arterial blood is only partially saturated with dioxygen (85, 673) (Figure 10-4).

The cooperativity of subunits facilitates the difficult task of O₂-transport. Another important point is the so-called haem–haem interaction, that is, the influence of haem-bound O₂ on the affinity of the haem groups in other subunits for O₂ (634). This effect enables hemoglobin to take up O₂ in the lungs at a free O_2-concentration of 130 μM and to release approximately 30% of it in the capillaries of O_2-consuming tissues (Figure 10-4). Here, the concentration of free O_2 is still as high as 50 μM (in order to guarantee the oxygen supply of the tissue cells by diffusion). The value of 30% is not greatly affected by a decrease of the O_2-concentration in the lungs, and it can be improved by reducing the average diffusion pathway of O_2 in the tissues. In contrast, an O_2-carrier with independent binding sites would release less than 20% of its O_2 under favorable conditions and would have no adaptation possibilities (Figure 10-4). Thus, cooperativity in oligomers allows much better adjustment to the given task.

Many concepts referring to protein action and interaction were introduced in the course of hemoglobin research. Many concepts and models concerning protein–ligand interaction and protein–protein interaction were developed during the history of hemoglobin research: sigmoid binding behavior (674–676); Hill coefficient (677); successive binding constants for ligands in oligomeric proteins (678); cooperativity based on conformational changes (679, 680); and allosteric control of proteins (92, 681, 682). It should be noted that many of these concepts were introduced and (mathematically) formalized before the structure of any protein was known. Consequently the actual meaning and usefulness of such concepts should be continuously reexamined. The example of diphosphoglycerate, whose influence on hemoglobin action and structure was ignored for decades, attests to the potential dangers of rigid formalizations in biology.

10.4 Nucleotide-Binding Sites

NAD-Binding Sites in Dehydrogenases

The structures of four NAD-dependent dehydrogenases are known: lactate dehydrogenase (LDH) (232), s-malate dehydrogenase (MDH) (233), liver alcohol dehydrogenase (ADH) (234), and D-glyceraldehyde-3-phosphate dehydrogenase (GAPDH) (230, 231). These enzymes catalyze the transfer of a hydride ion from the substrate, after which they are named, to the C4-atom of the nicotinamide ring of NAD. Each subunit of each enzyme contains one NAD-binding domain.

All NAD-binding domains contain parallel stranded β-sheets of identical topology. The domains are homologous (91) since they contain the same complicated sheet topology (section 9.6, page 204). In particular the binding region of NAD with respect to the sheet as sketched in Figure 7-7 has been conserved in great detail. This conservation has survived many amino acid changes; there is no detectable sequence homology among any two of these domains.

NAD is bound in extended conformation. When bound to any of these four dehydrogenases NAD assumes a characteristic, extended conformation, which may reflect the similarity of the NAD-binding domains. The adenine and nicotinamide rings are about 14 Å apart, and their planes are nearly perpendicular to each other. It is advantageous to consider the adenosine-containing half and the nicotinamide-containing half of NAD separately, one reason being that the NAD-binding domain itself might have arisen by duplication of a smaller unit, the Rossmann-fold (section 9.4, page 196). As sketched in Figure 7-7, each mononucleotide moiety of NAD attaches to one Rossmann-fold.

The binding pocket for adenine is not highly specific. The adenosine of NAD binds in a pocket which is not very specific for this ligand, but for aromatics in general. One invariant residue of the adenosine-binding site in the four dehydrogenases is the last residue of the first sheet strand of the Rossmann-fold (Figure 5-12b). It is a glycine; any larger residue would interfere with the binding of the ribosyl-moiety. The last residue of the second β-strand is also conserved. It is an aspartate which forms an H-bond to the O-2' of ribose. This bond may be crucial for the enzyme's ability to discriminate between NAD and NADP.

In LDH, ADH, and MDH the A-side of nicotinamide accepts the hydride ion; in GAPDH it is the B-side. The nicotinamide moiety of NAD is invariably bound in a cavity which is hydrophobic on one side and hydrophilic on the side which interacts with substrates such as lactate, ethanol, malate, or glyceraldehyde phosphate. The hydride ion is transferred stereospecifically to nicotinamide. In LDH, ADH, and MDH the so-called A-side (85) of the nicotinamide ring touches the substrates and accepts the hydride ion, whereas in GAPDH the nicotinamide ring has been rotated by 180° around the glycosidic bond which links ribose and nicotinamide so that the B-side is exposed to the substrate and accepts the hydride ion. This relative rotation is, of course, forced upon NAD by the geometry of the binding site.

Nucleotide-Binding Sites in Other Proteins

Many cofactors contain adenosine moieties. In the NAD-dependent dehydrogenases the subsite for the adenosine moiety is particularly well conserved. Many other cofactors of proteins also contain adenosine moieties, examples being ATP, NADP, FAD, coenzyme A, S-adenosylmethionine, PAPS, and deoxyadenosyl cobalamin. Thus it is important to determine whether the generalizations concerning NAD-binding by dehydrogenases can be extended to other nucleotide–protein interactions.

A variety of nucleotide binding proteins show similarities with NAD-dependent dehydrogenases. To some degree the findings with dehydrogenases are confirmed; all structurally known proteins which bind nucleotides show similarities with the dehydrogenases. Phosphoglycerate kinase (235, 310, 311) contains a domain which has the same sheet topology (section 9.6) as the NAD-binding domain, and the adenosine moiety of the cofactor ATP binds in a position corresponding to the adenosine moiety of NAD in the dehydrogenases.

In adenylate kinase the binding sites for the substrates ATP and AMP (665) are at geometrically similar positions relative to the β-sheet (Figure 7-7) as in the NAD-binding site in the dehydrogenases. In addition, the sheet topologies are similar (255). Topologically speaking, the position of ATP in adenylate kinase corresponds exactly to that of the adenosine-containing half of NAD in lactate dehydrogenase. The location of AMP in adenylate kinase, on the other hand, corresponds in a geometrical, but not in a topological, sense to the position of nicotinamide in lactate dehydrogenase. It is topologically nonequivalent because ATP and AMP touch each other below the connection between sheet strands A and B and not above the connection as shown in Figure 7-7. The adenine of ATP is probably H-bonded to the phenol OH-group of Tyr-95 in adenylate kinase; the isofunctional residue in LDH is Tyr-85. Similarities between LDH and adenylate kinase are also conspicuous at the (pyro) phosphate-binding subsite. Furthermore, in both proteins there are large conformational changes on nucleotide-binding at corresponding positions (665).

In flavodoxin the binding site for the prosthetic group flavinmononucleotide (FMN) corresponds to the site of nicotinamide moiety of NAD in dehydrogenases (229). Aligning FMN with nicotinamide superimposes almost the whole chain fold of flavodoxin with the NAD binding domain of LDH (91).

In dihydrofolate reductase and in glutathione reductase NADP is bound to domains containing β-sheets with different topologies. Other proteins which have been compared with the NAD-binding domain are described in Refs. (91, 254, and 683). Of particular interest in this respect are proteins which are specific for NADP. Whereas NAD is typical for enzymes of the catabolism, NADP is characteristic for enzymes of the anabolism. There are two structurally known enzymes with NADP as cofactor, dihydrofolate reductase (308) and glutathione reductase (124).

In dihydrofolate reductase helices and β-strands of the NADP-binding site show some geometrical resemblance to those forming the NAD-binding site of LDH. However, the topologies and chain folds of these two enzymes are quite different. In glutathione reductase there is an NADP-binding domain with a central parallel pleated sheet (124). This sheet contains a Rossmann-fold which binds the adenosine moiety of NADP in a position corresponding to the equivalent site in dehydrogenases. There is no similar Rossmann-fold for the nicotinamide moiety of NADP. Glutathione reductase contains FAD as a prosthetic group. Here too the adenosine moiety of FAD is bound to a structure resembling a Rossmann-fold.

Certain proteinases are directed against the apoprotein of nucleotide-binding enzymes. Long before common features in nucleotide-binding proteins were recognized by scientists (91), these features have been the targets of other naturally occurring agents. There are, for example, proteinases which specifically attack the apoenzymes of nucleotide-binding proteins (684). Another interesting example is salicylate, a compound occurring in certain plants, which is the active principle of aspirin. Both in alcohol dehydrogenase (685) and in adenylate kinase (665) salicylate attaches as a (weak) inhibitor to functional groups which are normally used for binding the adenosine moieties of NAD and ATP, respectively.

Induced Fit in Adenylate Kinase

An induced fit of the substrate increases the specificity. Many enzymes which transfer phosphoryl groups from ATP to acceptor molecules use an induced fit mechanism for increasing their specificity, in particular for excluding H_2O as an acceptor of the phosphoryl group. Adenylate kinase, a typical example (Figure 10-5), phosphorylates H_2O at a 10^5 times slower rate than its specific substrate AMP.

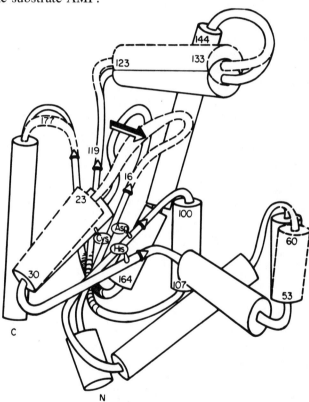

Figure 10-5. Movement of the phosphoryl-binding loop (residues 16–22) during the transition of conformation A (solid lines) to conformation B (dashed lines) of crystallized adenylate kinase (665, 688). Conformation B has an open adenosin pocket at the ATP site, a wide cleft for accomodating the phosphoryl groups, and an open AMP site. The A conformation also has an open ATP site, but a narrow phosphoryl cleft, and the adenosin pocket of the AMP site is closed. These findings suggest that conformation B is related to the free enzyme E before an AMP-induced fit, whereas conformation A is related to enzyme E' after such a change.

An enzyme for which the induced fit model (686, 687) applies exists almost quantitatively in an inactive state E. Only a small fraction of the molecules have the active conformation E'. According to the assumption of Jencks (631), the ratio of [E]/[E'] can be derived directly from the phosphorylation rates of H_2O and of the specific substrate. For adenylate kinase [E]/[E'], in the absence of substrate, would be 10^5. Only the binding of a specific substrate brings about an interaction with the active site which leads to the transition of the enzyme into the active form E' (Figure 10-5).

Most of the binding energy is used for changing the enzyme conformation. If Jenck's assumption is true the observed binding constant $K_{obs} = $ [E'-AMP]/[E]·[AMP] $ = 10^4 \ M^{-1}$ is 10^5 times smaller than the "total" binding constant $K_{total} = $ [E'-AMP]/[E']·[AMP]. Using the general formula $\Delta G = -R·T·\ln K$, we find that the total free energy of binding that is available from the interaction of AMP with the enzyme E' ($\Delta G_{total} = -12.5$ kcal/mol) is more than twice the free energy of binding as derived from the observed dissociation constant ($\Delta G_{obs} = -5.3$ kcal/mol). Thus most of the binding energy is used up to force a change in conformation of the enzyme from the inactive to the active form and the remainder appears as the observed binding energy. The total binding energy would be manifested as observed binding energy if the enzyme could be frozen into the active form E'. Unfortunately this has only been possible in crystallized adenylate kinase under conditions where meaningful binding constants could not be determined (665, 688).

10.5 Binding Sites for Phosphoryl Groups

There is no phosphoryl binding site in proteins which is both unique and common. Approximately 50% of the known proteins bind and/or process compounds possessing phosphoryl groups. The contrasting examples of ribonuclease (689), human hemoglobin [which binds 2,3-diphosphoglycerate, ATP, or inositol hexaphosphate at a site between the two β-chains (672)], and staphylococcal nuclease (242) illustrate that there is no phosphoryl-binding site in proteins which is both unique and common.

The loop between the first sheet strand of a Rossmann-fold and the following α-helix is one characteristic phosphoryl-binding site. There is a group of proteins which bind a phosphoryl or a pyrophosphoryl moiety at a characteristic site. This site is the loop of the polypeptide chain which connects the *first*[4] strand of a parallel pleated sheet and a helix running more or less antiparallel to this sheet strand. Usually it corresponds to the loop between the first β-strand of a Rossmann-fold and the following α-helix (Figure 5-12b).

---

[4]It should be recalled that sheet strands are numbered according to their appearance along the linear polypeptide, the residues of the first strand being closest to the N-terminus.

Phosphoryl groups are fixed to the backbone of the loop by hydrogen bonds. In s-malate dehydrogenase (691), D-glyceraldehyde-3-phosphate dehydrogenase (230) and lactate dehydrogenase (230) the pyrophosphate moiety of NAD is hydrogen-bonded to the backbone of a corresponding loop. For alcohol dehydrogenase (692), binding studies with an NAD analogue showed that the pyrophosphate is at the same position (within 3 Å) relative to this loop. In flavodoxin this loop wraps around the phosphoryl moiety of FMN (145, 237, 238). The backbone of the loop in adenylate kinase wraps around the phosphate of AMP, as derived from substrate-binding studies; in crystalline adenylate kinase this loop fixes a sulfate from the mother liquor (665). In phosphoglycerate kinase (310, 311) the phosphoryl groups of ATP bind at a similar position. In glutathione reductase (124) the pyrophosphoryl of NADP binds to the backbone of the corresponding loop of the Rossmann-fold. The same is true for the pyrophosphate of FAD. In triose phosphate isomerase (305) the phosphate is also located at loops connecting β-strands and following α-helices, but in the C-terminal part of the polypeptide chain (Figure 5-17e). In all these cases a loop between a carbonyl end of a β-strand and a following α-helix is used. This may be explained by a favorable electrostatic interaction between the negative charge of the phosphoryl group and the dipole of the α-helix (792), which originates from the superposition of hydrogen bond dipoles.

Phosphoryl-binding loops can be pliable and flexible. Some other generalizations are also apparent. In all cases the phosphoryl groups form hydrogen bonds to the backbone of the polypeptide chain. Thus phosphoryl-binding occurs at loops standing out of a rigid secondary or supersecondary structure such as the Rossmann-fold. As their base is strong, these loops can be flexible without endangering the protein structure as a whole. In porcine adenylate kinase, for example, the phosphate-binding loop has the sequence Gly–Pro–Gly–Ser–Gly–Lys–Gly (389); the glycyl residues at every second position allow for a large range of dihedral angles at the C_α-atoms (section 2.3). Consequently this chain segment can assume a wide range of conformations which may lead to pliability and flexibility even in the intact enzyme molecule. X-ray analysis (Figure 10-5) revealed that this loop undergoes a displacement of 6Å during the transition from conformation A to conformation B of adenylate kinase (688).

The similarities within this group of phosphoryl-binding proteins may not have a uniform basis. With respect to the dehydrogenases described in section 10.4 (page 220), the mode of (pyro)phosphoryl binding is probably a conserved feature, since these proteins are likely to be homologous (section 9.6, page 204). However, for the other proteins of this group, the mode of phosphoryl binding should not be accepted as a trait which indicates homology, since it may simply reflect a physico-chemically favorable interaction.

A number of proteins which are modified by covalent phosphorylation have typical properties in common with nucleotide-binding proteins. There is a large number of proteins whose function is modified or tuned by reversible phosphorylation (175). The amino acid sequences at the phosphorylation sites of various proteins show some similarities (Table 10-2). The phosphorylated side chain is either Ser, Thr, or His. Often this residue is preceded or followed by a Gly; in most cases the residue two positions further along the chain is a positively charged Lys or Arg. These sequences bear some resemblance to the phosphoryl-binding loop of adenylate kinase (Table 10-2). Furthermore, the hypothesis that reversible phosphorylation of a given residue triggers conformational changes of a (phospho)protein (175, 693) correlates well with the fact that in lactate dehydrogenase (232) and in adenylate kinase (688) the loop is the epicenter of a large conformational change. Consequently nucleotide-binding proteins may help to clarify static and dynamic aspects of protein phosphorylation.

Table 10-2. Phosphate Sites in Nucleotide-Binding Proteins and in Phosphoproteins

Protein	Sequence			Comment
Myelin basic protein	G–Ser(P)–G–K–D			Phosphoproteins; (P) indi-
Histone H2a	Ac–Ser(P)–G–R–G			cates the residue which is
Histone H4	Ac–Ser(P)–G–R–G			reversibly phosphoryl-
Histone H1	G–Ser(P)–F–K–L			ated; the data are taken
Glycogen synthetase	I–Ser(P)–V–R–X			from Ref. (693)
Troponin I	I–Thr(P)–A–R–R			
HPr-factor (720)	I–His(P)–A–R–P			HPr is involved in vectorial phosphorylation. A typical phosphate loop topology has been predicted from the sequence
Flavodoxin (145)	G–Thr	$-\overset{10}{G}$–N–T		This loop wraps around the phosphate moiety of FMN
Adenylate kinase (665)	G–Ser	$-\overset{20}{G}$–K–G		This loop wraps around the P-moiety of AMP; it is the epicenter of a conformational change
Glyceraldehyde-3-phosphate dehydrogenase (145)	R–Ile	–G–R–L		The backbone amides of Arg-10 and of Ile-11 form hydrogen bonds with the pyrophosphate moiety of NADH

10.6 Interactions of Proteins With Other Macromolecules

Proteins can interact specifically with other macromolecules such as nucleic acids or polysaccharides. Lipids are also regarded as macromolecules since they form large aggregates in aqueous solutions. In nucleoproteins, glycoproteins, or lipoproteins the protein part can be less than 50%, and the bulk properties of the complexes are often dominated by the nonprotein moieties. Furthermore both structure formation and stability of the proteins may depend on their partners in the complexes. This is most obvious for those membrane proteins which span the hydrocarbon moiety of the lipid bilayer.

Lipoproteins

Proteins penetrate the interfaces of aqueous and apolar phases. Two major biological functions are based on bulk interactions of proteins and lipids: lipid transport in aqueous solutions and membrane activities. The transport capacity in the blood plasma for otherwise poorly soluble lipids such as cholesterol (2 mg are soluble in 1 liter of 150 mM NaCl, pH 7.4) or triacylglycerides is increased 1000-fold through the effects of specific apolipoproteins. Such lipid-transporting proteins may play a role in atherosclerosis and other diseases of affluent societies; consequently much work is devoted to their characterization (145, 694). Of great interest are the secondary and tertiary structures of lipoproteins. Prediction attempts (chapter 6) were made on the basis of the correlation between secondary structure and amino acid sequence of normal globular proteins (694). Since lipoproteins do not seem to crystallize readily more exact structural data will not be available for some time.

In membranes, lipids form biological barriers and compartments whereas the specific membrane functions such as transport, signal mediation, and energy transduction are carried out by proteins (19, 695–697). Sequence information on intramembranous parts of proteins is scarce and indicates that relatively long stretches of apolar residues are present (698). The most detailed information on the tertiary structure is available for the purple membrane protein of *Halobacterium halobium* (699, 700). The subunit of this protein consists essentially of seven parallel or antiparallel α-helices, which extend from one side of the membrane to the other. Another well-studied system is discussed below.

The calcium pump is a typical, well-analyzed membrane protein. The sarcoplasmic reticulum of muscle (701, 702) is a tubular system with a highly specialized membrane; its only function seems to be the release and accumulation of calcium ions (703, 704). This is reflected in the fact that one protein of 100,000 daltons, the so-called Ca^{2+}-transport ATPase or Ca^{2+}

pump, constitutes more than 50% of the membrane mass and 80% of the membrane protein mass. This protein, a cylinder 50 Å in diameter and 80 Å in length, spans the membrane the thickness of which is 60 to 90 Å. The membrane contains 90 (phospho)lipid molecules per molecule of the Ca^{2+} pump.

It is obvious that membrane and Ca^{2+} pump form a functional unit; the membrane serves as the barrier which holds back the accumulated calcium. Furthermore the membrane is a two-dimensional solvent for the protein; as shown by lipid-binding to the purified protein, the Ca^{2+}-transport ATPase has a surface with a higher affinity for a nonpolar environment than for water. There is only one type of water but many types of lipid molecules and this wide spectrum of lipid molecules makes possible a wide variety of specific interactions with membrane proteins. The role of individual lipid classes for a given membrane process is tested in so-called reconstitution experiments (705, 706). In the case of the calcium pump, a functionally active membrane system can be formed by adding only phospholipid(s) to the purified pump protein (701, 702).

Proteins move in the plane of the membrane. A bulky antibody attached to the outer part of the Ca^{2+}-transport ATPase does not interfere with the translocation of calcium ions, indicating that the protein does not rotate around an axis parallel to the membrane surfaces for Ca^{2+} translocation (707). This seems to be a general rule for membrane proteins. On the other hand, proteins can rotate (708) and move laterally within the plane of membranes, the actual degree of mobility depending on physico-chemical properties of the membrane and on steering effects of protein networks attached to either side of a membrane. Lateral diffusion is important for the interactions between the constituents of multicomponent membrane-bound systems since functionally linked membrane proteins are not always in physical contact with each other (709, 710).

Glycoproteins

Covalently bound carbohydrate contributes to size, viscosity, and localization of many glycoproteins. The most diverse group of biologically occurring macromolecules is represented by the glycoproteins which consist of carbohydrate moieties covalently linked to polypeptides via the side chains of Thr, Ser, or Asn (711, 712). In the only structurally known glycoprotein, myeloma protein IgG Kol (543), the carbohydrate moiety has a fixed position and presumably a structural role. This is unlikely to be true for the majority of glycoproteins.

The carbohydrate component can represent from less than 1 to over 80% of the total weight of a glycoprotein. As the carbohydrate moieties attached at a specific position to a protein can differ, a glycoprotein rarely represents a homogeneous molecular species. The first examples of Table 10-3 show that the viscous and hydrophilic properties of carbohydrate moieties are

Table 10-3. Contributions of the Carbohydrate Moiety to the Properties and Effects of Glycoproteins[a]

Property or effect of glycoprotein	Examples	Biological aspects
Size	Blood plasma proteins (145, 721)	The loss of the proteins through the kidney is prevented
Viscosity	Mucins (glycoproteins of mucous) (722)	Mucins serve as protective linings, as lubricants, and as transport media
Antifreeze effect	Antifreeze glycoproteins in Antarctic fish species (723, 724)	The growth, not the formation, of ice crystals in body fluids is prevented. The glycoproteins lower the freezing point of aqueous solutions to the same extent as equal weights of NaCl
Orientation in membranes	Glycoproteins of cell membranes in mammals; the carbohydrate moieties are located at the outer surface (709)	As there is a high energy barrier to the passage of a carbohydrate moiety through the hydrophobic part of the membrane, glycoproteins help to establish and maintain an asymmetric distribution of membrane proteins
Interaction with specific surfaces	Prothrombin (725) and other proteins of the blood plasma in mammals (173)	The survival time of plasma proteins is regulated in the following way. Loss of the terminal nonreducing sugar of protein-linked carbohydrate side chains leads to the exposure of terminal D-galactose. This residue is recognized by a receptor at the surface of liver cells. The whole glycoprotein is then taken up and degraded by the cell
Action as membrane-bound receptor	Lectin-receptors in cell membranes; e.g., the receptor for concanavalin A (726)	The lectin (e.g., concanavalin A) has binding sites for a specific sugar (α-methyl mannoside) of the receptor, a glycoprotein. Lectin-binding triggers cell division. Lectins are valuable tools for elucidating this process. Their true function is not known

[a]The examples given in lines 1 to 4 mainly represent bulk properties of carbohydrates. The last two examples illustrate the importance of specific sugar residues for specific recognition.

important for the function of many glycoproteins. In other cases [the standard examples are pancreatic ribonucleases from different mammals (145)] the carbohydrate moiety may be neither advantageous nor disadvantageous to a protein and only reflect the fact that a nonspecific sugar-attaching enzyme recognizes a structural feature of a protein as a substrate (712). Artificial carbohydrate–protein conjugates are more stable against heat, denaturants, and proteases than the corresponding proteins. As the catabolism and other biological features of such conjugates differ from those of the proteins they have potential as novel therapeutic agents (796).

Specific sugar residues serve recognition functions. The last two examples of Table 10-3 show that sugars play an important role in specific interactions between cell surfaces and soluble macromolecules. Intercellular recognition, such as during tissue formation from different cell types, is also based on the structural diversity of glycoproteins (709, 713). Sugars are indeed well suited for creating specific structural features (85). Whereas only six different peptides can be made from three different given amino acids (all permutations), at least ten times as many primary structures can be formed from three sugar residues because many of the possibilities of joining monosaccharides together are used *in vivo*. However, the recognition mechanisms involving sugar residues may often be based on stochastic rather than stoichiometric processes as the synthesis of complex carbohydrates lacks the precision of protein synthesis.

Nucleoproteins

Under this term we include all proteins which interact with nucleic acids and not only with mono- or dinucleotides such as the enzymes mentioned in section 10.4.

Ribosomes are complexes of proteins and nucleic acids. A well-analyzed complex between proteins and nucleic acids is the ribosome which catalyzes the formation of polypeptide chains (714, 715). A ribosome contains several RNA and various protein molecules. It consists of two particles of unequal size which bury the messenger RNA between themselves. In *E. coli* the larger particle contains two RNA molecules (named "23S" and "5S" according to their sedimentation constants) and 34 proteins (named "L1" through "L34," L standing for large). The smaller particle contains the "16S" RNA and 21 proteins ("S1" through "S21," S standing for small).

Many of these proteins are tightly attached to the nucleic acid. Some proteins are so elongated (7) that external contacts with nucleic acid and/or other ribosomal proteins far outnumber the internal contacts within the polypeptide chain. Thus, the polypeptide chain cannot assume a defined structure on its own. Accordingly, it is very unlikely that such proteins can be crystallized and structurally analyzed. So far, it has only been possible to crystallize fragments of the protein L12 (and/or protein L7) which dissociates easily from the ribosome and thus is likely to form a defined structure on its own. An X-ray structure analysis is under way (716).

Elongation factor Tu · GDP has a tadpole structure. Elongation factor Tu · GDP, which forms a ternary complex with the ribosome and tRNA, could also be crystallized (717). Its structure is known at 6 Å resolution (718). The shape of the molecule is quite different from the shape of usual globular proteins. It resembles a tadpole, consisting of a globular domain (the "head") and an elongated thin domain (the "tail") separated by a scissure in overall density. The head contains several α-helices and seems to have a rather rigid structure. In contrast the tail is rather fluffy and seems to consist of β-structure. Interestingly enough, head and tail have a second connection, so that a ring structure is formed. Probably, tRNA is bound near the hole of this ring in a large groove between domains. Relative movements of the domains during the elongation step on the ribosome are easy to visualize. The separation into a flexible and a rigid domain is also known for the nucleoproteins L7/L12 (716) and lac repressor (719). In L7/L12 and flexible domain is at the N-terminal side.

Aminoacyl-tRNA synthetases are also involved in protein synthesis. However, these enzymes do not interact with the ribosome but only with tRNA. The chain fold of one synthetase is known (221). There is no striking difference to other globular proteins. The enzyme is a symmetric dimer. It was suggested that the tRNA binding site is formed by both subunits, resembling the situation observed in some enzymes with smaller substrates (124, 313).

Proteins which control gene expression by associating with DNA are of high interest for molecular biologists. One of these proteins, the lac repressor, was crystallized. An X-ray structure analysis is in progress (719). In conclusion, we find that the interactions between proteins and nucleic acids are poorly understood. As mentioned earlier, twisted β-sheets may play a major role in these interactions (206–208). Examples such as ribosomal proteins or elongation factor Tu · GDP indicate that some nucleoproteins have quite peculiar structures.

Summary

The smooth operation of living organisms is possible only because a natural protein does not take advantage of the intrinsic capacity of polypeptides to bind all kinds of small molecules, but specializes in a few specific ones. Only one type of protein, the immunoglobulins, provides as many different binding sites as possible. Thus, a systematic analysis of immunoglobulin–ligand systems may yield general principles governing protein–ligand interactions. Furthermore, the immunoglobulin architecture may serve as a basis for the *in vitro* synthesis of polypeptides with desired binding properties.

Although living organisms contain an immense variety of low molecular weight compounds, a limited group of metabolites including among others amino acids, adenosine derivatives, and porphyrins plays a dominant role; most proteins interact specifically with one or more of these distinguished metabolites. One rule is, for example, that adenosine nucleotides and other phosphoryl-containing compounds are often bound at a characteristic site of a Rossmann-fold.

Haem proteins illustrate how the properties of a versatile chemical compound, the porphyrin–iron complex, are exploited for biological functions. The protein moiety ensures that in each case only one specific reaction occurs at the haem iron and that intrinsic chemical properties of the haem are modified by physiologically important parameters.

Proteins which form complexes with lipid bilayers, saccharides, and nucleic acids are biologically most important. Some information on these proteins can be derived from known protein–ligand interactions. For further insight the three-dimensional structures of such complexes should be elucidated.

In conclusion, for studying the interactions of proteins with any other compound the immunoglobulins are the proteins of choice. With respect to the interactions between proteins and the group of distinguished metabolites structural predictions can already be made which concern the majority of proteins.

Chapter 11

The Structural Basis of Protein Mechanism, Action, and Function

11.1 Definitions

Most if not all activities of an organism are based on the activities of proteins. Atomic or even subatomic detail is responsible for the overall answer of a whole organism to problems of homeostasis and to challenges by the environment (728). This chapter summarizes aspects of protein structure which can be subsumed in the categories of mechanism, action or activity, and function.

A *mechanism* describes the order, in time and space, of fundamental processes involved in an action or reaction. Detailed atomic geometry and molecular orbitals have to be known before a mechanism of protein action can be formulated. At present this is possible only in rare cases, hemoglobin (section 10.3, page 217) and chymotrypsin (section 11.2, page 235) being the classical examples. The question concerning the mechanism of a protein is: *How does it work?*

Action or *activity* refers to the most obvious effects of a protein. For example, the activity of chymotrypsin is the cleavage of a peptide or polypeptide. The question concerning the activity of a protein is: *What does it do?*

Biological function is the contribution of a part to the performance of a whole system. This means that the function of a protein has to be studied in relation to higher levels of the functional hierarchy (729). In the case of hemoglobin, for example, these higher levels are circulation and metabolism. Chymotrypsin has a function in the digestive system, processing nutritional proteins, and a function in the body's defense system, inactivating harmful polypeptides (certain hormones, toxins, and viral proteins) before they can attack the epithelial barrier. The question concerning the function of a protein is: *What is its purpose?*

The following sections discuss known mechanisms of protein action using enzymes as examples. Finally the performance of one organ, skeletal muscle, is dealt with in terms of protein action and function.

11.2 Enzyme Catalysis

Enzymes are a billion times more efficient than man-made catalysts although they operate within the limits of biological conditions. A fascinating property of enzymes is their catalytic power. Enzyme-catalyzed reactions proceed at rates that are from 10^8 to 10^{20} times faster than the corresponding uncatalyzed reaction. Traditionally, enzymes are compared with man-made catalysts which as a rule are 10^8 to 10^9 times less effective in accelerating a given reaction than the corresponding enzyme (730–732). Furthermore, in contrast to the rather extreme conditions often required to accelerate chemical reactions in the organic laboratory, enzymes achieve their catalytic effects in aqueous solution at biological pH values and moderate temperatures and pressure.

The specificity of enzyme action prevents pollution of an organism with by-products. Another attribute of an enzyme, which a chemical catalyst possesses only in rare cases, is specificity of action. Only one or a few compounds, the substrate(s), are acted upon and only a single type of reaction takes place. Side reactions or by-products do not occur, a reflection of the fact that uncontrolled pollution cannot be tolerated in a living cell. From this observation it may be deduced that not only the catalytic power but also (and even more so) the specificity of enzymes should be exploited for industrial purposes.

The free energy of substrate-binding is utilized for catalysis and for specificity. Using chymotrypsin activity as an example, the tradition is followed of separating specific binding interactions between substrate and enzyme (section 10.2, page 209) and the catalytic process itself which is described in terms of chemical "model" reactions (see below).

As emphasized by Jencks (631), the concept of "utilization" may reconcile the two main aspects of enzyme action. The noncovalent interaction of a specific substrate with the active site of an enzyme can be quantitated as free energy of binding ("binding energy"). The so-called intrinsic binding energy is of the order of 10 to 20 kcal/mol for most enzyme–substrate pairs. However, most of this energy is utilized for the catalytic process; catalysis is driven at the expense of binding forces which can be distant from the catalytic site. The observed binding energy for enzyme–substrate pairs represents only what is left over after this utilization and amounts to approximately 5 to 7 kcal/mol. Binding forces can also be utilized in order to pay for an induced-fit mechanism (section 10.4, page 223) or for nonproductive binding (section 10.2, page 210), two mechanisms which increase the specificity of enzyme action.

The Catalytic Mechanism of Chymotrypsin

The catalytic mechanism of chymotrypsin, an enzyme which cleaves peptide bonds, is understood in more detail than that of any other enzyme (537). Chymotrypsin has features which simplify such studies. It is a monomeric enzyme which shows no allosteric effects; the structural transitions which occur in the process of peptide bond cleavage are very small; and it can transfer acyl groups from a variety of donors, such as peptides or esters, to a variety of acceptors, such as water, alcohols, or amines. Comparisons among different donors and among different acceptors greatly facilitated the analysis of the individual catalytic steps (733, 734).

On the basis of X-ray crystallographic, spectroscopic, and chemical data [for a review see Ref. (537)] a plausible mechanism of chymotrypsin action has been proposed (Figure 11-1). After the Michaelis complex between enzyme and substrate has been formed (section 10.2, page 210), the oxygen atom of the hydroxyl group of Ser-195 attacks the carbonyl carbon atom of the scissile bond in the substrate. A transient tetrahedral intermediate is formed (735). This reaction is facilitated by the charge-relay network (628, 736) which serves to draw a proton away from the hydroxyl group of Ser-195, thereby making it a powerful nucleophile. His-57 then donates a proton to the nitrogen atom of the scissile peptide bond; as a result the bond is cleaved. At this stage, the amine component is hydrogen-bonded to His-57 whereas the acyl group is bound to Ser-195 in an ester linkage. The acylation stage of the hydrolytic reaction is reached.

The second half of the process is deacylation. The amine component of the substrate diffuses away, and a water molecule takes its place at the active site. In principle, deacylation is the reverse of acylation with H_2O substituting for the amine component (85). First the charge-relay network draws a proton away from water; the resulting OH^- ion simultaneously attacks the carbonyl atom of the acyl group that is attached to Ser-195. A transient tetrahedral intermediate is formed and His-57 donates a proton to the oxygen atom of Ser-195, leading to the release of the substrate's acid component which diffuses away. The enzyme is then ready for another round of catalysis.

Chymotrypsin, glyceraldehyde-3-phosphate dehydrogenase, and papain use analogous reaction mechanisms. The mechanism described above is valid not only for the phylogenetic relatives of chymotrypsin such as trypsin or thrombin; it may also be relevant for other families. For instance, a step-by-step analogy has been traced for the reaction catalyzed by enzymes as different as chymotrypsin, papain, and glyceraldehyde-3-phosphate dehydrogenase (737) (Figure 11-1).

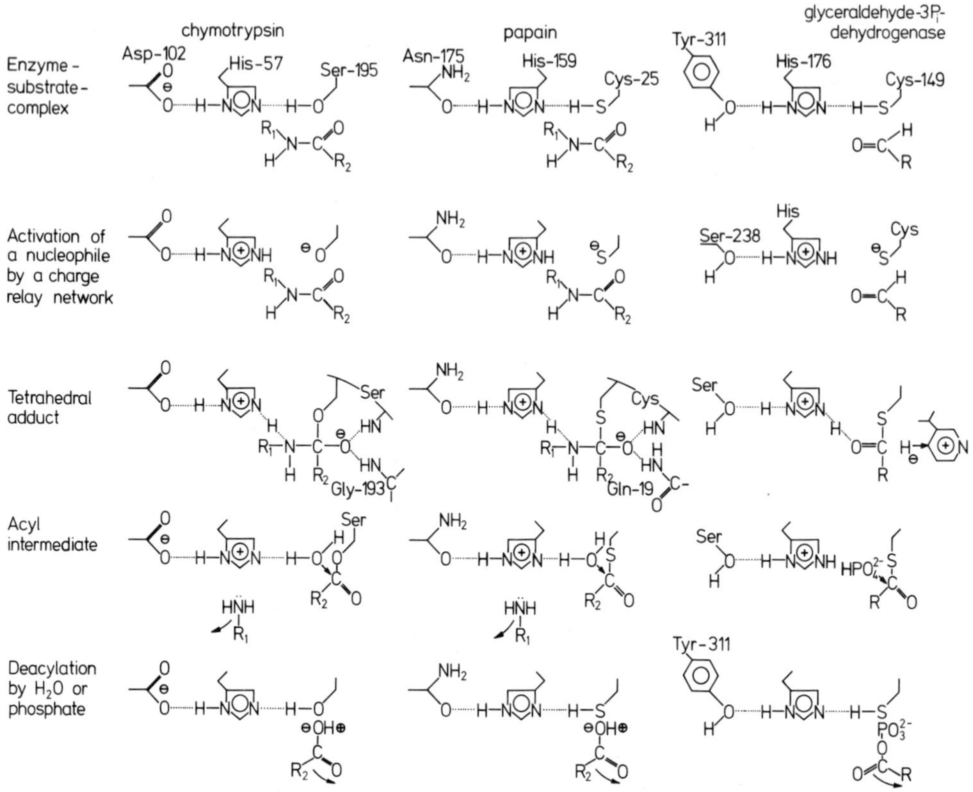

Figure 11-1. The analogous catalytic mechanisms of chymotrypsin, papain, and glyceraldehyde-3-phosphate dehydrogenase [from Ref. (737)]. The individual steps designated in the first column are valid for all three enzymes. The tetrahedral intermediates (tetrahedral adducts) probably represent transition states. Chymotrypsin (left) is represented by its charge-relay network Asp–His–Ser. Of the substrate, R_1–NH–CO–R_2, only the scissile peptide bond is shown. Notice that the peptide chain goes from right to left as common in sketches of protease mechanisms. The individual steps of the reaction catalyzed by chymotrypsin are discussed in the text; the second tetrahedral intermediate mentioned there is shown in the figure only in the nascent state: H_2O attacks the acyl intermediate. Papain is a so-called sulfhydryl protease; the nucleophilic group analogous to the OH group of Ser-195 in chymotrypsin is an SH group in papain. The enzyme shown in the fourth column catalyzes the reaction glyceraldehyde-3-phosphate + NAD + inorganic phosphate $\rightleftharpoons$ 1,3-diphosphoglycerate + NADH (the three substrates are represented as O=CHR, as a nicotinamide ring and as HPO_4^{2-}). In this case the reaction leading from the tetrahedral adduct to the acyl intermediate is the elimination of a hydride ion, that is an oxidation step. The acyl intermediate is not attacked by H_2O as in the two proteases but by inorganic phosphate so that a mixed acid anhydride is the final reaction product.

Transition State Theory

In each step of a chemical reaction, reactants proceed from one relatively stable state to another through a state of higher energy (Figure 11-2). The transition state is the state of highest energy through which the reactants must pass to get from one stable state to another (738–740). The additional free energy needed to reach the transition state is $\Delta G\ddagger$, the free energy of activation ("activation energy"). According to the Arrhenius equation $k =$

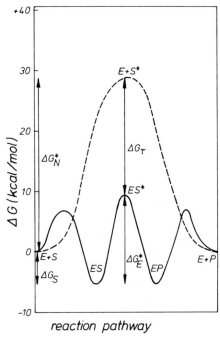

Figure 11-2. Transition state theory and enzyme catalysis [from Ref. (740)]. The free energy profiles of an uncatalyzed reaction (dashed line) and the corresponding enzyme-catalyzed reaction (solid line) are compared. Only one compound (substrate) is involved; the reaction in question could be an isomerization. Furthermore it is assumed that the reaction mechanism is the same in the catalyzed reaction as it is in the uncatalyzed reaction. The following symbols are used: E= enzyme; S = substrate; P = product; S$\ddagger$ = transition state of the substrate. The free energy changes refer to the following reactions: ΔG_s, formation of ES from E and S; ΔG_τ, formation of ES$\ddagger$ from E and S$\ddagger$; $\Delta G_N\ddagger$ formation of S$\ddagger$ from S; $\Delta G_E\ddagger$ formation of ES$\ddagger$ from ES. The values which were assumed in order to calculate the free energy changes were: K_S, association constant of ES, 10^4 M^{-1}; K_P, association constant of EP, 10^4 M^{-1}; k_d (the first-order rate constant for dissociation of ES to E and S, and of EP to E and P), 10^4 sec^{-1}; k_E (the first-order rate constant for the conversion of ES to EP), 10^2 sec^{-1}; k_N (first-order rate constant for the corresponding noncatalyzed reaction), 10^{-8} sec^{-1}; the concentrations of S and of P, 1 M; the equilibrium constant for the formation of P from S, 1.0.

$s \cdot \exp(-\Delta G\ddagger/RT) = s \cdot \exp(-\Delta H\ddagger/RT + T\Delta S\ddagger/RT)$, the rate k of a chemical reaction depends on the activation energy (697).

The activation energy is partly entropic and represents the additional order which needs to be imposed on the system to reach the transition state; for example, the chance that two atoms would approach sufficiently close to react. Another part of the activation energy is enthalpic, the work which needs to be done to bring two atoms close enough so that a covalent bond can form between them. Furthermore, enzymes may decrease the observed enthalpy of activation (compared with a solution reaction) by fixing specific substrates in such a way as to select a favorable reaction pathway from several pathways that are followed by the solution reaction.

Effects Which Contribute to High Reaction Rates

Proximity, Orientation, Orbital Steering, and Other Entropic Effects

An enzyme can accelerate a chemical reaction by orienting the reactants at the active site in the optimal position for reaction (741, 742). This may even include orbital steering, the precise orientation of binding orbitals of the atomic reaction partner relative to each other (743, 744).

The binding of two separate molecules at the active site of an enzyme serves to convert a bimolecular reaction to a monomolecular, intramolecular reaction. Intramolecular model reactions provide the most direct means of estimating the magnitude of the rate acceleration that may be obtained from bringing the reactants together (631, 745). In other words, the entropic contribution of an enzyme leads to an increase in effective substrate concentration. Since chemical reactions proceed at rates proportional to the concentrations of the reactants, a rate enhancement by a factor of up to 10^8 can be expected in such a local area of high concentration and order (631, 744).

Destabilization of Transformable Groups of Substrates

The transformable group of a substrate can be destabilized at the moment when the substrate is bound to the enzyme; the cost of this destabilization (631) must be paid for by the overall binding energy. Destabilizing mechanisms include change of solvent, charge–charge interactions, and geometric strain of bond lengths and bond angles. Destabilization can be relieved in the transition state so that the free energy of activation that is required to reach the transition state is reduced.

Forcing chemical groups into a nonpolar environment can increase rates by a factor of 50,000. Pyruvate decarboxylase is a thiamine-pyrophosphate-containing enzyme, for which a simple mechanism has been suggested (746) on the basis of two observations. One is that the adduct of pyruvate and an analogue of the cofactor (Figure 11-3) is decarboxylated in organic solvents

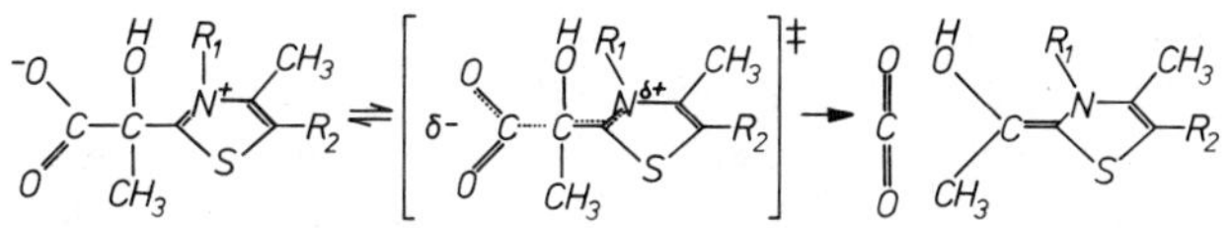

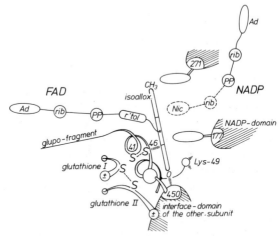

Figure 11-3. Catalysis of decarboxylation by solvent change. A thiazole group is the reactive moiety of thiamine pyrophosphate, a cofactor of decarboxylating enzymes. In thiamine pyrophosphate R_1 is a pyrimidine and R_2 is a pyrophosphate moiety; in the model compound discussed here, $R_1 = CH_3$ and $R_2 = H$. The model compound loses CO_2 in organic solvents 50,000 times faster than in water (746). It is believed that organic solvents favor the transition state shown in brackets.

10^4 to 10^5 times more rapidly than in water; the other is that the binding site for pyruvate in pyruvate decarboxylase is hydrophobic. The increase in decarboxylation rate that is expected from the transfer of the pyruvate-coenzyme adduct into such an environment is sufficient to account for the observed enzymic rate. The pyrophosphate group and the pyrimidine ring of the coenzyme could provide the binding energy to fix the charged substrate in this unfavorable environment and bring about the rate increase.

The ADP moiety of coenzymes such as ATP, FAD, NAD(P), or coenzyme A may play a role similar to the pyrimidine and pyrophosphate groups of thiamine pyrophosphate. In all known cases (section 10.4) the cofactors bind to their enzymes in extended conformations, thereby forming numerous interactions with the enzyme (Figure 11-4). In each case the

Figure 11-4. The active site of glutathione reductase [from Ref. (124)]. There are two identical active sites in the dimeric enzyme. The reduction equivalents are transferred along the following pathway (from right to left): from the nicotinamide moiety (Nic) of NADPH → isoalloxazine ring of FAD (shown edge-on) → redox active disulfide (of the so-called glupo fragment)→ disulfide bond of glutathione. Both subunits contribute residues to the glutathione-binding site and to the catalytic site. Note that both NADPH and FAD are bound to the enzyme in extended conformations.

ADP moiety provides binding energy which can be utilized for specific tasks of the other moiety of the coenzyme.

Charged groups of the enzyme may become more reactive when solvating water molecules are replaced by substrate. Destabilization based on charge desolvation may contribute to the catalytic effect of the bound Zn^{2+}-ion in carboxypeptidase A (747). The replacement of a solvating water molecule by the substrate decreases the dielectric constant surrounding the metal and increases its activity in polarizing the acyl group of the substrate for nucleophilic attack. A similar effect contributes to the events at the D subsite of lysozyme. The interaction of the desolvated carboxylate group of Asp-42 with substrate leads to a local destabilization of the enzyme–substrate complex; this destabilization is relieved upon formation of the oxocarbonium ion-like transition state (531).

Bond lengths and bond angles of the substrate can be distorted when it is bound to the enzyme. A chemical group of the substrate can be forced by the enzyme to resemble the transition state; the mechanisms of this geometric destabilization include deformation of bond angles, compression of reacting atoms to an internuclear distance that is smaller than the sum of their van der Waals radii, and stretching of a covalent bond to a length that is longer than the sum of the covalent radii of the bound atoms (631, 739, 740). The best known example of geometric destabilization is found in an N-acetyl-amino sugar residue bound to the subsite D of lysozyme (748). The tetrahedral (sp^3) carbon atom-1 is distorted and this strain is relieved in the transition state where the carbon atom-1 is planar (sp^2). Taking into account the desolvation of Asp-42 on substrate binding, the destabilization energy ΔG is at least $+8.6$ kcal/mol. This energy is compensated for by the overall binding energy of the oligosaccharide substrate.

Binding interactions that are far from the catalytic site stabilize the trypsin-inhibitor complex. This is illustrated by comparing trypsin-substrate interaction with trypsin-inhibitor interaction (269, 536). The complex of trypsin with its pseudosubstrate, the inhibitor, is characterized by an abnormal distance of 2.6 Å between carbon of the carbonyl group-15 of the pseudosubstrate and Ser-195 of the enzyme. The normal substrates of chymotrypsin and trypsin which make few favorable interactions need $+5$ to $+15$ kcal/mol of activation energy for reaching the acylation stage. During the formation of the trypsin-inhibitor complex, on the other hand, numerous interactions are optimized and the apparent ΔG is -18 kcal/mol despite the strained C–O^γ adduct (Table 5-6). Thus, the difference of stabilization energy can be explained by the different contact areas in the complexes which trypsin forms with the inhibitor and an ordinary substrate, respectively (749).

Events at the catalytic site itself can also stabilize the transition state. So far the point has been stressed that remote interactions provide free energy for activating groups at the catalytic site of an enzyme–substrate complex; however, events at the catalytic site itself can also stabilize the transition state and thereby contribute to the efficiency of enzyme catalysis. In chymotrypsin, the energy gained by the formation of two hydrogen bonds between the activated substrate and backbone nitrogens to the enzyme, as well as by the partial compensation of the buried charge of Asp-102 (Figure 11-1), helps to balance the energy of formation of the strained bond between enzyme and substrate in the tetrahedral intermediate (537).

Chemical Teams as Parts of Enzymes

The effects subsumed in the category of destabilization represent the strained work of the enzyme–substrate complex on the transformable groups in the substrates. However, an enzyme also provides functional teams which elegantly act upon the transformable groups. General acid or general base catalysis is quite common in enzymes; this effect can increase the reaction rate by a factor of 1000. In chymotrypsin it is the charge-relay system which facilitates, by polarizing hydrogen bonds, the proton transfers occurring at several steps of the reaction (Figure 11-1). In other enzymes, such as glutathione reductase, the protein provides active groups (FAD and a redox-active cysteine pair) for the transport of electrons across the enzyme molecule (Figure 11-4).

Separation of the Rate-Accelerating Effects in Enzyme Catalysis

Attempts have been made to quantitate the contributions of diverse effects (such as proximity, orbital steering, destabilization, or general acid catalysis, etc.) to the rate acceleration brought about by an enzyme. This may be possible in rare cases; the example of chymotrypsin, however, illustrates that the "effects" describe the same active site in different ways and cannot really be separated from one another. On the other hand, the concepts developed for the analysis of enzyme action have furthered the understanding of chemical catalysis and have helped in the design of enzyme analogues based on polymers other than polypeptides (745, 750).

11.3 Biological and Medical Aspects of Protein Action and Function

Proteins as Functional Parts of an Organism

Proteins *in situ* must possess a variety of properties which allow them to operate in an environment of high complexity (586). These properties include rapid spontaneous folding, appropriate physico-chemical properties such as solubility and isoelectric point, and good manners [an intracellular protein must not interact with the overwhelming majority of cell constituents (cf. section 10.1)].

Many proteins are present *in situ* in a much higher concentration than is usually needed for their specific function, for example, for the catalysis of a certain chemical reaction. This indicates that proteins may serve as a storage form of amino acids and/or that they may have functions of which we are not aware. For example, the high concentration of creatine kinase in muscle [20 mg per g of tissue (751)] was a mystery until it was found that this protein functions not only as an enzyme but also, in the form of an M-disc constituent (Figure 11-5), as structural support of the contractile apparatus (752).

One difficulty in the study of biological function is that functions may remain unnoticed as long as they operate normally. Consequently the elucidation of function is often based on the study of diseases and mutations and on the effects of poisons, toxins, and drugs. As a rule a defective or disturbed function can be traced to a missing, defective, or disturbed protein structure. This indicates that protein function is the interface at which molecular detail is responsible for the overall performance of an organ or an organism (753).

Development of New Functions on the Basis of Old Proteins

The structure of a protein can be older than its function. Lactalbumin, for example, which was developed as a functional part of the mammary gland some 100 million years ago, existed long before this function was required, probably as a lysozyme-like protein. It is conceivable that any thermodynamically possible function can evolve on the basis of existing protein structures (cf. section 9.3).

A new function can also be developed by using preexisting proteins in a new functional context (754). As illustrated in Table 9-4, the serine protease is the prototype of a functional unit which was repeatedly employed in the development of complex physiological systems. Another example are the proteins actin and myosin which are widely employed for moving cells and their contents (755, 756). In higher organisms, functions as diverse as muscle contraction, release of transmitter compounds in the nervous system, amoeboid movement of white blood cells, or sealing of injured blood vessels by clot retraction are based on actin–myosin systems. Furthermore, the property of actin to form a variety of structures by reversible polymerization plays a role in several of biological processes in which the shape of cells must be stabilized or changed (757).

The fact that diversity in function is not paralleled by diversity in structures (586) has a positive consequence for practical medicine as it shows that molecular physiology may not be as inscrutable as assumed until now. To some extent, however, this positive aspect is muted because structure-specific drugs such as inhibitors of serine proteases or acetylsalicylate may have a variety of effects and thus act nonspecifically (84).

11.4 Skeletal Muscle, a System in Which Protein Action Can be Related to the Overall Performance of an Organ

Three properties of skeletal muscle made this organ a favorite for structural and functional analysis: (a) it is relatively insensitive to manipulations; (b) the typical proteins are present in large amounts; a muscle cell contains 10% myosin, 3.5% actin, 2% tropomyosin plus troponin, and 2% Ca^{2+}-transport-ATPase; and (c) the contractile proteins are arranged in highly ordered geometrical patterns. Techniques such as low angle X-ray diffraction analysis and high resolution electron microscopy, in combination with biochemical and physiological methods, made it possible to relate molecular details to the overall performance of the organ (755, 758).

Recognition, response, and regulation are cell-biological aspects of protein structure. Although a muscle cell is highly specialized it contains most features typical of a living system (Table 11-1). It has the capacity to act and the capacity to control its actions (687). A signal to this system (a nerve impulse) gives rise to a massive response (movement or tension) which is strictly controlled in time, space, and extent, and which is coordinated with functionally related systems, for example, with processes yielding chemical energy. In this context, the roles of these proteins may be subsumed under categories of cell biology, namely, recognition (With which molecules does a protein interact?), response (How does a protein react to a stimulus or signal?), and regulation (How is a protein's activity controlled or which process does a protein control?). However, all these terms describe different aspects of protein structure and they therefore cannot be clearly differentiated.

Structural and Functional Organization of the Contractile Proteins

The central structural and functional unit of a muscle cell is the sarcomere (Figure 11-5), a cylinder of 1.5 μm in diameter and 2 μm in length which contains about 2 × 2000 thin protein filaments and 1000 thick protein filaments. The formation of a thick filament from some 200 myosin molecules is shown on the left-hand side of Figure 11-5. Thin filaments are formed by the association of 2 × 175 monomers of globular actin, 2 × 25 tropomyosin molecules, and 2 × 25 units of the three-component protein

Table 11-1. The Mammalian Muscle Cell as a System in which Signals (Nerve Impulses or Hormone Binding) Lead to Controlled Responses[a]

Stimulus (signal) to the cell membrane	Nerve impulses	Binding of a hormone (epinephrine)
Mediator	Ca^{2+}-ions → Release of Ca^{2+}-ions from the sarcoplasmic reticulum	Ca^{2+}-ions Cyclic AMP → Synthesis of cyclic AMP by adenylate cyclase
Amplification mechanism	None	A cascade of enzyme phosphorylations
Effector	The contractile apparatus; the binding of calcium to troponin C results in cooperative protein interactions which switch on the contractile apparatus	Glycogen phosphorylase; phosphorylation of this enzyme leads to the transition from the b-form to the more active a-form
Effector action (response)	Cross-bridge cycles based on the interactions of myosin, actin, and Mg-ATP	Breakdown of glycogen to glucose-1 phosphate units
Fate of the mediator after stimulation stops	Uptake of Ca^{2+}-ions into the sarcoplasmic reticulum	Enzymatic hydrolysis of cyclic AMP
Time lag between cessation of stimulus and cessation of effector action	Milliseconds	Seconds (The enzyme cascade is dephosphorylated enzymatically)

[a]The main functions of the cell are movement, based on contraction and relaxation, and force development. Auxiliary functions concern the supply of chemical energy for the chemomechanical energy transduction. In this context the mobilization of glycogen, a storage form of energy, is an important step.

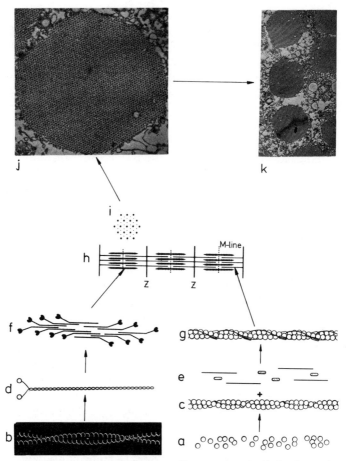

Figure 11-5. Structural organization of the contractile apparatus in skeletal muscle (754). As sketched at the lower left the myosin molecule (b and d) consists of a long coiled-coil α-helix, the tail, and two globular heads. The tail is formed from the heavy chains of the molecule whereas these chains and four light chains contribute to the two heads. Bipolar aggregates of myosin molecules, so-called thick filaments (f), are deposited in the sarcomeres (h). A thick filament is 1200 nm long and 10 nm in diameter (1nm = 10Å); its molecular weight is 10^8. Polymerization of actin (a) leads to long helical filaments (F-actin). Two other proteins, troponin and the cable-like tropomyosin, are bound to F-actin; by this process [an example of self-assembling (754)] the thin filament (1000 nm long and 5 nm in diameter; 2.5×10^7 daltons) is formed (g). The thin filaments attach themselves to the Z-discs which separate sarcomeres from each other. Thin filaments emerging from different sides of a sarcomere have opposite structural polarity. In the sarcomere thin and thick filaments are arranged in hexagonal lattices (i). As revealed in electron micrographs (courtesy of W. Hofmann) the order of thick and thin filaments is almost crystalline (j). Myofibrils (k), the next level of the structural hierarchy, consist of sarcomeres. The sum of myofibrils form the contractile apparatus of a muscle cell.

troponin. The actin filament is usually described as a double stranded helix (section 5.1, page 68) with a helical repeat distance of 360 to 370 Å (755, 758). The sarcomere is one of the few *in vivo* assemblies of macromolecules which have space group symmetry (page 99).

The cross-bridge cycle is the elementary process of muscle contraction. Contraction occurs as the result of a process in which the thin filaments slide past the thick filaments toward the center of the sarcomere (759, 760). The force between the filaments is generated by cross-bridges which project from the thick filaments. The cross-bridges, also called myosin heads, are the biochemically active portions of the myosin molecule which interact with Mg-ATP and actin in so-called cross-bridge cycles (Figure 11-6). One cross-bridge undergoes 10 to 100 cycles per second. As the cycles of the myosin heads (one million per sarcomere!) are not synchronized, a steady force and/or a smooth movement result.

Contraction involves chemo-mechanical energy transduction. When a muscle cell is activated its consumption of chemical energy in the form of ATP rises by one to two orders of magnitude (758). One proposal for the underlying mechanism is the following. In the resting state of muscle actin and myosin do not interact and the myosin head is a slow ATPase, the release of products, in particular of Mg-ADP, being the rate-limiting step. In the working muscle actin activates the ATPase activity of myosin about 100-fold by displacing Mg-ADP. It is unclear as yet whether this is a direct or an allosteric effect of actin on the ADP-binding site of myosin.

The response of the myosin ATPase shows that protein action and control of this action are equally important. The ATPase activity of myosin functions only for muscle contraction; in the absence of actin–myosin interaction, a high ATPase activity would be a waste of chemical energy.

The long half-lives of Mg^{2+}-nucleotide complexes may explain their widespread occurrence as substrates. Kinetic considerations might be relevant to the question of why Mg-ATP and not Ca-ATP is the physiological substrate of the myosin ATPase and numerous other enzymes. The stability constants of Mg-nucleotide complexes and of Ca-nucleotide complexes are nearly identical; however, the half-lives of Ca^{2+}-complexes are 1000 times shorter than the corresponding Mg^{2+}-complexes (762). As the half-lives of Mg-ATP- and Mg-ADP-complexes are in the order of milliseconds, magnesium, but not calcium, recommends itself for inhibiting the ATPase activity of myosin in the resting state and for playing a part in relatively slow conformational changes ($t_{1/2} > 1$ msec) which are steps in the catalytic mechanism of the actin-activated myosin ATPase as well as of other enzymes (758).

Such considerations illustrate that both equilibrium and kinetic properties of complexes and of structural isomers must be known in order to evaluate their significance for protein action. This in turn underlines the importance of techniques (763–766) used to determine the kinetics of fast

$(10^{-7}$ sec) and very fast $(10^{-9}$ sec) reactions whose rates are close to those of diffusion-controlled kinetics $(10^{-10}$ sec).

Analysis of Responses to Nerve Impulses and to Hormone Action in the Muscle Cell

In the resting state of a muscle cell, the attachment sites for cross-bridges at the thin filaments are blocked by tropomyosin molecules (767–769), sturdy two-stranded cables (214). Tropomyosin, the prototype of an α-helical coiled coil (section 5.2, page 79), is 400 Å long and covers seven globular actin units (768, 769). The off-position of tropomyosin appears to be determined by troponin, more precisely speaking by the overall conformation of the Ca^{2+}-free troponin-complex (Figure 11-6).

The chain of events which leads to a displacement of tropomyosin from the off-position starts at the cell membrane. When nerve impulses activate a muscle cell of 1 μl in volume, 10^{14} Ca^{2+}-ions are released from the sarcoplasmic reticulum (770) into the cytoplasm which raises the concentration of free Ca^{2+} by two orders of magnitude to values above 1 μM (Figure 11-7). This leads to a saturation of troponin C, the calcium-sensitive component of the thin filament (771); in actual fact 90% of the 10^{14} ions are bound to troponin C molecules. Ca^{2+} binding leads to a structural transition of the whole troponin complex (772). The altered troponin structure can no longer hold tropomyosin in the off-position. The tropomyosin cable rolls sideways to a new position closer to the center of the groove. In this

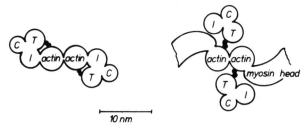

Figure 11-6. Protein interactions in the contractile apparatus [based on Ref. (615)]. To the left a cross-section of the thin filament in the resting state is shown. C, I, and T are constituent proteins of the troponin complex. Tropomyosin blocks the binding sites for myosin heads. When Ca^{2+} is bound to troponin C the thin filament switches to the active state (right). Tropomyosin moves toward the center of the groove of the actin helix, and myosin heads (cross-bridges) can interact with actin. A cycle starts with the attachment of a myosin head to the actin filament (755). Then the myosin tilts over drawing the actin filament toward the center of a sarcomere (in the figure out of the paper plane) by about 80 Å. ATP then dissociates the thin filament from the myosin head, which can start another cross-bridge cycle further down on the actin filament.

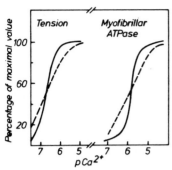

Figure 11-7. Effect of the Ca^{2+}-concentration on muscle tension and on the ATPase activity of myosin (785). pCa^{2+} is an expression analogous to pH. The observed cooperativity (solid lines) is probably based on the fact that it is the binding of the fourth Ca^{2+}-ion to troponin C which induces the structural and enzymatic transitions of the contractile apparatus (615, 773, 785). The dashed curves show the hypothetical activities of the contractile apparatus in the absence of structural cooperativity.

manner one tropomyosin molecule switches on seven actin monomers for their interaction with myosin (767, 769, 785).

Activation of the contractile apparatus by Ca^{2+}-ions is a model of a cooperative process (772, 773). The activation of the contractile apparatus by Ca^{2+} (Figure 11-7) is a suitable process for illustrating cooperative behavior (92, 678, 682, 774) and allosteric control (92, 681, 775, 776) of proteins. Cooperativity refers to the modes in which the components of a protein or a supramolecular assembly, e.g., the thin filament, act together to switch from one stable state, the off-state, to another, the on-state. The typical switching process within structural units, as in the components of troponin, is often a conformational change. This structural isomerization is in turn induced by a specific allosteric ligand, Ca^{2+} in our example. The behavior of tropomyosin illustrates that not all dynamic protein–protein interactions are based on structural isomerization within individual proteins; tropomyosin exerts its effects by moving as a whole structure (Figure 11-6).

Metabolic, nerval, and hormonal stimuli, often amplified by enzyme cascades, correlate energy demand and energy supply. What mechanisms provide the chemical energy for the active contractile machine? An important role in the immediate coupling of energy supply to energy demand is played by ADP, the major product of the myosin ATPase. ADP, as a metabolic signal, triggers ATP production by activating proteins in key positions of the glycolytic pathway and of oxidative phosphorylation (19). The supply of energy for a longer time period is triggered by Ca^{2+}-ions which stimulate the breakdown of glycogen (777). In this case, however, the activation mechanism by calcium involves several steps which contrasts with the one-step activation of the contractile apparatus by this ion (Table 11-1).

Mobilization of glycogen can also be effected by the hormone epinephrine which functions independently of or parallel to the stimulation by nerve impulses. The cascade of membrane and cytoplasmic events that leads from the binding of epinephrine to the phosphorylation of the glucose moieties of glycogen functions as a highly effective kinetic amplification mechanism (Table 11-1). The concentration of the hormone in the extracellular space is approximately $10^{-11}M$; cyclic AMP which is formed from ATP as the first effector of the hormone's action activates, at a concentration of as high as 10^{-8} to 10^{-7} M, a protein kinase. A further 10-fold amplification occurs when the phosphorylase b kinase is stimulated and a final 20- to 50-fold amplification is produced by the conversion of phosphorylase b to phosphorylase a.

By analogy to the events in muscle, it is believed that in other physiological processes cyclic AMP, the second messenger in the action of many hormones, acts by modulating a system of protein phosphorylation–dephosphorylation reactions (777, 778).

The fast dissociation of calcium–protein complexes is a mechanism for switching off protein activities. Finally it is necessary to consider the intracellular events when the stimuli (nerve impulses or hormone binding) cease. The metabolic processes will come to an end with a time lag of seconds after the levels of cyclic AMP or Ca^{2+} have decreased below a critical value (Table 11-1). In contrast, the contractile apparatus is turned off within 10–20 msec by removal of calcium from the thin filament. The speed of this process attests to the efficiency of the sarcoplasmic reticulum as a calcium pump (701). On the other hand, the presence of this complex system which is exclusively devoted to Ca^{2+}-uptake and Ca^{2+}-release suggests that Ca^{2+} must have unique properties for switching functions. This is also indicated by the fact that Ca^{2+}-ions play a role in many other physiological processes as mediators between incoming stimuli and cellular responses (615), such as in light-sensitive cells (779), or in the ovum where a Ca^{2+}-dependent mechanism seems to restrict fertilization to one sperm cell (780). In such processes protein activities must be turned on or off rapidly. This is also true for the contractile apparatus; fast contraction *and* fast relaxation is essential for many movements.

Calcium might have been selected for the control of rapid biological processes on the basis of its coordination chemistry (615), which permits Ca^{2+}-containing complexes to disintegrate under physiological conditions with an acceptable rate constant (781). For a trigger system based on a signal molecule S and a target molecule T, association and dissociation occur according to $S + T \underset{k_{-1}}{\overset{k_{+1}}{\rightleftharpoons}} ST$; k_{+1} and k_{-1} are the rate constants of complex formation and disintegration, respectively. The stability constant (association constant) is $K = k_{+1}/k_{-1} = [ST]/[S] \cdot [T]$. Optimal values for K range from 10^5 to 10^7 M^{-1} in trigger systems which operate within the physiological concentration range (782). k_{+1} as measured for Ca^{2+} substitution in water is 10^8 M^{-1} sec^{-1} and for Mg^{2+} is 10^5 M^{-1} sec^{-1} (762). Now, for $k_{-1} = k_{+1}/K$ the value is up to 1000 per second in the case of Ca^{2+}-complexes, and 1 per second in the case of Mg^{2+}-complexes. The shortest half-lives ($\ln 2/k_{-1}$) of calcium and magnesium complexes under these conditions would be 0.7 msec and 0.7 sec, respectively.

Thus, whenever the required time resolution for protein action is in the order of milliseconds to seconds Ca^{2+} is well suited as an on-*and*-off switch; this is not the case for Mg^{2+} and other compounds which are readily available to biological systems.

As a final point, it should be emphasized that it is probably the coordination chemistry and the biochemistry of calcium, and not protein action, which sets a limit to the speed of control in vertebrate muscles. The fast external muscles of the human eye perform up to 50 contraction–relaxation cycles per second. Insect flight muscles, on the other hand, can oscillate with (audible!) frequencies above 1000 Hz. These muscles exploit structural properties of proteins, such as the response of the myosin ATPase to tension, and not Ca^{2+}, for their rapid oscillations (783, 784).

Summary

Since our ultimate goal is the explanation of biological function any structure analysis of a protein should be followed by studies on its mechanism, its action, and its function. The proteins for which the mechanisms are known best are hemoglobin and chymotrypsin. The catalytic power of chymotrypsin (as well as of other enzymes) has been ascribed to a number of effects which can accelerate chemical reactions in model systems. Such effects include orientation of the substrate(s) at the enzyme's active site in optimal position for reaction, destabilization of the transformable groups in the substrates relative to their transition states, and general base or general acid catalysis. The energy for the rate-accelerating effects is believed to stem from the free energy of enzyme–substrate complex formation.

Many properties of proteins can be explained only in light of their function, that is, their contribution to a larger entity. One of the few systems in which protein functions can be correlated with organ function is skeletal muscle. The muscle cell is activated by nerve impulses (membrane-directed signals). The molecular basis of muscle contraction is the cross-bridge cycle consisting of periodic interactions between myosin, actin, and Mg-ATP. Ca^{2+}-ions and calcium-binding proteins serve as mediators between nerve impulses and the effector proteins. Mediation by Ca^{2+} limits the speed of the response to on/off signals but it prevents contraction without signal efficiently. In contrast, the individual oscillations of insect flight muscle are not controlled by Ca^{2+} or similar low molecular weight compounds but by the contractile proteins themselves. This makes possible very fast periodic contractions which, once they are initiated (by Ca^{2+}-ions), run by themselves. In conclusion, the studies on muscle indicate that protein function is indeed the interface at which molecular detail is responsible for the overall performance of an organism.

Appendix

Statistical Mechanics of the Helix-Coil Transition

A.1 Partition Function

The partition function is the most rigorous description of an ensemble of molecules in equilibrium. The distribution of conformational states of a polypeptide chain of N residues is rigorously described by the partition function

$$Z = \sum_{\text{all states } \kappa} \exp \frac{-E_\kappa}{RT} \tag{A-1}$$

where E_κ is the energy of the whole system, chain plus surrounding solvent, in state κ. The term $\exp \dfrac{-E_\kappa}{RT}$ is called the "statistical weight" of state κ. If the solvent contribution and strain energies in bond angles and lengths are neglected, E_κ reduces to a function of the dihedral angles (ϕ_i, ψ_i) in the backbone and of the dihedral angles $(\chi_i^1, \chi_i^2, \ldots) \equiv \chi_i$ of the side chain of residue i (Figure 2-2). Since conformational space is continuous, the sum becomes an integral

$$Z = \text{const} \cdot \int \exp \frac{-E(\phi_1 \cdots \chi_N)}{RT} \, d\phi_1 d\psi_1 d\chi_1 d\phi_2 d\psi_2 \cdots d\chi_N. \tag{A-2}$$

A.2 Probability of a Residue Being in a Certain Conformation

If the solvent and all residue–residue interactions are neglected, the partition function becomes the product of the statistical weights of the constituting residues. These statistical weights can be split into weights for α_R, α_L, and ϵ regions. If all interactions *between* residues are neglected, the energy can be separated into single residue contributions, and the integral can be factored:

$$E(\phi_1 \cdots \chi_N) = E_1(\phi_1\psi_1\chi_1) + E_2(\phi_2\psi_2\chi_2) + \cdots + E_N(\phi_N\psi_N\chi_N)$$

$$Z = const \cdot \int \exp\frac{-E_1}{RT} d\phi_1 d\psi_1 d\chi_1 \cdot \ldots \cdot \int \exp\frac{-E_N}{RT} d\phi_N d\psi_N d\chi_N. \tag{A-3}$$

The resulting single residue integrals can be approximated by restricting the integration area in $(\phi_i\psi_i)$ to those regions in which no steric hindrance occurs, i.e., where E_i is not too large. As shown in Figure 2-3b, these are the regions of right-handed and left-handed α-helix as well as extended chain, designated as α_R, α_L, and ϵ.

$$\int \exp\frac{-E_i}{RT} d\phi_i d\psi_i d\chi_i = \int_{\alpha_R} \exp\frac{-E_i}{RT} d\phi_i d\psi_i d\chi_i + \int_{\alpha_L} \cdots + \int_{\epsilon} \cdots$$

$$= z_i^{\alpha_R} + z_i^{\alpha_L} + z_i^\epsilon.$$

$$Z = const \prod_{i=1}^{N} (z_i^{\alpha_R} + z_i^{\alpha_L} + z_i^\epsilon). \tag{A-4}$$

Here $z_i^{\alpha_R}$, $z_i^{\alpha_L}$, z_i^ϵ are the statistical weights for the α_R-, α_L-, ϵ-regions, respectively. They are specific for a given residue type. Thus, $z_i^{\alpha_R}$, $z_i^{\alpha_L}$, z_i^ϵ are the probabilities that the backbone moiety of residue i assumes any of the conformations in the α_R-, α_L-, ϵ-regions, respectively. In other words $z_i^{\alpha_R}$, $z_i^{\alpha_L}$, z_i^ϵ reflect the propensity of residue i to be in α_R-, α_L-, ϵ-conformation, respectively. These probabilities depend on the energy E_i in this region and on the area of this region. Thus, in a comparison of α_R- and ϵ-probabilities, a lower energy in α_R (better binding) may well be compensated by the larger area of ϵ (Figure 2-3).

Empirical statistical weights can be derived from observed frequencies. In principle these propensities can be calculated from energy maps as shown in Figure 2-5 for each type of residue. However, these maps have to be extended to include the dependence on side chain conformation χ_i. Moreover, in order to obtain reasonable results the solvent has to be taken in account, which of course is very difficult. On the other hand approximations to these probabilities $z_i^{\alpha_R}$, $z_i^{\alpha_L}$, z_i^ϵ can be obtained by equating them to the frequency of states α_R, α_L, ϵ observed in globular proteins of known structure for the residue type in question. This procedure has been implicitly adopted in most prediction methods.

A general empirical energy map was derived from frequencies. Using the approximation of Eq. (A-3), a general empirical energy map $E(\phi, \psi)$ can also be obtained from the observed (ϕ, ψ)-angles of all residue types in globular proteins. For this purpose the frequencies of residues (irrespective of residue type) found in a given (ϕ, ψ)-area (see Figure 2-4) are equated to the statistical weights $\exp\dfrac{-E(\phi, \psi)}{RT}$ in this area. Consequently, the statistical weight and therefore E can be derived as a function of the (ϕ, ψ)-angles. The empirical $E(\phi, \psi)$ fits the calculated one (Figure 2-5) quite well (786).

A.3 Ising Model

A linear array with interactions between nearest neighbors was first described by Ising. The simplifications of the partition function beyond Eq. (A-2) are based on the assumption of no interaction between different residues. This is certainly wrong for α-helices, because there are hydrogen bonds between residues i and $i + 3$ (Figure 5-4). Moreover, helix–coil transitions of synthetic polypeptides (328,787) have a sigmoidal shape, indicating cooperativity. To allow for this fact, other approximations of the partition function are necessary. For a similar case, namely, a linear array of ferromagnets with nearest neighbor interactions, such an approximation was described by Ising (788).

A.4 Zimm–Bragg Model for Helix–Coil Transition

Basic Formula

A nearest neighbor interaction term is reintroduced in the simplified formula. Zimm and Bragg (789) adapted the Ising model to the helix–coil transition of homopolymeric peptide chains. For this purpose they divided conformation space into two regions, α_R or "α" and non-α_R or "coil." Furthermore, they used the approximation Eq. (A-4) based on no residue–residue interaction and subsequently introduced an interaction term between nearest neighbors. For a chain of N residues of a given type Eq. (A-4) becomes

$$Z = \text{const}\,(z_1^\alpha + z_1^c) \cdot (z_2^\alpha + z_2^c) \cdots (z_N^\alpha + z_N^c).$$

After multiplication one gets a sum of 2^N terms

$$Z = \text{const} \cdot \sum_{\substack{\text{all } 2^N \text{ chain} \\ \text{conformations}}} z_1^\alpha \cdot z_2^\alpha \cdot z_3^\alpha \cdot z_4^c \cdot z_5^c \cdot z_6^\alpha \cdot z_7^c \cdots z_N^\alpha. \qquad (A\text{-}5)$$

For each helix–coil and coil–helix junction the probability or the statistical weight is reduced by applying a factor $\sqrt{\sigma}$, $\sigma \ll 1$:

$$Z = \text{const} \cdot \sum_{\substack{\text{all } 2^N \text{ chain} \\ \text{conformations}}} z_1^\alpha \cdot z_2^\alpha \cdot z_3^\alpha \cdot \sqrt{\sigma} \cdot z_4^c \cdot z_5^c \cdot \sqrt{\sigma} \cdot z_6^\alpha \cdot \sqrt{\sigma} \cdot z_7^c \cdots z_N^\alpha.$$

This can be converted to

$$Z = \text{const} \cdot \sum_{\substack{\text{all } 2^N \text{ chain} \\ \text{conformations}}} z_1^\alpha \cdot z_2^\alpha \cdot z_3^\alpha \cdot \sigma \cdot z_4^c \cdot z_5^c \cdot z_6^\alpha \cdot \sigma \cdot z_7^c \cdots z_N^\alpha \qquad (A\text{-}6)$$

by pairing $\sqrt{\sigma}$ and applying it only to helix–coil junctions. This is possible because there are as many α-segments as coil-segments, if the fact that one type may exceed the other by one segment is neglected.

This σ is a penalty term for too many junctions. Consequently, it favors long segments of identical conformation. This is an alternative description of "cooperativity between residues in identical conformation" or, more particularly, "cooperativity between residues in helical conformation." It can also be described as favoring "nuclei of coil or helix conformation." Therefore, σ is called a "cooperativity or nucleation parameter."

Matrix Representation

The partition function can be written as the result of a matrix multiplication. As proposed by Kramers and Wannier (790), Eq. (A-6) can be written in a compact matrix form, which enables a simple evaluation of Z.

$$Z = \text{const} \cdot (1,\,1) \cdot \begin{pmatrix} z_1^c & 0 \\ 0 & z_1^\alpha \end{pmatrix} \cdot \begin{pmatrix} z_2^c & z_2^\alpha \cdot \sigma \\ z_2^c & z_2^\alpha \end{pmatrix} \cdot \ldots \cdot \begin{pmatrix} z_N^c & z_N^\alpha \cdot \sigma \\ z_N^c & z_N^\alpha \end{pmatrix} \cdot \begin{pmatrix} 1 \\ 1 \end{pmatrix}. \quad \text{(A-7)}$$

Let us check the correspondence of this expression with Eq. (A-7) for small N:

$$\begin{aligned}
N = 1 \quad & Z = \text{const} \cdot (z_1^c + z_1^\alpha) \\
N = 2 \quad & Z = \text{const} \cdot (z_1^c \cdot z_2^c + z_1^c \cdot z_2^\alpha + z_1^\alpha \cdot \sigma \cdot z_2^c + z_1^\alpha \cdot z_2^\alpha) \\
N = 3 \quad & Z = \text{const} \cdot (z_1^c \cdot z_2^c \cdot z_3^c + z_1^c \cdot z_2^\alpha \cdot \sigma \cdot z_3^c + z_1^\alpha \cdot \sigma \cdot z_2^c \cdot z_3^c \\
& \quad + z_1^\alpha \cdot z_2^\alpha \cdot \sigma \cdot z_3^c + z_1^c \cdot z_2^c \cdot z_3^\alpha + z_1^c \cdot z_2^\alpha \cdot z_3^\alpha \\
& \quad + z_1^\alpha \cdot \sigma \cdot z_2^c \cdot z_3^\alpha + z_1^\alpha \cdot z_2^\alpha \cdot z_3^\alpha).
\end{aligned}$$

For larger N the relation can be proved by induction.

The simple case of a homopolymer is essentially described by a matrix to the power of N. This allows a comprehensive description in terms of the eigenvalues. When Eq. (A-7) is applied to *homo*polymers, all z_i^α and all z_i^c are equal. As a further simplification the relative statistical weight s is defined by

$$s_i = z_i^\alpha / z_i^c \equiv s. \quad \text{(A-8)}$$

Include $(z_i^c)^N$ in the constant, and omit this constant completely, because only the change of Z during the helix–coil transition and not its absolute value is of interest.

$$Z = (1,\,1) \cdot \begin{pmatrix} 1 & 0 \\ 0 & s \end{pmatrix} \cdot \begin{pmatrix} 1 & s\sigma \\ 1 & s \end{pmatrix}^{N-1} \begin{pmatrix} 1 \\ 1 \end{pmatrix}. \quad \text{(A-9)}$$

By diagonalizing the matrix, this can be converted to

$$Z = (1,\,1) \begin{pmatrix} 1 & 0 \\ 0 & s \end{pmatrix} A(A^{-1}UA)^{N-1}A^{-1} \begin{pmatrix} 1 \\ 1 \end{pmatrix} \quad \text{(A-10)}$$

$$\text{with } U = \begin{pmatrix} 1 & s\sigma \\ 1 & s \end{pmatrix},$$

$$A^{-1}UA = \begin{pmatrix} \lambda_1 & 0 \\ 0 & \lambda_2 \end{pmatrix},$$

$$\lambda_{1(2)} = \frac{1}{2}(1 + s_{(\pm)} \sqrt{(1-s)^2 + 4s\sigma}),$$

$$A = \begin{pmatrix} 1 - \lambda_2 & 1 - \lambda_1 \\ 1 & 1 \end{pmatrix},$$

$$A^{-1} = \begin{pmatrix} 1 & \lambda_1 - 1 \\ -1 & 1 - \lambda_2 \end{pmatrix},$$

$$(A^{-1}UA)^{N-1} = \begin{pmatrix} \lambda_1^{N-1} & 0 \\ 0 & \lambda_2^{N-1} \end{pmatrix},$$

so that Z can be given explicitly as a function of the eigenvalues.

$$Z = \lambda_1^N \frac{\lambda_1}{\lambda_1 - \lambda_2} - \lambda_2^N \frac{\lambda_2}{\lambda_1 - \lambda_2}. \tag{A-11}$$

As $\lambda_1 > 1 > \lambda_2$ for all s and σ, the partition function can be approximated for large N by

$$Z = \lambda_1^N. \tag{A-12}$$

Fraction of Helical Residues in a Homopolymer

The helix content varies with relative statistical weight and thus with temperature. Z can be used to determine the average number of helical residues $<n>$ as a function of s and σ. Using Eq. (A-7) and introducing Eq. (A-8) for chains containing ν helical segments:

$$Z = \sum_{\substack{\text{all } 2^N \text{ chain} \\ \text{conformations}}} s^n \sigma^\nu \tag{A-13}$$

$$<n> = \sum_{\substack{\text{all } 2^N \text{ chain} \\ \text{conformations}}} n \cdot \frac{s^n \sigma^\nu}{Z}, \tag{A-14}$$

i.e., the number of helical residues in each chain conformation is weighted with the relative statistical weight of this conformation, which is normalized by $1/Z$.

$$<n> = \frac{s}{Z} \frac{\partial Z}{\partial s} = \frac{\partial \ln Z}{\partial \ln s} \simeq \frac{\partial \ln \lambda_1^N}{\partial \ln s} = \frac{N \cdot s}{\lambda_1} \cdot \frac{\partial \lambda_1}{\partial s} = N \frac{\lambda_1 - 1}{\lambda_1 - \lambda_2}. \tag{A-15}$$

$<n>$ is the experimentally observed variable during helix–coil transitions. Theoretical curves for $<n>$ as a function of s are plotted for several σ-values in Figure A-1a. At $s = 1$ half the chain residues are helical for all σ-

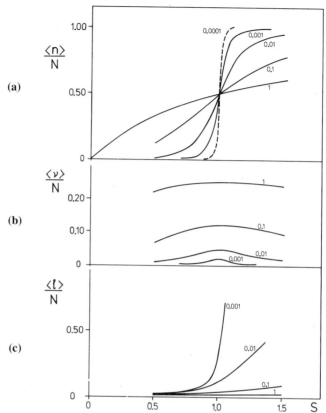

Figure A-1. Zimm–Bragg model of helix–coil transition in a polypeptide chain of length N. σ-Values are written onto the curves. (a) Average helix fraction as a function of the relative statistical weight s. (b) Average number of helical segments related to N as a function of s. (c) Average length of helical segments related to N as a function of s.

values. The smaller σ is the steeper is the transition at $s = 1$, i.e., the stronger is the cooperativity.

Number and Length of Helical Segments

The largest number of helical segments is found at the transition point. Equation (A-13) introduced ν as the number of helical segments in a given chain conformation. The average number of helical segments $<\nu>$ is given by

$$<\nu> = \sum_{\text{all } 2^N \text{ chain conformations}} \nu \, \frac{s^n \sigma^\nu}{Z} \tag{A-16}$$

$$= \frac{\partial \ln Z}{\partial \ln \sigma} \simeq \frac{\partial \ln \lambda_1^N}{\partial \ln \sigma} = N \frac{(\lambda_1 - 1)(1 - \lambda_2)}{\lambda_1(\lambda_1 - \lambda_2)}.$$

This function is shown in Figure A-1b; it assumes its maximum at $s = 1$. The number of segments decreases with decreasing σ. The average length of a helical segment $<l>$ is derived from

$$<l> = \frac{<n>}{<\nu>} = \frac{\lambda_1}{1 - \lambda_2}$$

$$<l>_{s=1} = \frac{1 + \sqrt{\sigma}}{\sqrt{\sigma}} \simeq \frac{1}{\sqrt{\sigma}}.$$

(A-17)

As shown in Figure A-1c, the helix length increases monotonically with s.

It should be mentioned that the splitting of the chain into several segments of helical and coil conformation is in accordance with a more general proof given by Landau and Lifshitz (791). The authors state that in one dimension a separation into two distinct phases is impossible, unless the boundary tension is infinite. Infinite boundary tensions correspond to $\sigma = 0$, where the Zimm and Bragg model also results in a splitting into two phases.

Relation between s and Temperature

Around the transition point the relative statistical weight has a linear relationship to the temperature. This relation is important because helix–coil transitions are often observed by temperature variation. It can be derived by returning to Eq. (A-4) but restricting it to two states

$$
\begin{aligned}
\frac{1}{s}\frac{\partial s}{\partial T} &= \frac{\partial \ln s}{\partial T} \\
&= \frac{\partial \ln z^\alpha}{\partial T} - \frac{\partial \ln z^c}{\partial T} \\
&= \frac{1}{z^\alpha}\frac{\partial z^\alpha}{\partial T} - \frac{1}{z^c}\frac{\partial z^c}{\partial T} \\
&= \frac{\displaystyle\int_\alpha \frac{E}{RT^2}\exp\left(\frac{-E}{RT}\right)d\phi\, d\psi\, d\chi}{\displaystyle\int_\alpha \exp\left(\frac{-E}{RT}\right)d\phi\, d\psi\, d\chi} - \frac{\displaystyle\int_{coil}\frac{E}{RT^2}\exp\left(\frac{-E}{RT}\right)d\phi\, d\psi\, d\chi}{\displaystyle\int_{coil}\exp\left(\frac{-E}{RT}\right)d\phi\, d\psi\, d\chi} \\
&= \frac{<E>_{\alpha-region}}{RT^2} - \frac{<E>_{coil}}{RT^2} \equiv \frac{\Delta H}{RT^2}.
\end{aligned}
$$

(A-18)

Thus, in the transition region ($s \simeq 1$) the relative change of s with T is roughly constant, i.e., the change of s is approximately proportional to the change of temperature (Figure A-2).

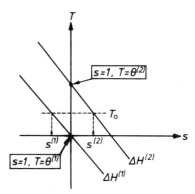

Figure A-2. Relation between relative statistical weight s, temperature T, and enthalpy difference ΔH around the transition point ($s = 1$, $T = \theta$) for two different polymers. This is a graph of Eqs. (A-18) and (A-19). Both $\Delta H^{(1)}$ and $\Delta H^{(2)}$ are negative and $\Delta H^{(2)} < \Delta H^{(1)}$. The transition temperature difference $\theta^{(2)} - \theta^{(1)}$ is proportional to $\Delta H^{(2)} - \Delta H^{(1)}$. For a given temperature T_0 both polymers have different relative statistical weights for helix, $s^{(1)}$ and $s^{(2)}$. These s-values are the helix propensities at this temperature.

The Transition Temperature

The transition temperature is roughly proportional to the change in enthalpy. The transition occurs at $s = 1$, i.e., at $z^\alpha = z^c$ or

$$\int_\alpha \exp\left(\frac{-E}{RT}\right) d\phi\, d\psi\, d\chi = \int_{coil} \exp\left(\frac{-E}{RT}\right) d\phi\, d\psi\, d\chi.$$

Assuming that $E = E_\alpha$ = constant over the α-region of size Ω_α and $E = E_c$ = constant over the coil-region of size Ω_c, then

$$\Omega_\alpha \exp\left(\frac{-E_\alpha}{RT}\right) = \Omega_c \exp\left(\frac{-E_c}{RT}\right)$$

$$T_{s=1} \equiv \theta = \frac{-(E_\alpha - E_c)}{R \ln \dfrac{\Omega_c}{\Omega_\alpha}} = \frac{-\Delta H}{R \ln \dfrac{\Omega_c}{\Omega_\alpha}}. \qquad (A\text{-}19)$$

Transitions occur only if the state with lower energy occupies the smaller conformational region. For $\Omega_c > \Omega_\alpha$ the transition temperature becomes negative if the energy of the α-conformation is less favorable than the energy of the coil-conformation, $E_\alpha > E_c$. Therefore, transitions occur only if the state with the lower energy E (stronger binding) occupies the smaller conformational region Ω. In our case with $\Omega_c > \Omega_\alpha$ (see Figure 2-3), the helical conformation is assumed only if it is energetically favorable. The transition direction is such that at temperatures below θ the state with the favorable energy wins, and at temperatures above θ the state with the larger conformational region wins (see Figure A-2). The transition temperature θ is higher, the smaller the difference in conformational region and the more energetically favorable the helix conformation.

A.5 Comparison With Experimental Data

Helix–Coil Transition Curves and Zimm–Bragg Model

Experimental helix–coil transition curves can be fitted using the Zimm–Bragg model. In order to confirm the Zimm–Bragg model, the theoretical helix–coil transition curves [Eq. (A-15)] shown in Figure A-1a should be compared to experimental ones. When measuring helix content $<n>$ as a function of the temperature T, sigmoidal curves similar to those of Figure A-1a are obtained. For a comparison temperature T has to be converted to relative statistical weight s. In the transition region T is roughly a linear function of s (section A.4, page 258 and Figure A-2). According to Eq. (A-18) T can be replaced by s if the enthalpy difference of the transition, ΔH, is known. The resulting experimental transition curve, $<n>$ as a function of s, can then be fitted by a theoretical curve with appropriate nucleation parameter σ. In most cases the fit is reasonably good. Thus, the Zimm–Bragg model is essentially confirmed.

Approximate σ-Values for Synthetic Polymers

For a temperature-induced helix–coil transition the nucleation parameter can be estimated from the slope at the inflection point. As shown above a temperature-induced experimental helix–coil transition curve can be fitted by a theoretical curve if ΔH is known. This fit yields the nucleation parameter. However, in many cases ΔH is not known. Nevertheless, a good estimate for σ can be obtained. From Eq. (A-15) using Eq. (A-10) the slope of the transition curve at the inflection point ($s = 1$) is found to be

$$\frac{1}{N} \left.\frac{\partial <n>}{\partial s}\right|_{s=1} = \frac{1}{4s\sigma} = \frac{1}{4\sigma}. \tag{A-20}$$

ΔH is always of the order 1 kcal/mol. This value can be used to convert the observed $\partial<n>/\partial T$ to $\partial<n>/\partial s$ (Figure A-2), which in turn yields σ according to Eq. (A-20). The observed σ-values for synthetic polymers are of the order 10^{-4}. This means that each junction between segments of helix and coil reduces the statistical weight of the corresponding chain conformation by a factor of 100 [Eq. (A-6)]. Furthermore, at the transition temperature the average length of helical segments is 100 residues [Eq. (A-17)]. Note that these values are derived with the assumption that N is very large.

Transition Temperature and Helix Propensity

The transition temperatures of synthetic polymers can be converted to relative statistical weights at a given temperature. The relation between s, T, and ΔH around the transition point is shown in Figure A-2. In a first approximation it is linear (section A.4, page 258). These relations are now compared for two different homopolymers with enthalpy differences $\Delta H^{(1)}$ and $\Delta H^{(2)}$ and transition temperatures $\theta^{(1)}$ and $\theta^{(2)}$. If the ΔH- and θ-values are reasonably close both linear relationships can be combined in the same graph (Figure A-2). This graph can be used to extract at a given temperature T_0 relative statistical weights $s^{(1)}$ and $s^{(2)}$ for both types of homopolymers. These statistical weights are the helix propensities [Eqs. (A-8) and (A-4)] at this particular temperature.

Usually, ΔH is not known. However, it can be estimated from the respective θ using Eq. (A-19). Thus, whenever θ can be measured for a homopolymer of a given residue type,[1] this θ can be converted to an approximate helix propensity.

Solvent Contribution

After Eq. (A-1) the solvent in all formulae was neglected for the sake of clarity. To be correct the translational and rotational parameters of all surrounding solvent molecules should be added to the chain parameters (ϕ, ψ, χ). However, since this is virtually impossible, the solvent contribution is always considered separately.

Solvent changes can induce helix–coil transitions without temperature changes. Furthermore, in a given solvent, ΔH of Eq. (A-19) can even be positive (789) and $\Omega_c < \Omega_\alpha$ so that the helix forms at high temperatures. Here, of course, Ω_c and Ω_α include the solvent contribution and can no longer be derived from Figure 2-3.

---

[1] For a number of residue types homopolymers do not form helices readily. In such cases mixed polymers with the residue type in question as a "guest" and a good helix former (e.g., poly-γ-benzyl L-glutamate) as the "host" can be used. Analysis of the θ-change then yields the transition temperature of the "guest" residue type.

References

1. W. Kühne (1876), Über das Verhalten verschiedener organisirter und sogenannter ungeformter Fermente. Über das Trypsin (Enzym des Pankreas). *Verhandlungen des Heidelb. Naturhist.-Med. Vereins N.S.* **I**, 2–5. Reproduced in *FEBS Lett.* **62**, E3–E7 (1976).
2. F. Hoppe-Seyler (1864), Über die chemischen und optischen Eigenschaften des Blutfarbstoffs. *Virchows Arch.* **29**, 233–235.
3. M. Calvin (1969), *Chemical Evolution.* Clarendon Press, Oxford.
4. F. H. C. Crick (1968), The origin of the genetic code. *J. Mol. Biol.* **38**, 367–379.
5. L. E. Orgel (1968), Evolution of the genetic apparatus. *J. Mol. Biol.* **38**, 381–393.
6. M. Eigen (1971), Selforganization of matter and the evolution of biological macromolecules. *Naturwissenschaften* **58**, 465–523.
7. G. W. Tischendorf, H. Zeichardt, and G. Stöffler (1975), Architecture of the *Escherichia coli* ribosome as determined by immune electron microscopy. *Proc. Natl. Acad. Sci. USA* **72**, 4820–4824.
8. S. L. Miller (1954), Production of some organic compounds under possible primitive Earth conditions. *J. Amer. Chem. Soc.* **77**, 2351–2361.
9. W. A. Bonner, M. A. van Dort, and M. R. Yearian (1975), Asymmetric degradation of DL-leucin with longitudinally polarised electrons. *Nature* **258**, 419–421.
10. J. Tze-Fei Wong (1975), A co-evolution theory of the genetic code. *Proc. Natl. Acad. Sci. USA* **72**, 1909–1912.
11. K. E. Richards, H. Guilley, G. Jonard, and L. Hirth (1974), A specifically encapsidated fragment from the RNA of tobacco mosaic virus: Sequence homology with the coat protein cistron. *FEBS Lett.* **43**, 31–32.
12. F. Sanger, G. M. Air, B. G. Barrell, N. L. Brown, A. R. Coulson, J. C. Fiddes, C. A. Hutchison III, P. M. Slocombe, and M. Smith (1977), Nucleotide sequence of bacteriophage ϕX174 DNA. *Nature* **265**, 687–695.
13. W. Fiers, R. Contreras, F. Duerinck, G. Haegeman, D. Inserentant, J.

Merregaert, W. Min Jou, F. Molemans, A. Raeymaekers, A. Van den Berghe, G. Volckaert, and M. Ysebaert (1976), Complete nucleotide sequence of bacteriophage MS2 RNA: Primary and secondary structure of the replicase gene. *Nature* **260**, 500–507.

14. B. G. Barrell, G. M. Air, and C. A. Hutchison III (1976), Overlapping genes in bacteriophage φX174. *Nature* **264**, 34–41.

15. W. Kauzmann (1959), Some factors in the interpretation of protein denaturation. *Adv. Prot. Chem.* **14**, 1–63.

16. Y. Nozaki and C. Tanford (1971), The solubility of amino acids and two glycine peptides in aqueous ethanol and dioxane solutions. *J. Biol. Chem.* **246**, 2211–2217.

17. C. Chothia (1974), Hydrophobic bonding and accessible surface area in proteins. *Nature* **248**, 338–339.

18. J. J. Birktoft and D. M. Blow (1972), Structure of crystalline α-chymotrypsin. *J. Mol. Biol.* **68**, 187–240.

19. A. L. Lehninger (1975), *Biochemistry,* 2nd ed. Worth, New York.

20. M. O. Dayhoff (1972), *Atlas of Protein Sequence and Structure.* Natl. Biom. Res. Foundation, Washington D.C.

21. IUPAC–IUB Commission on Biochemical Nomenclature 1969 (1970), Abbreviations and symbols for the description of the conformation of polypeptide chains. *Biochemistry* **9**, 3471–3479.

22. J. D. Watson (1976), *Molecular Biology of the Gene,* 3rd ed. Benjamin, Menlo Park, California.

23. L. Pauling (1960), *The Nature of the Chemical Bond,* p. 281. Cornell University Press, Ithaca, New York.

24. A. L. Lehninger (1971), *Bioenergetics,* p. 173. W. A. Benjamin, Menlo Park, California.

25. L. Pauling, R. B. Corey, and H. R. Branson (1951), The Structure of proteins: Two hydrogen-bonded helical configurations of the polypeptide chain. *Proc. Nat. Acad. Sci. USA* **37**, 205–211.

26. Tables of Interatomic Distances and Configuration in Molecules and Ions (1965), Suppl. Spec. Publ. No. 18, The Chemical Society, London.

27. J. T. Edsall, P. J. Flory, J. C. Kendrew, A. M. Liquori, G. Némethy, G. N. Ramachandran, and H. A. Scheraga (1966), A proposal of standard conventions and nomenclature for the description of polypeptide conformations. *J. Mol. Biol.* **15**, 399–407.

28. G. N. Ramachandran, C. Ramakrishnan, and V. Sasisekharan (1963), Stereochemistry of polypeptide chain configurations. *J. Mol. Biol.* **7**, 95–99.

29. G. N. Ramachandran and V. Sasisekharan (1968), Conformation of polypeptides and proteins. *Adv. Prot. Chem.* **23**, 283–437.

30. M. Levitt (1976), A simplified representation of protein conformations for rapid simulation of protein folding. *J. Mol. Biol.* **104**, 59–116.

31. A. W. Burgess, P. K. Ponnuswamy, and H. A. Scheraga (1974), Analysis of conformations of amino acid residues and prediction of backbone topography in proteins. *Israel J. Chem.* **12**, 239–286.

32. D. A. Brant, W. G. Miller, and P. J. Flory (1967), Conformational energy estimates for statistically coiling polypeptide chains. *J. Mol. Biol.* **23**, 47–65.

33. G. Némethy and H. A. Scheraga (1977), Protein folding. *Quart. Rev. Biophys.* **10**, 239–352.

34. G. N. Ramachandran, C. M. Venkatachalam, and S. Krimm (1966), Stereo-chemical criteria for polypeptide and protein chain conformations. *Biophys. J.* **6**, 849–872.

35. A. Gieren, B. Dederer, and F. Schanda (1978), Some new aspects concerning conformation of polypeptide chains in proteins. *Z. Naturforsch.*, in press.

36. F. K. Winkler and J. D. Dunitz (1971), The non-polar amide group. *J. Mol. Biol.* **59**, 169–182.

37. T. Wieland and C. Birr (1976), Homodetic cyclic peptides. *International Review of Science, Organic Chemistry Series Two*, Vol. 6, pp. 183–218. Butterworth, London.

38. I. L. Karle, J. Karle, T. Wieland, W. Burgermeister, H. Faulstich, and B. Witkop (1973), Conformations of the Li-antamanide complex and Na-/Phe^4, Val^6/antamanide complex in the crystalline state. *Proc. Nat. Acad. Sci. USA* **70**, 1836–1840.

39. H. W. Wyckoff, D. Tsernoglou, A. W. Hanson, J. R. Knox, B. Lee, and F. M. Richards (1970), The three-dimensional structure of ribonuclease-S. *J. Biol. Chem.* **245**, 305–328.

40. C. Schubert-Wright, R. A. Alden, and J. Kraut (1969), Structure of subtilisin BPN' at 2.5 Å resolution. *Nature* **221**, 235–242.

41. R. S. Cahn, C. K. Ingold, and V. Prelog (1956), The specification of asymmetric configuration in organic chemistry. *Experentia* **12**, 81–94.

42. O. Epp, E. E. Lattman, M. Schiffer, R. Huber, and W. Palm (1975), The molecular structure of a dimer composed of the variable portions of the Bence–Jones protein REI refined at 2.0 Å resolution. *Biochemistry* **14**, 4943–4952.

43. F. A. Quiocho and W. N. Lipscomb (1971), Carboxypeptidase A: A protein and an enzyme. *Adv. Prot. Chem.* **25**, 1–78.

44. E. Fischer (1894), Einfluß der Configuration auf die Wirkung der Enzyme. *Chem. Ber.* **27**, 2985–2993.

45. F. London (1930), Über einige Eigenschaften und Anwendungen der Molekularkräfte. *Z. Phys. Chem. Abt. B* **11**, 222–251.

46. J. C. Slater and J. G. Kirkwood (1931), The van der Waals forces in gases. *Phys. Rev.* **37**, 682–697.

47. J. E. Jones (1924), On the determination of molecular fields. *Proc. Roy. Soc. London,* Ser. **A 106**, 441–462.

48. J. C. Slater (1928), The normal state of helium. *Phys. Rev.* **32**, 349–360.

49. R. A. Buckingham (1938), The classical equation of state of gaseous helium, neon and argon. *Proc. Roy. Soc. London,* Ser. **A 168**, 264–283.

50. *Handbook of Chemistry and Physics* (1959) 41st ed., p. 2523. Chemical Rubber, Cleveland, Ohio.

51. D. A. Brant (1972), Conformational analysis of biopolymers: Conformational energy calculations. *Annu. Rev. Biophys. Bioeng.* **1**, 369–408.

52. F. A. Momany, R. F. McGuire, A. W. Burgess, and H. A. Scheraga (1975), Energy parameters in polypeptides. VII. Geometric parameters, partial atomic charges, nonbonded interactions, and intrinsic torsional potentials for the naturally occurring amino acids. *J. Phys. Chem.* **79**, 2361–2381.

53. A. T. Hagler and A. Lapiccirela (1976), Spatial electron distribution and population analysis of amides, carboxylic acids, and peptides, and their relation to empirical potential functions. *Biopolymers* **15**, 1167–1200.

54. A. T. Hagler, E. Huler, and S. Lifson (1974), Energy functions for peptides

and proteins. I. Derivation of a consistent force field including the hydrogen bond from amide crystals. *J. Am. Chem. Soc.* **96**, 5319–5327.

55. M. Levitt (1974), Energy refinement of hen egg-white lysozyme. *J. Mol. Biol.* **82**, 393–420.
56. B. Jeziorski and M. van Hermet (1976), Variation–perturbation treatment of the hydrogen bond between water molecules. *Mol. Phys.* **31**, 713–729.
57. L. Pauling (1960), *The Nature of the Chemical Bond*, p. 461. Cornell University Press, Ithaca, New York.
58. G. C. Pimentel and A. L. McClellan (1960), *The Hydrogen Bond*, pp. 242, 282–288. Freeman & Co, London.
59. P. C. Moews and R. H. Kretsinger (1975), Refinement of the structure of carp muscle calcium-binding parvalbumin by model building and difference Fourier analysis. *J. Mol. Biol.* **91**, 201–228.
60. M. F. Perutz (1970), Stereochemistry of cooperative effects of haemoglobin. *Nature* **228**, 726–739.
61. R. H. Kretsinger and C. E. Nockolds (1973), Carp muscle calcium-binding protein. *J. Biol. Chem.* **248**, 3313–3326.
62. W. Bode and P. Schwager (1975), The refined crystal structure of bovine β-trypsin at 1.8 Å resolution. *J. Mol. Biol.* **98**, 693–717.
63. F. M. Richards (1974), The interpretation of protein structures: Total volume, group volume distributions and packing density. *J. Mol. Biol.* **82**, 1–14.
64. F. M. Richards (1977), Areas, volumes, packing and protein structure. *Annu. Rev. Biophys. Bioeng.* **6**, 151–176.
65. J. F. Brandts, R. J. Oliveira, and C. Westort (1970), Thermodynamics of protein denaturation. Effect of pressure on the denaturation of ribonuclease A. *Biochemistry* **9**, 1038–1047.
66. d'Ans-Lax (1967), *Taschenbuch für Chemiker und Physiker,* Vol. 1. Springer-Verlag, Berlin.
67. H.W. Wyckoff (1968), Compensating nature of substitutions in pancreatic ribonucleases. *Brookhaven Symp. Biol.* **21**, 252–258.
68. J.L. Finney (1975), Volume occupation, environment and accessibility in proteins. The problem of the protein surface. *J. Mol. Biol.* **96**, 721–732.
69. F.A. Momany, L.M. Carruthers, R.F. McGuire, and H.A. Scheraga (1974), Intermolecular potentials from crystal data. III. Determination of empirical potentials and application to the packing configurations and lattice energies in crystals of hydrocarbons, carboxilic acids, amines, and amides. *J. Phys. Chem.* **78**, 1595–1620.
70. S. Lifson and A. Warshel (1968), Consistent force field for calculations of conformations, vibrational spectra, and enthalpies of cycloalkane and *n*-alkane molecules. *J. Chem. Phys.* **49**, 5116–5129.
71. A. Warshel, M. Levitt, and S. Lifson (1970), Consistent force field for calculation of vibrational spectra and conformations of some amides and lactam rings. *J. Mol. Spectrosc.* **33**, 84–99.
72. E. Adman, K. D. Watenpaugh, and L. H. Jensen (1975), NH–S hydrogen bonds in *Peptococcus aerogenes* ferredoxin, *Clostridium pasteurianum* rubredoxin, and chromatium high potential iron protein. *Proc. Nat. Acad. Sci. USA* **72**, 4854–4858.
73. A. Bondi (1964), Van der Waals volumes and radii. *J. Phys. Chem.* **68**, 441–451.

74. S. Benzer (1957), (W. D. McElroy and B. Glass, Eds.), *The Chemical Basis of Heredity*, pp. 70–93. John Hopkins University Press, Baltimore.

75. G. M. Edelman, B. A. Cunningham, W. E. Gall, P. D. Gottlieb, U. Rutishauser, and M. J. Waxdal (1969), The covalent structure of an entire G immunoglobulin molecule. *Proc. Nat. Acad. Sci. USA* **63**, 78–85.

76. K. Kirschner and H. Bisswanger (1976), Multifunctional proteins. *Annu. Rev. Biochem.* **45**, 143–166.

77. M. E. Goldberg (1969), Tertiary structure of *Escherichia coli* β-D-galactosidase. *J. Mol. Biol.* **46**, 441–446.

78. W. Gilbert (1978), Why genes in pieces? (editorial) *Nature* **271**, 501.

79. E. H. Davidson and R. J. Britten (1973), Organization, transcription and regulation in the animal genome. *Quart. Rev. Biol.* **48**, 565–613.

80. D. K. Stammers and H. Muirhead (1977), Three-dimensional structure of cat muscle pyruvate kinase at 3.1 Å resolution. *J. Mol. Biol.* **112**, 309–316.

81. I. M. Klotz, D. W. Darnall, and N. R. Langerman (1975), Quaternary structure of proteins. In: *The Proteins* (H. Neurath, Ed.), 3rd ed., Vol. I, pp. 293–411. Academic Press, New York.

82. J. R. Brown (1976), Structural origins of mammalian albumin. *Fed. Proc.* **35**, 2141–2144.

83. L. Noda (1973), Adenylate kinase. *The Enzymes* **8**, 279–305.

84. I. von Zabern, B. Wittmann-Liebold, R. Untucht-Grau, R. H. Schirmer, and E. Pai (1976), Primary and tertiary structure of the principal human adenylate kinase. *Eur. J. Biochem.* **68**, 281–290.

85. L. Stryer (1975), *Biochemistry*. W. H. Freeman, San Francisco.

86. P. A. Peterson, L. Rask, K. Sege, L. Klareskog, H. Anundi, and L. Ostberg (1975), Evolutionary relationship between immunoglobulins and transplantation antigens. *Proc. Nat. Acad. Sci. USA* **72**, 1612–1616.

87. F. H. Bach (1976), Genetics of transplantation: The major histocompatibility complex. *Annu. Rev. Genet.* **10**, 281.

88. P. A. Peterson, B. A. Cunningham, I. Berggård, and G. M. Edelman (1972), β_2-Microglobulin—A free immunoglobulin domain. *Proc. Nat. Acad. Sci. USA* **69**, 1697–1701.

89. E. Frieden (1971), Protein–protein interaction and enzymatic activity. *Annu. Rev. Biochem.* **40**, 653–696.

90. W. H. C. Chan (1976), Relationship between quaternary structure and enzyme activity. *Trends Biol. Sci.* **1**, 258–260.

91. M. G. Rossmann, A. Liljas, C.-I. Brändén, and L. J. Banaszak (1975), Evolutionary and structural relationships among dehydrogenases. *The Enzymes* **11**, 61–102.

92. J. Monod, J. Wyman, and J.-P. Changeux (1965), On the nature of allosteric transitions: A plausible model. *J. Mol. Biol.* **12**, 88–118.

93. D. E. Koshland, Jr. (1976), The evolution of function in enzymes. *Fed. Proc.* **35**, 2104–2111.

94. C. B. Anfinsen and H. A. Scheraga (1975), Experimental and theoretical aspects of protein folding. *Adv. Prot. Chem.* **29**, 205–300.

95. M. Sela, F. H. White, Jr., and C. B. Anfinsen (1957), Reductive cleavage of disulfide bridges in ribonuclease. *Science* **125**, 691–692.

96. F. H. White, Jr. (1961), Regeneration of native secondary and tertiary structures by air oxidation of reduced ribonuclease. *J. Biol. Chem.* **236**, 1353–1360.

97. C. B. Anfinsen and E. Haber (1961), Studies on the reduction and reformation of protein disulfide bonds. *J. Biol. Chem.* **236**, 1361–1363.

98. D. E. Olins and G. M. Edelman (1964), Reconstitution of 7 S molecules from L and H polypeptide chains of antibodies and γ-globulins. *J. Exp. Med.* **119**, 789–815.

99. M. H. Freedman and M. Sela (1966), Recovery of antigenic activity upon reoxydation of completely reduced polyalanyl rabbit immunoglobulin G. *J. Biol. Chem.* **241**, 2383–2396.

100. J. G. L. Petersen and K. I. Dorrington (1974), An *in vitro* system for studying the kinetics of interchain disulfide bond formation in immunoglobulin G. *J. Biol. Chem.* **249**, 5633–5641.

101. D. Givol, F. DeLorenzo, R. F. Goldberger, and C. B. Anfinsen (1965), Disulfide interchange and the three-dimensional structure of proteins. *Proc. Nat. Acad. Sci. USA* **53**, 676–684.

102. E. Della Corte and R. M. E. Parkhouse (1973), Biosynthesis of immunoglobulin A (IgA) and immunoglobulin M (IgM). Requirement for J-chain and a disulfide-exchanging enzyme for polymerization. *Biochem. J.* **136**, 597–606.

103. D. M. Ziegler and L. L. Poulsen (1977), Protein disulfide bond synthesis, a possible intracellular mechanism. *Trends Biol. Sci.* **2**, 79–81.

104. P. Venetianer and F. B. Straub (1963), The enzymic reactivation of reduced ribonuclease. *Biochim. Biophys. Acta* **67**, 166–167.

105. R. F. Goldberger, C. J. Epstein, and C. B. Anfinsen (1963), Acceleration of reactivation of reduced bovine pancreatic ribonuclease by a microsomal system from rat liver. *J. Biol. Chem.* **238**, 628–635.

106. F. DeLorenzo, R. F. Goldberger, E. Steers, D. Givol, and C. B. Anfinsen (1966), Purification and properties of an enzyme from beef liver which catalyzes sulfhydryl–disulfide interchange in proteins. *J. Biol. Chem.* **241**, 1562.

107. S. Fuchs, F. DeLorenzo, and C. B. Anfinsen (1967), Studies on the mechanism of the enzymic catalysis of disulfide interchange in proteins. *J. Biol. Chem.* **242**, 398–402.

108. P. J. Flory (1953), *Principles of Polymer Chemistry,* Chap. 14, Cornell University Press, Ithaca, New York.

109. F. B. Straub (1967), SH groups and SS bridges in the structure of enzymes. *Proc. 7th Int. Congress Biochem. (Tokyo),* 42–50.

110. J. S. Wall (1971), Disulfide bonds: Determination, location and influence on molecular properties of proteins. *J. Agr. Food Chem.* **19**, 619–625.

111. C. H. Williams, Jr. (1976), Flavin-containing dehydrogenases. *The Enzymes* (3rd ed.) **13**, 89–173.

112. G. W. Hatfield and R. O. Burns (1970), Threonine deaminase from *Salmonella typhimurium. J. Biol. Chem.* **245**, 787–791.

113. S. Pontremoli and B. L. Horecker (1971), Fructose-1,6-diphosphatases. *The Enzymes* (3rd ed.) **4**, 611–646.

114. M. J. Ernest and K. H. Kim (1974), Regulation of rat liver glycogen synthetase D. *J. Biol. Chem.* **249**, 5011–5018.

115. M. J. Ernest and K. H. Kim (1973), Regulation of rat liver glycogen synthetase. *J. Biol. Chem.* **248**, 1550–1555.

116. F. DeLorenzo, S. Fuchs, and C. B. Anfinsen (1966), Characterization of a peptide fragment containing the essential half-cystine residue of a microsomal disulfide interchange enzyme. *Biochemistry* **5**, 3961–3965.

117. H. Sakai (1967), A ribonucleoprotein which catalyzes thiol-disulfide exchange in the sea urchin egg. *J. Biol. Chem.* **242**, 1458–1461.
118. P. Cuatrecasas (1972), *Proc. Symp. Insulin Action, Toronto, Canada*, p. 137. Academic Press, New York.
119. M. Czech (1977), Molecular basis of insulin action. *Annu. Rev. Biochem.* **46**, 359–384.
120. S. Bhakdi, H. Knüfermann, R. Schmidt-Ullrich, H. Fischer, and D. F. Wallach (1974), Interaction between erythrocyte membrane proteins and complement components. *Biochim. Biophys. Acta* **363**, 39–53.
121. Y. Morino and E. E. Snell (1967), The subunit structure of tryptophanase. *J. Biol. Chem.* **242**, 5602–5610.
122. G. W. Hatfield and R. O. Burns (1970), Specific binding of leucyl transfer RNA to an immature form of L-threonine deaminase: Its implications in repression. *Proc. Nat. Acad. Sci. USA* **66**, 1027–1035.
123. E. T. Jones and C. H. Williams, Jr. (1975), The sequence of amino acid residues around the oxidation–reduction active disulfide in yeast glutathione reductase. *J. Biol. Chem.* **250**, 3779–3784.
124. G. E. Schulz, R. H. Schirmer, W. Sachsenheimer, and E. F. Pai (1978), The structure of the flavoenzyme glutathione reductase. *Nature* **273**, 120–124.
125. P. Bornstein (1974), The biosynthesis of collagen. *Annu. Rev. Biochem.* **43**, 567–604.
126. M. L. Tanzer, (1978), The biological diversity of collagenous proteins. *Trends Biochem. Sci.* **3**, 15–17.
127. J. Uitto and J. R. Lichtenstein (1976), Defects in the biochemistry of collagen in diseases of connective tissue. *J. Invest. Dermatol.* **66**, 59–79.
128. P. H. Byers, E. M. Click, E. Harper, and P. Bornstein (1975), Interchain disulfide bonds on procollagen are located in a large nontriple-helical COOH-terminal domain. *Proc. Nat. Acad. Sci. USA* **72**, 3009–3013.
129. A. Veis and A. G. Brownell (1974), Collagen biosynthesis. *CRC Crit. Rev. Biochem.* **2**, 417–453.
130. M. L. Tanzer, R. L. Church, J. A. Yaeger, D. E. Wampler, and E. D. Park (1974), Procollagen: Intermediate forms containing several types of peptide chains and non-collagen peptide extensions at NH$_2$ and COOH ends. *Proc. Nat. Acad. Sci. USA* **71**, 3009–3013.
131. G. R. Martin, P. H. Byers, and K. A. Piez (1975), Procollagen. *Adv. Enzymol.* **42**, 167–191.
132. K. Narita, H. Matsuo, and T. Nakajima (1975), End group determination. In: *Protein Sequence Determination* (S. B. Needleman, Ed.), 2nd ed., pp. 30–103. Springer-Verlag, Heidelberg.
133. G. H. Jones (1974), Cell-free synthesis of amino-terminal L-pyroglutamic acid. *Biochemistry* **13**, 854–860.
134. Y. Burstein, F. Kantor, and I. Schechter (1976), Partial amino-acid sequence of the precursor of an immunoglobulin light chain containing NH$_2$-terminal pyroglutamic acid. *Proc. Nat. Acad. Sci. USA* **73**, 2604–2608.
135. R. Chen, J. Brosius, and B. Wittmann-Liebold (1977), Occurrence of methylated amino acids as N-termini of proteins from *Escherichia coli* ribosomes. *J. Mol. Biol.* **111**, 173–181.
136. R. G. Gillen and A. Riggs (1973), Structure and function of the isolated

hemoglobins of the American eel, *anguilla rostrata*. *J. Biol. Chem.* **248**, 1961–1969.

137. M. Ottesen (1967), Induction of biological activity by limited proteolysis. *Annu. Rev. Biochem.* **36**, 55–76.

138. H. Neurath and K. A. Walsh (1976), Role of proteolytic enzymes in biological regulation (a review). *Proc. Nat. Acad. Sci. USA* **73**, 3825–3832.

139. C. de Haen, H. Neurath, and D. C. Teller (1975), The phylogeny of trypsin-related serine proteases and their zymogens. New methods for the investigation of distant evolutionary relationships. *J. Mol. Biol.* **92**, 225–259.

140. A. Hershko and M. Fry (1975), Post-translational cleavage of polypeptide chains: Role in assembly. *Annu. Rev. Biochem.* **44**, 775–798.

141. B. N. Apte and D. Zipser (1976), Authors' statement on polypeptide splicing. *Proc. Nat. Acad. Sci. USA* **73**, 661.

142. R. W. Hendrix and S. R. Casjens (1974), Protein fusion: A novel reaction in bacteriophage λ head assembly. *Proc. Nat. Acad. Sci. USA* **71**, 1451–1455.

143. S. Maroux, J. Baratti, and P. Desnuelle (1971), Purification and specificity of porcine enterokinase. *J. Biol. Chem.* **246**, 5031–5039.

144. J. L. King and T. H. Jukes (1969), Non-Darwinian evolution. *Science* **164**, 788–798.

145. M. O. Dayhoff (1976), *Atlas of Protein Sequence and Structure,* Vol. 5, Suppl. 2 and earlier volumes. Nat. Biom. Res. Foundation, Washington, D. C.

146. D. F. Steiner (1974), Proteolytic processing in the biosynthesis of insulin and other proteins. *Fed. Proc.* **33** (10) 2105.

147. H. J. Müller-Eberhard (1975), Complement. *Annu. Rev. Biochem.* **44**, 697–724.

148. E. W. Davie and K. Fujikawa (1975), Basic mechanisms in blood coagulation. *Annu. Rev. Biochem.* **44**, 799–829.

149. G. Blobel and B. Dobberstein (1975), Transfer of proteins across membranes. *J. Cell Biol.* **67**, 835–851.

150. C. Milstein, G. G. Brownleee, T. M. Harrison, and M. B. Mathews (1972), A possible precursor of immunoglobulin light chains. *Nature New Biol.* **239**, 117–120.

151. J. Schechter, D. J. McKean, R. Guyer, and W. Terry (1974), Partial amino acid sequence of the precursor of immunoglobulin light chain programmed by messenger RNA *in vitro*. *Science* **188**, 160–162.

152. B. Kemper, J. F. Habener, R. C. Mulligan, J. T. Potts, and A. Rich (1974), Pre-proparathyroid hormone: A direct translational product of parathyroid messenger RNA. *Proc. Nat. Acad. Sci. USA* **71**, 3731–3735.

153. A. Devillers-Thiery, T. Kindt, G. Scheele, and G. Blobel (1975), Homology in amino terminal sequence of precursors to pancreatic secretory proteins. *Proc. Nat. Acad. Sci. USA* **72**, 5016–5020.

154. A. L. Goldberg and A. C. St. John (1976), Intracellular protein degradation in mammalian and bacterial cells. *Annu. Rev. Biochem.* **45**, 747–804.

155. N. Katunuma (1977), New intracellular proteases and their role in intracellular enzyme degradation. *Trends Biol. Sci.* **2**, 122–124.

156. O. A. Scornik and V. Botbol (1976), Role of changes in protein degradation in the growth of regenerating livers. *J. Biol. Chem.* **251**, 2891–2897.

157. R. T. Schimke and D. Doyle (1970), Control of enzyme levels in animal tissues. *Annu. Rev. Biochem.* **39**, 929–976.

158. H. G. Williams-Ashman, A. C. Notides, S. S. Pabalan, and L. Lorand (1972), Transamidase reactions involved in the enzymic coagulation of semen: Isolation of γ-glutamyl-ε-lysine dipeptide from clotted secretion protein of guinea pig seminal vesicle. *Proc. Nat. Acad. Sci. USA* **69**, 2322–2325.

159. L. Lorand and T. Gotoh (1970), Fibrinoligase—The fibrin stabilizing factor system. *Meth. Enzymol.* **19**, 770–782.

160. K. Decker (1976), Structure and synthesis of a flavoprotein with covalently bound FAD. *Trends Biochem. Sci.* **1**, 184–185.

161. A.S. Acharya and H. Taniuchi (1976), A study of renaturation of reduced hen egg white lysozyme. Enzymatically active intermediate formed during oxidation of the reduced protein. *J. Biol. Chem.* **251**, 6934–6946.

162. A. S. Acharya and H. Taniuchi (1977), Formation of the four isomers of hen egg white lysozyme containing three native disulfide bonds and one open disulfide bond. *Proc. Nat. Acad. Sci. USA* **74**, 2362–2366.

163. M. F. Jacobson and D. Baltimore (1968), Polypeptide cleavages in the formation of poliovirus proteins. *Proc. Nat. Acad. Sci. USA* **61**, 77–84.

164. D. Baltimore (1971), Polio is not dead. *Perspect. Virol.* **7**, 1–12.

165. A. M. Pappenheimer, Jr. (1977), Diphtheria toxin. *Annu. Rev. Biochem.* **46**, 69–94.

166. H. Holzer (1976), Catabolite inactivation in yeast. *Trends Biol. Sci.* **1**, 178–181.

167. H. S. Tager and D. F. Steiner (1974), Peptide hormones. *Annu. Rev. Biochem.* **43**, 509–538.

168. E. R. Stadtman (1973), Adenylyl transfer reactions. *The Enzymes* **8**, 2–49.

169. O. Hagaishi and K. Ueda (1977), Poly(ADP-ribose) and ADP-ribosylation of proteins. *Annu. Rev. Biochem.* **46**, 95–116.

170. J. Stenflo and J. W. Suttie (1977), Vitamin K-dependent formation of γ-carboxyglutamic acid. *Annu. Rev. Biochem.* **46**, 157–172.

171. P. A. Price, J. W. Poser, and N. Raman (1976), Primary structure of the γ-carboxyglutamic acid-containing protein from bovine bone. *Proc. Nat. Acad. Sci. USA* **73**, 3374–3375.

172. J. Lunney and G. Ashwell (1976), A hepatic receptor of avian origin capable of binding specifically modified glycoproteins. *Proc. Nat. Acad. Sci. USA* **73**, 341–343.

173. G. Ashwell and A. Morell (1974), The role of surface carbohydrates in the hepatic recognition and transport of circulating glycoproteins. *Adv. Enzymol.* **41**, 99–128.

174. W. K. Paik and S. Kim (1975), Protein methylation: Chemical, enzymological and biological significance. *Adv. Enzymol.* **42**, 227–286.

175. C. S. Rubin and O. M. Rosen (1975), Protein phosphorylation. *Annu. Rev. Biochem.* **44**, 831–887.

176. K. U. Linderström-Lang and J. A. Schellman (1959), Protein structure and enzyme activity. *The Enzymes,* (P. D. Boyer, Ed.), Vol. 1, 2nd ed., pp. 443–510. Academic Press, New York.

177. C. B. Anfinsen (1973), Principles that govern the folding of protein chains. *Science* **181**, 223–230.

178. R. Moorhouse, W. T. Winter, and S. Arnott (1977), Conformation and molecular organization in fibers of the capsular polysaccharide from *Escherichia coli* M41 mutant. *J. Mol. Biol.* **109**, 373–391.

179. S. Arnott (1970), The geometry of nucleic acids. *Progr. Biophys. Mol. Biol.* **21**, 265–319.

180. G. Stubbs, S. Warren, and K. Holmes (1977), Structure of RNA and RNA binding site in tobacco mosaic virus from 4 Å map calculated from X-ray fibre diagrams. *Nature* **267**, 216–221.

181. C. Cohen, S. C. Harrison, and R. E. Stephens (1971), X-ray diffraction from microtubules. *J. Mol. Biol.* **59**, 375–380.

182. M. F. Moody (1973), Sheath of Bacteriophage T4. *J. Mol. Biol.* **80**, 613–635.

183. M. F. Perutz (1951), New X-ray evidence on the configuration of polypeptide chains. *Nature* **167**, 1053–1054.

184. R. E. Dickerson (1964), X-ray analysis and protein structure. *The Proteins* (H. Neurath, Ed.), 2nd ed., Vol. 2, pp. 603–778. Academic Press, New York.

185. J. C. Kendrew, R. E. Dickerson, B. E. Strandberg, R. G. Hart, D. R. Davies, D. C. Phillips, and V. C. Shore (1960), Structure of myoglobin. *Nature* **185**, 422–427.

186. G. E. Schulz, M. Elzinga, F. Marx, and R. H. Schirmer (1974), Three-dimensional structure of adenylate kinase. *Nature* **250**, 120–123.

187. J. Donohue (1953), Hydrogen-bonded helical configurations of the polypeptide chain. *Proc. Nat. Acad. Sci. USA* **39**, 470–478.

188. W. A. Hendrickson and W. E. Love (1971), Structure of lamprey haemoglobin. *Nature New Biol.* **232**, 197–203.

189. W. A. Hendrickson, W. E. Love, and J. Karle (1973), Crystal structure analysis of sea lamprey hemoglobin at 2 Å resolution. *J. Mol. Biol.* **74**, 331–361.

190. B. W. Low and H. J. Grenville-Wells (1953), Generalized mathematical relationships for polypeptide chain helices. The coordinates of the π-helix. *Proc. Nat. Acad. Sci. USA* **39**, 785–801.

191. G. N. Ramachandran and G. Kartha (1955), Structure of collagen. *Nature* **176**, 593–595.

192. A. Rich and F. H. C. Crick (1961), The molecular structure of collagen. *J. Mol. Biol.* **3**, 483–506.

193. A. Yonath and W. Traub (1969), Polymers of tripeptides as collagen models. *J. Mol. Biol.* **43**, 461–477.

194. W. Traub and K. A. Piez (1971), The chemistry and structure of collagen. *Adv. Prot. Chem.* **25**, 243–352.

195. G. Balian, E. M. Click, M. Hermodsen, and P. Bornstein (1972), Structure of rat skin collagen α1-CB8. Amino acid sequence of the hydroxylamine-produced fragment HA2. *Biochemistry* **11**, 3798–3806.

196. P. P. Fietzek, F. Rexrodt, K. Hopper, and K. Kühn (1973), The covalent structure of collagen. *Eur. J. Biochem.* **38**, 396–400.

197. D. J. S. Hulmes, A. Miller, D. A. D. Parry, K. A. Piez, and J. Woodhead-Galloway (1973), Analysis of the primary structure of collagen for the origins of molecular packing. *J. Mol. Biol.* **79**, 137–148.

198. K. B. M. Reid and R. R. Porter (1976), Subunit composition and structure of subcomponent C1q of the first component of human complement. *Biochem. J.* **155**, 19–23.

199. C. M. Venkatachalam (1968), Stereochemical criteria for polypeptides and proteins.. V. Conformation of a system of three linked peptide units. *Biopolymers* **6**, 1425–1436.

200. J. L. Crawford, W. N. Lipscomb, and C. G. Schellman (1973), The reverse turn as a polypeptide conformation in globular proteins. *Proc. Nat. Acad. Sci. USA* **70**, 538–542.

201. P. Y. Chou and G. D. Fasman (1974), Conformational parameters for amino acids in helical, β-sheet, and random coil regions calculated from proteins. *Biochemistry* **13**, 211–222.

202. P. N. Lewis, F. A. Momany, and H. A. Scheraga (1973), Chain reversals in proteins. *Biochim. Biophys. Acta* **303**, 211–229.

203. I. D. Kuntz (1972), Protein folding. *J. Amer. Chem. Soc.* **94**, 4009–4012.

204. L. Pauling and R. B. Corey (1951), Configurations of polypeptide chains with favored orientations around single bonds: Two new pleated sheets. *Proc. Nat. Acad. Sci. USA* **37**, 729–740.

205. C. Chothia (1973), Conformation of twisted β-pleated sheets in proteins. *J. Mol. Biol.* **75**, 295–302.

206. C. C. F. Blake and S. J. Oatley (1977), Protein–DNA and protein–hormone interactions in prealbumin: A model of the thyroid hormone nuclear receptor? *Nature* **268**, 115–120.

207. C. Carter and J. Kraut (1974), A proposed model for the interaction of polypeptides with RNA. *Proc. Nat. Acad. Sci. USA* **71**, 283–287.

208. G. M. Church, J. L. Sussmann, and S. H. Kim (1977), Secondary structural complementarity between DNA and proteins. *Proc. Nat. Acad. Sci. USA* **74**, 1458–1462.

209. J. S. Richardson, E. D. Getzoff, and D. C. Richardson (1978), The β-bulge: A common small unit of non-repetitive protein structure. *Proc. Nat. Acad. Sci. USA*, **75**, 2574–2578.

210. F. H. C. Crick (1953), The packing of α-helices: Simple coiled coils. *Acta Crystallogr.* **6**, 689–697.

211. R. D. B. Fraser and T. P. MacRae (1971), Structure of α-keratin. *Nature* **233**, 138–140.

212. D. A. D. Parry, W. G. Crewther, R. D. B. Fraser, and T. P. MacRae (1977), Structure of α-keratin: Structural implication of the amino acid sequences of the type I and type II chain segments. *J. Mol. Biol.* **113**, 449–454.

213. D. L. D. Caspar, C. Cohen, and W. Longley (1969), Tropomyosin: Crystal structure, polymorphism and molecular interactions. *J. Mol. Biol.* **41**, 87–107.

214. C. Cohen and K. C. Holmes (1963), X-ray diffraction evidence for α-helical coiled-coils in native muscle. *J. Mol. Biol.* **6**, 423–432.

215. H. E. Huxley (1969), The mechanism of muscular contraction. *Science* **164**, 1356–1366.

216. I. M. Klotz, G. L. Klippenstein, and W. A. Hendrickson (1976), Hemerythrin: Alternative oxygen carrier. *Science* **192**, 335–344.

217. R. E. Stenkamp, L. C. Sieker, L. H. Jensen, and J. S. Loehr (1976), Structure of methemerythrin at 5 Å resolution. *J. Mol. Biol.* **100**, 23–34.

218. J. N. Champness, A. C. Bloomer, C. Bricogne, P. J. G. Butler, and A. Klug (1976), The structure of the protein disk of tobacco mosaic virus to 5 Å resolution. *Nature* **259**, 20–24.

219. R. Henderson and P. N. T. Unwin (1975), Three-dimensional model of purple membrane obtained by electron microscopy. *Nature* **257**, 28–32.

220. D. A. Marvin and E. J. Wachtel (1976), Structure and assembly of filamentous bacterial viruses. *Phil. Trans. Roy. Soc. London* **B 276**, 81–98.

221. M. J. Irwin, J. Nyborg, B. R. Reid, and D. M. Blow (1976), The crystal structure of tyrosyl-transfer RNA synthetase at 2.7 Å resolution. *J. Mol. Biol.* **105**, 577–586.

222. D. Stone, J. Sodek, P. Johnson, and L. B. Smillie (1975), Tropomyosin: Correlation of amino acid sequence and structure. *Proc. 9th FEBS Meeting (Budapest)* **31**, 125–136.

223. M. Stewart (1975), Tropomyosin: Evidence for no stagger between chains. *FEBS Lett.* **53**, 5–7.

224. M. Stewart and A. D. McLachlan (1976), Structure of magnesium paracrystals of α-tropomyosin. *J. Mol. Biol.* **103**, 251–269.

225. C. Chothia, M. Levitt, and D. Richardson (1977), Structure of proteins: Packing of α-helices and pleated sheets. *Proc. Nat. Acad. Sci. USA* **74**, 4130–4134.

226. J. S. Richardson (1976), Handedness of crossover connections in β-sheets. *Proc. Nat. Acad. Sci. USA* **73**, 2619–2623.

227. M. J. E. Sternberg and J. M. Thornton (1976), On the conformation of proteins: The handedness of the β-strand–α-helix–β-strand unit. *J. Mol. Biol.* **105**, 367–382.

228. K. Nagano (1977), Logical analysis of mechanism of protein folding. IV. Supersecondary structures. *J. Mol. Biol.* **109**, 235–250.

229. S. T. Rao and M. G. Rossmann (1973), Comparison of super-secondary structures in proteins. *J. Mol. Biol.* **76**, 241–256.

230. D. Moras, K. W. Olsen, M. N. Sabesan, M. Buehner, G. C. Ford, and M. G. Rossmann (1975), Studies of asymmetry in the three-dimensional structure of lobster D-glyceraldehyde-3-phosphate dehydrogenase. *J. Biol. Chem.* **250**, 9137–9162.

231. G. Biesecker, J. I. Harris, J. C. Thierry, J. E. Walker, and A. J. Wonacott (1977), Sequence and structure of D-glyceraldehyde 3-phosphate dehydrogenase from *Bacillus stearothermophilus*. *Nature* **266**, 328–333.

232. M. G. Rossmann, M. J. Adams, M. Buehner, G. C. Ford, M. L. Hackert, P. J. Lentz, Jr., A. McPherson, Jr., R. W. Schevitz, and I. E. Smiley (1971), Structural constraints of possible mechanisms of lactate dehydrogenase as shown by high resolution studies of the apoenzyme and a variety of enzyme complexes. *Cold Spring Harbor Symp. Quant. Biol.* **36**, 179–191.

233. E. Hill, D. Tsernoglou, L. Webb, and L. J. Banaszak (1972), Polypeptide conformation of cytoplasmic malate dehydrogenase from an electron density map at 3.0 Å resolution. *J. Mol. Biol.* **72**, 577–591.

234. H. Eklund, B. Nordström, E. Zeppezauer, G. Söderlund, I. Ohlsson, T. Boiwe, B. O. Söderberg, O. Tapia, and C. I. Brändén (1976), Three-dimensional structure of horse liver alcohol dehydrogenase at 2.4 Å resolution. *J. Mol. Biol.* **102**, 27–59.

235. C. C. F. Blake (1975), X-ray studies of glycolytic enzymes. *Essays Biochem.* **11**, 37–79.

236. R. J. Fletterick, J. Sygusch, H. Semple, and N. B. Madsen (1976), Structure of glycogen phosphorylase *a* at 3.0 Å resolution and its ligand binding sites at 6 Å. *J. Biol. Chem.* **251**, 6142–6146.

237. R. M. Burnett, G. D. Darling, D. S. Kendall, M. E. LeQuesne, S. G. Mayhew, W. W. Smith, and M. L. Ludwig (1974), The structure of the oxidized form of clostridial flavodoxin at 1.9 Å resolution. *J. Biol. Chem.* **249**, 4383–4392.

238. K. D. Watenpaugh, L. C. Sieker, L. H. Jensen, J. Legall, and M. Dubourdieu (1972), Structure of the oxidized form of a flavodoxin at 2.5 Å resolution: Resolution of the phase ambiguity by anomalous scattering. *Proc. Nat. Acad. Sci. USA* **69**, 3185–3188.

239. J. Drenth, W. G. J. Hol, J. N. Jansonius, and R. Koekoek (1971), A comparison of the three-dimensional structures of subtilisin BPN' and subtilisin novo. *Cold Spring Harbor Symp. Quant. Biol.* **36**, 107–116.

240. C. S. Wright, R. A. Alden, and J. Kraut (1969), Structure of subtilisin BPN' at 2.5 Å resolution. *Nature* **221**, 235–242.

241. B. W. Matthews and S. J. Remington (1974), The three-dimensional structure of the lysozyme from bacteriophage T4. *Proc. Nat. Acad. Sci. USA* **71**, 4178–4182.

242. A. Arnone, C. J. Bier, F. A. Cotton, V. W. Day, E. E. Hazen, Jr., D. C. Richardson, J. S. Richardson, and A. Yonath (1971), A high resolution structure of an inhibitor complex of the extracellular nuclease of *Staphylococcus aureus*. *J. Biol. Chem.* **246**, 2302–2316.

243. D. M. Shotton and H. C. Watson (1970), Three-dimensional structure of tosyl-elastase. *Nature* **225**, 811–816.

244. R. M. Stroud, L. M. Kay, and R. E. Dickerson (1974), The structure of bovine trypsin: Electron density maps of the inhibited enzyme at 5 Å and at 2.7 Å resolution. *J. Mol. Biol.* **83**, 185–208.

245. H. Fehlhammer, W. Bode, and R. Huber (1977), Crystal structure of bovine trypsinogen at 1.8 Å resolution. *J. Mol. Biol.* **111**, 415–438.

246. L. T. J. Delbaere, W. L. B. Hutcheon, M. N. G. James, and W. E. Thiessen (1975), Tertiary structural differences between microbial serine proteases and pancreatic serine enzymes. *Nature* **257**, 758–763.

247. J. S. Richardson (1977), β-sheet topology and the relatedness of proteins. *Nature* **268**, 495–500.

248. D. E. Wetlaufer (1973), Nucleation, rapid folding, and globular intrachain regions in proteins. *Proc. Nat. Acad. Sci. USA* **70**, 697–701.

249. M. Levitt and C. Chothia (1976), Structural patterns in globular proteins. *Nature* **261**, 552–557.

250. G. E. Schulz (1977), Structural rules for globular proteins. *Angew. Chem. Int. Edit.* **16**, 23–33.

251. M. Hamermesh (1962), *Group Theory,* Addison-Wesley, London.

252. H. Weyl (1952), *Symmetry,* Princeton University Press, Princeton, New Jersey.

253. C. S. Wright (1978), Multi-domain structure of the dimeric lectin wheat germ agglutinin. International Symposium of Biomolecular Structure, Conformation, Function and Evolution, Madras.

254. M. G. Rossmann, D. Moras, and K. W. Olsen (1974), Chemical and biological evolution of a nucleotide-binding protein. *Nature* **250**, 194–199.

255. G. E. Schulz and R. H. Schirmer (1974), Topological comparison of adenylate kinase with other proteins. *Nature* **250**, 142–144.

256. H. A. Zappe, G. Krohne-Ehrich, and G. E. Schulz (1977), Low resolution structure of human erythrocyte glutathione reductase. *J. Mol. Biol.* **113**, 141–152.

257. J. Bergsma, W. G. J. Hol, J. N. Jansonius, K. H. Kalk, J. H. Ploegman, and J. D. G. Smit (1975), The double domain structure of rhodanese. *J. Mol. Biol.* **98**, 637–643.

258. J. C. S. Clegg and S. I. T. Kennedy (1975), Initiation of synthesis of the structural proteins of semliki forest virus. *J. Mol. Biol.* **97**, 401–411.

259. T. Blundell, G. Dodson, D. Hodgkin, and D. Mercola (1972), Insulin: The structure in the crystal and its reflection in chemistry and biology. *Adv. Prot. Chem.* **26**, 279–402.

260. J. Moult, A. Yonath, W. Traub, A. Smilansky, A. Podjarny, D. Rabinovich, and A. Saya (1976), The structure of triclinic lysozyme at 2.5 Å resolution. *J. Mol. Biol.* **100**, 179–195.

261. T. A. Steitz, R. J. Fletterick, W. F. Anderson, and C. M. Anderson (1976), High resolution X-ray structure of yeast hexokinase, an allosteric protein exhibiting a non-symmetric arrangement of subunits. *J. Mol. Biol.* **104**, 197–222.

262. D. L. D. Caspar and A. Klug (1962), Physical principles in the construction of regular viruses. *Cold Spring Harbor Symp. Quant. Biol.* **27**, 1–24.

263. F. K. Winkler, C. E. Schutt, S. C. Harrison, and G. Bricogne (1977), Tomato bushy stunt virus at 5.5 Å resolution. *Nature* **265**, 509–513.

264. A. Liljas and M. G. Rossmann (1974), X-ray studies of protein interactions. *Annu. Rev. Biochem.* **43**, 475–507.

265. M. G. Rossmann, M. J. Adams, M. Buehner, G. C. Ford, M. L. Hackert, A. Liljas, S. T. Rao, L. J. Banaszak, E. Hill, D. Tsernoglou, and L. Webb (1973), Molecular symmetry axes and subunit interfaces in certain dehydrogenases. *J. Mol. Biol.* **76**, 533–537.

266. C. Chothia and J. Janin (1975), Principles of protein–protein recognition. *Nature* **256**, 705–708.

267. J. Janin and C. Chothia (1976), Stability and specificity of protein–protein interactions: The case of the trypsin–trypsin inhibitor complexes. *J. Mol. Biol.* **100**, 197–211.

268. C. Chothia, S. Wodak, and J. Janin (1976), Role of subunit interfaces in the allosteric mechanism of hemoglobin. *Proc. Nat. Acad. Sci. USA* **73**, 3793–3797.

269. R. Huber, D. Kukla, W. Bode, P. Schwager, K. Bartels, J. Deisenhofer, and W. Steigemann (1974), Structure of the complex formed by bovine trypsin and bovine pancreatic trypsin inhibitor. *J. Mol. Biol.* **89**, 73–101.

270. C. Levinthal, S.J. Wodak, P. Kahn, and A.K. Dadivanian (1975), Hemoglobin interaction in sickle cell fibers. I: Theoretical approaches to the molecular contacts. *Proc. Nat. Acad. Sci. USA* **72**, 1330–1334.

271. R. Timkovich and R.E. Dickerson (1973), Recurrence of the cytochrome fold in a nitrate-respiring bacterium. *J. Mol. Biol.* **79**, 39–56.

272. F. R. Salemme, S. T. Freer, NG. H. Xuong, R. A. Alden, and J. Kraut (1973), The structure of oxidized cytochrome c_2 of *Rhodospirillum rubrum*. *J. Biol. Chem.* **248**, 3910–3921.

273. T. Takano, O. B. Kallai, R. Swanson, and R. E. Dickerson (1973), The structure of ferrocytochrome c at 2.45 Å resolution. *J. Biol. Chem.* **248**, 5234–5255.

274. J. A. Frier and M. F. Perutz (1977), Structure of human foetal deoxyhaemoglobin. *J. Mol. Biol.* **112**, 97–112.

275. B. K. Vainshtein, E. H. Harutyunyan, I. P. Kuranova, V. V. Borisov, N. I. Sosfenov, A. G. Pavlovsky, A. I. Grebenko, and N. V. Konareva (1975), Structure of leghaemoglobin from lupin root nodules at 5 Å resolution. *Nature* **254**, 163–164.

276. W. E. Love, P. A. Klock, E. E. Lattmann, E. A. Padlan, K. B. Ward, Jr., and W. A. Hendrickson (1971), The structures of lamprey and bloodworm hemoglobins in relation to their evolution and function. *Cold Spring Harbor Symp. Quant. Biol.* **36**, 349–357.

277. R. Huber, O. Epp, W. Steigemann, and H. Formanek (1971), The atomic structure of erythrocruorin in the light of the chemical sequence, and its comparison with myoglobin. *Eur. J. Biochem.* **19**, 42–50.

278. K. Sasaki, S. Dockerill, D. A. Adamiak, I. J. Tickle, and T. Blundell (1975), X-ray analysis of glucagon and its relationship to receptor binding. *Nature* **257**, 751–757.

279. J. Drenth, J. N. Jansonius, R. Koekoek, L. A. A. Sluyterman, and B. G. Wolthers (1970), The structure of the papain molecule. *Phil. Trans. Roy. Soc. London* **B 257**, 231–236.

280. P. M. Colman, J. N. Jansonius, and B. W. Matthews (1972), The structure of thermolysin: An electron density map at 2.3 Å resolution. *J. Mol. Biol.* **70**, 701–724.

281. K. D. Hardman and C. F. Ainsworth (1972), Structure of concanavalin A at 2.4 Å resolution. *Biochemistry* **11**, 4910–4919.

282. G. N. Reeke, Jr., J. W. Becker, and G. M. Edelman (1975), The covalent and three-dimensional structure of concanavalin A. *J. Biol. Chem.* **250**, 1525–1547.

283. C. C. F. Blake, M. J. Geisow, I. D. Swan, C. Rerat, and B. Rerat (1974), Structure of human plasma prealbumin at 2.5 Å resolution. *J. Mol. Biol.* **88**, 1–12.

284. K. D. Watenpaugh, L. C. Sieker, J. R. Herriott, and L. H. Jensen (1971), The structure of a non-heme iron protein: Rubredoxin at 1.5 Å resolution. *Cold Spring Harbor Symp. Quant. Biol.* **36**, 359–367.

285. K. D. Watenpaugh, L. C. Sieker, J. R. Herriott, and L. H. Jensen (1973), Refinement of the model of a protein: Rubredoxin at 1.5 Å resolution. *Acta Crystallogr.* **B29**, 943–956.

286. J. S. Richardson, K. A. Thomas, B. H. Rubin, and D. C. Richardson (1975), Crystal structure of bovine Cu,Zn superoxide dismutase at 3 Å resolution: Chain tracing and metal ligands. *Proc. Nat. Acad. Sci. USA* **72**, 1349–1353.

287. B. W. Low, H. S. Preston, A. Sato, L. S. Rosen, J. S. Searl, A. D. Rudko, and J. S. Richardson (1976), Three-dimensional structure of erabutoxin *b* neurotoxic protein: Inhibitor of acetylcholine receptor. *Proc. Nat. Acad. Sci. USA* **73**, 2991–2994.

288. E. Subramanian, I. D. A. Swan, M. Liu, D. R. Davies, J. A. Jenkins, I. J. Tickle, and T. L. Blundell (1977), Homology among acid proteases: Comparison of crystal structures at 3 Å resolution of acid proteases from *Rhizopus chinensis* and *Endothia parasitica*. *Proc. Nat. Acad. Sci. USA* **74**, 556–559.

289. I. N. Hsu, L. T. J. Delbaere, M. N. G. James, and T. Hofmann (1977), Penicillopepsin from *Penicillum janthinellum* crystal structure at 2.8 Å and sequence homology with porcine pepsin. *Nature* **266**, 140–145.

290. J. Tang, M. N. G. James, I. N. Hsu, J. A. Jenkins, and T. L. Blundell (1978), Structural evidence for gene duplication in the evolution of acid proteases. *Nature* **271**, 618–621.

291. R. J. Poljak, L. M. Amzel, B. L. Chen, R. P. Phizackerley, and F. Saul (1974), The three-dimensional structure of the Fab' fragment of a human myeloma

immunoglobulin at 2.0 Å resolution. *Proc. Nat. Acad. Sci. USA* **71**, 3440–3444.

292. M. Schiffer, R. L. Girling, K. R. Ely, and A. B. Edmundson (1973), Structure of a λ-type Bence–Jones protein at 3.5 Å resolution. *Biochemistry* **12**, 4620–4631.

293. D. M. Segal, E. A. Padlan, G. H. Cohen, S. Rudikoff, M. Potter, and D. R. Davies (1974), The three-dimensional structure of a phosphorylcholine-binding mouse immunoglobulin Fab and the nature of the antigen binding site. *Proc. Nat. Acad. Sci. USA* **71**, 4298–4302.

294. R. Huber, J. Deisenhofer, P. M. Colman, M. Matsushima, and W. Palm (1976), Crystallographic structure studies of an IgG molecule and an F_c fragment. *Nature* **264**, 415–420.

295. A. Liljas, K. K. Kannan, P. C. Bergstén, I. Waara, K. Fridborg, B. Strandberg, U. Carlbom, L. Järup, S. Lövgren, and M. Petef (1972), Crystal structure of human carbonic anhydrase C. *Nature New Biol.* **235**, 131–137.

296. K. K. Kannan, B. Notstrand, K. Fridborg, S. Lövgren, A. Ohlsson, and M. Petef (1975), Crystal structure of human erythrocyte carbonic anhydrase B. Three dimensional structure at a nominal 2.2 Å resolution. *Proc. Nat. Acad. Sci. USA* **72**, 51–55.

297. F. S. Mathews, P. Argos, and M. Levine (1971), The structure of cytochrome b_5 at 2.0 Å resolution. *Cold Spring Harbor Symp.* **36**, 387–395.

298. F. S. Mathews, M. Levine, and P. Argos (1972), Three-dimensional Fourier synthesis of calf liver cytochrome b_5 at 2.8 Å resolution. *J. Mol. Biol.* **64**, 449–464.

299. C. W. Carter, Jr., J. Kraut, S. T. Freer, N. H. Xuong, R. A. Alden, and R. G. Bartsch (1974), Two-Ångstrom crystal structure of oxidized *Chromatium* high potential iron protein. *J. Biol. Chem.* **249**, 4212–4225.

300. C. C. F. Blake, D. F. Koenig, G. A. Mair, A. C. T. North, D. C. Phillips, and V. R. Sarma (1965), Structure of hen egg-white lysozyme. *Nature* **206**, 757–763.

301. G. Kartha, J. Bello, and D. Harker (1967), Tertiary structure of ribonuclease. *Nature* **213**, 862–865.

302. C. H. Carlisle, R. A. Palmer, S. K. Mazumdar, B. A. Gorinsky, and D. G. R. Yeates (1974), The structure of ribonuclease at 2.5 Å resolution. *J. Mol. Biol.* **85**, 1–18.

303. R. E. Fenna and B. W. Matthews (1975), Chlorophyll arrangement in a bacteriochlorophyll protein from *Chlorobium limicola*. *Nature* **258**, 573–577.

304. A. Holmgren, B. O. Söderberg, H. Eklund, and C. I. Brändén (1975), Three-dimensional structure of *Escherichia coli* thioredoxin-S_2 to 2.8 Å resolution. *Proc. Nat. Acad. Sci. USA* **72**, 2305–2309.

305. D. W. Banner, A. C. Bloomer, G. A. Petsko, D. C. Phillips, C. I. Pogson, I. A. Wilson, P. H. Corran, A. J. Furth, J. D. Milman, R. E. Offord, J. D. Priddle, and S. G. Waley (1975), Structure of chicken muscle triose phosphate isomerase determined crystallographically at 2.5 Å resolution using amino acid sequence data. *Nature* **255**, 609–614.

306. I. M. Mavridis and A. Tulinsky (1976), The folding and quaternary structure of trimeric 2-keto-3-deoxy-6-phosphogluconic aldolase at 3.5 Å resolution. *Biochemistry* **15**, 4410–4417.

307. J. W. Campbell, H. C. Watson, and G. I. Hodgson (1974), Structure of yeast phosphoglycerate mutase. *Nature* **250**, 301–303.

308. D. A. Matthews, R. A. Alden, J. T. Bolin, S. T. Freer, R. Hamlin, N. Xuong, J. Kraut, M. Poe, M. Williams, and K. Hoogsteen (1977), Dihydrofolate reductase: X-ray structure of the binary complex with methotrexate. *Science* **197**, 452–455.

309. F. A. Quiocho, G. L. Gilliland, and G. N. Phillips, Jr. (1977), The 2.8 Å resolution structure of the L-arabinose binding protein from *E. coli. J. Biol. Chem.* **252**, 5142–5149.

310. C. C. F. Blake and P. R. Evans (1974), Structure of horse muscle phosphoglycerate kinase. *J. Mol. Biol.* **84**, 585–601.

311. T. N. Bryant, H. C. Watson, and P. L. Wendell (1974), Structure of yeast phosphoglycerate kinase. *Nature* **247**, 14–17.

312. M. Levine, H. Muirhead, D. K. Stammers, and D. I. Stuart (1978), Structure of pyruvate kinase and similarities with other enzymes: Possible implications for protein taxonomy and evolution. *Nature* **271**, 626–630.

313. P. J. Shaw and H. Muirhead (1977), Crystallographic structure analysis of glucose 6-phosphate isomerase at 3.5 Å resolution. *J. Mol. Biol.* **109**, 475–485.

314. E. T. Adman, L. C. Sieker, and L. H. Jensen (1973), The structure of a bacterial ferredoxin. *J. Biol. Chem.* **248**, 3987–3996.

315. J. Drenth, C. M. Enzing, K. H. Kalk, and J. C. A. Vessies (1976), Structure of porcine pancreatic prephospholipase A_2. *Nature* **264**, 373–377.

316. C. S. Wright (1977), The crystal structure of wheat germ agglutinine at 2.2 Å resolution. *J. Mol. Biol.* **111**, 439–457.

317. J. R. Knox and H. W. Wyckoff (1973), A crystallographic study of alkaline phosphatase at 7.7 Å resolution. *J. Mol. Biol.* **74**, 533–545.

318. M. J. Adams, J. R. Helliwell, and C. E. Bugg (1977), Structure of 6-phosphogluconate dehydrogenase from sheep liver at 6 Å resolution. *J. Mol. Biol.* **112**, 183–197.

319. S. G. Warren, B. F. P. Edwards, D. R. Evans, D. C. Wiley, and W. N. Lipscomb (1973), Aspartate transcarbamoylase from *Escherichia coli:* Electron density at 5.5 Å resolution. *Proc. Nat. Acad. Sci. USA* **70**, 1117–1121.

320. R. J. Hoare, P. M. Harrison, and T. G. Hoy (1975), Structure of horse-spleen apoferritin at 6 Å resolution. *Nature* **255**, 653–654.

321. M. F. Perutz, H. Muirhead, J. M. Cox, and L. C. G. Goaman (1968), Three-dimensional Fourier synthesis of horse oxyhaemoglobin at 2.8 Å resolution: The atomic model. *Nature* **219**, 131–139.

322. W. Bolton and M. F. Perutz (1970), Three dimensional Fourier synthesis of horse deoxyhaemoglobin at 2.8 Å resolution. *Nature* **228**, 551–552.

323. R. Srinivasan (1976), Helical length distribution from protein cyrstallographic data. *Ind. J. Biochem. Biophys.* **13**, 192–193.

324. C. N. Johnson (1976), Oak Ridge Nat. Lab Report-5138 ORTEP-II: A Fortran Thermal-Ellipsoid Plot Program for Crystal Structure Illustrations.

325. B. L. Trus and K. A. Piez (1976), Molecular packing of collagen: Three-dimensional analysis of electrostatic interactions. *J. Mol. Biol.* **108**, 705–732.

326. P. N. Lewis, F. A. Momany, and H. A. Scheraga (1971), Folding of polypeptide chains in proteins: A proposed mechanism for folding. *Proc. Nat. Acad. Sci. USA* **68**, 2293–2297.

327. M. J. E. Sternberg and J. M. Thornton (1977), On the conformation of proteins: An analysis of β-pleated sheets. *J. Mol. Biol.* **110**, 285–296.

328. E. R. Blout, C. de Lozé, S. M. Bloom, and G. D. Fasman (1960), The dependence of the conformations of synthetic polypeptides on amino acid composition. *J. Amer. Chem. Soc.* **82**, 3787–3789.

329. D. R. Davies (1964), A correlation between amino acid composition and protein structure. *J. Mol. Biol.* **9**, 605–609.

330. B. Jirgensons (1969), *Optical Rotary Dispersion of Proteins and Other Macromolecules*. Springer-Verlag, Heidelberg.

331. A. V. Guzzo (1965), The influence of amino acid sequence on protein structure. *Biophys. J.* **5**, 809–822.

332. J. W. Prothero (1966), Correlation between the distribution of amino acids and alpha helices. *Biophys. J.* **6**, 367–370.

333. Havensteen (1966), A study of the correlation between the amino acid composition and the helical content of proteins. *J. Theor. Biol.* **10**, 1–10.

334. D. A. Cook (1967), The relation between amino acid sequence and protein conformation. *J. Mol. Biol.* **29**, 167–171.

335. P. F. Periti, G. Guagliarotti, and A. M. Liquori (1967), Recognition of α-helical segments in proteins of known primary structure. *J. Mol. Biol.* **24**, 313–322.

336. P. Dunnill (1968), The use of helical net-diagrams to represent protein structures. *Biophys. J.* **8**, 865–874.

337. B. W. Low, F. M. Lovell, and A. D. Rudko (1968), Prediction of α-helical regions in proteins of known sequence. *Proc. Nat. Acad. Sci. USA* **60**, 1519–1526.

338. J. Dirkx (1972), Une méthode semi-empirique de prédiction des régions L-hélicoidales des chaînes polypeptidiques d'après leur structure primaire. *Arch. Int. Physiol. Biochim.* **80**, 185–187.

339. F. Beghin and J. Dirkx (1975), Une méthode statistique simple de prédiction des conformations protéiques. *Arch. Int. Physiol. Biochim.* **83**, 167–168.

340. P. Y. Chou and G. D. Fasman (1974), Prediction of protein conformation. *Biochemistry* **13**, 222–244.

341. O. B. Ptitsyn and A. V. Finkelstein (1970), Prediction of helical portions of globular proteins according to their primary structure. *Dokl. Akad. Nauk. SSSR* **195**, 221–224 (*Dokl. Biochem.* **195**, 322–325).

342. O. B. Ptitsyn and A. V. Finkelstein (1970), Connection between the secondary and primary structures of globular proteins. *Biofizika (USSR)* **15**, 757–768 (*Biophysics* **15**, 785–796).

343. A. V. Finkelstein and O. B. Ptitsyn (1971), Statistical analysis of the correlation among amino acid residues in helical, β-structural and non-regular regions of globular proteins. *J. Mol. Biol.* **62**, 613–624.

344. P. F. Periti (1974), A Bayesian approach to the recognition of discrete patterns with an application to a problem of protein molecular structure. *Boll. Chim. Farm.* **113**, 187–218.

345. R. H. Pain and B. Robson (1970), Analysis of the code relating sequence to secondary structure in proteins. *Nature* **227**, 62–63.

346. B. Robson and R. H. Pain (1971), Analysis of the code relating sequence to conformation in proteins: Possible implications for the mechanism of formation of helical regions. *J. Mol. Biol.* **58**, 237–259.

347. B. Robson (1974), Analysis of the code relating sequence to conformation in globular proteins. Theory and application of expected information. *Biochem. J.* **141**, 853–867.

348. B. Robson and R. H. Pain (1974), Analysis of the code relating sequence to

conformation in globular proteins. Development of a stereochemical alphabet on the basis of intra-residue information. *Biochem. J.* **141**, 869–882.

349. B. Robson and R. H. Pain (1974), Analysis of the code relating sequence to conformation in globular proteins. An informational analysis of the role of the residue in determining the conformation of its neighbours in the primary sequence. *Biochem. J.* **141**, 883–897.

350. B. Robson and R. H. Pain (1974), Analysis of the code relating sequence to conformation in globular proteins. The distribution of residue pairs in turns and kinks in the backbone chain. *Biochem. J.* **141**, 899–904.

351. B. Robson and E. Suzuki (1977), Conformational properties of amino acid residues in globular proteins. *J. Mol. Biol.* **107**, 327–356.

352. E. Suzuki and B. Robson (1977), Relationship between helix–coil transition parameters for synthetic polypeptides and helix conformation parameters for globular proteins. A simple model. *J. Mol. Biol.* **107**, 357–367.

353. K. Nagano (1973), Logical analysis of mechanism of protein folding. I. Predictions of helices, loops and β-structures from primary structure. *J. Mol. Biol.* **75**, 401–420.

354. K. Nagano (1974), Logical analysis of mechanism of protein folding. II. The nucleation process. *J. Mol. Biol.* **84**, 337–372.

355. K. Nagano and K. Hasegawa (1975), Logical analysis of mechanism of protein folding. III. Prediction of the strong long-range interactions. *J. Mol. Biol.* **94**, 257–281.

356. K. Nagano (1977), Triplet information in helix prediction applied to the analysis of supersecondary structures. *J. Mol. Biol.* **109**, 251–274.

357. E. A. Kabat and T. T. Wu (1973), The influence of neighbor amino acids on the conformation of the middle amino acid in proteins. Comparison of predicted and experimental determination of β-sheets in concanavalin A. *Proc. Nat. Acad. Sci. USA* **70**, 1473–1477.

358. E. A. Kabat and T. T. Wu (1973), The influence of nearest-neighboring amino acid residues on aspects of secondary structure of proteins. Attempts to locate α-helices and β-sheets. *Biopolymers* **12**, 751–774.

359. E. A. Kabat and T. T. Wu (1974), Further comparison of predicted and experimentally determined structure of adenylate kinase. *Proc. Nat. Acad. Sci. USA* **71**, 4217–4220.

360. T. T. Wu and E. A. Kabat (1971), An attempt to locate the non-helical and permissively helical sequences of proteins: Application to the variable regions of immunoglobulin light and heavy chains. *Proc. Nat. Acad. Sci. USA* **68**, 1501–1506.

361. E. A. Kabat and T. T. Wu (1972), Construction of a three-dimensional model of the polypeptide backbone of the variable region of kappa immunoglobulin light chains. *Proc. Nat. Acad. Sci. USA* **69**, 960–964.

362. T. T. Wu and E. A. Kabat (1973), An attempt to evaluate the influence of neighboring amino acids $(n-1)$ and $(n + 1)$ on the backbone conformation of amino acid (n) in proteins. Use in predicting the threedimensional structure of the polypeptide backbone of other proteins. *J. Mol. Biol.* **75**, 13–31.

363. D. Kotelchuck and H. A. Scheraga (1968), The influence of short-range interactions on protein conformation. I. Side chain–backbone interactions within a single peptide unit. *Proc. Nat. Acad. Sci. USA* **61**, 1163–1170.

364. D. Kotelchuck and H. A. Scheraga (1969), The influence of short-range

interactions on protein conformation. II. A model for predicting the α-helical regions of proteins. *Proc. Nat. Acad. Sci. USA* **62**, 14–21.

365. R. Leberman (1971), Secondary structure of tobacco mosaic virus protein. *J. Mol. Biol.* **55**, 23–30.

366. A. W. Burgess and H. A. Scheraga (1975), Assessment of some problems associated with prediction of the three-dimensional structure of a protein from its amino-acid sequence. *Proc. Nat. Acad. Sci. USA* **72**, 1221–1225.

367. P. K. Ponnuswamy, P. K. Warme, and H. A. Scheraga (1973), Role of medium-range interactions in proteins. *Proc. Nat. Acad. Sci. USA* **70**, 830–833.

368. P. N. Lewis, N. Gō, M. Gō, D. Kotelchuck, and H. A. Scheraga (1970), Helix probability profiles of denatured proteins and their correlation with native structures. *Proc. Nat. Acad. Sci. USA* **65**, 810–815.

369. P. N. Lewis and H. A. Scheraga (1971), Predictions of structural homologies in cytochrome *c* proteins. *Arch. Biophys. Biochem.* **144**, 576–583.

370. O. B. Ptitsyn, A. I. Denesyuk, A. V. Finkelstein, and V. I. Lim (1973), Prediction of the secondary structure of the L7, L12 proteins of the *E. coli* ribosome. *FEBS Lett.* **34**, 55–57.

371. A. V. Finkelstein and O. B. Ptitsyn (1976), A theory of protein molecule self-organization. IV. Helical and irregular local structures of unfolded protein chains. *J. Mol. Biol.* **103**, 15–24.

372. A. V. Finkelstein and O. B. Ptitsyn (1977), Theory of protein molecule self-organization. I. Thermodynamic parameters of local secondary structures in the unfolded protein chain. *Biopolymers* **16**, 469–495.

373. A. V. Finkelstein, O. B. Ptitsyn, and S. A. Kozitsyn (1977), Theory of protein molecule self-organization. II. A comparison of calculated thermodynamic parameters of local secondary structures with experiments. *Biopolymers* **16**, 497–524.

374. A. V. Finkelstein (1977), Theory of protein molecule self-organization. III. A calculating method for the probabilities of the secondary structure formation in an unfolded polypeptide chain. *Biopolymers* **16**, 525–529.

375. S. Lifson and A. Roig (1961), On the theory of helix-coil transitions in polypeptides. *J. Chem. Phys.* **34**, 1963–1974.

376. M. Schiffer and A. B. Edmundson (1967), Use of helical wheels to represent the structures of protein and to identify segments with helical potential. *Biophys. J.* **7**, 121–135.

377. M. Schiffer and A. B. Edmundson (1968), Correlation of amino acid sequence and conformation in tobacco mosaic virus. *Biophys. J.* **8**, 29–39.

378. J. Palau and P. Puigdoménech (1974), The structural code for proteins: Zonal distribution of amino acid residues and stabilisation of helices by hydrophobic triplets. *J. Mol. Biol.* **88**, 457–469.

379. V. I. Lim (1974), Structural principles of the globular organization of protein chains. A stereochemical theory of globular protein secondary structure. *J. Mol. Biol.* **88**, 857–872.

380. V. I. Lim (1974), Algorithms for prediction of α-helical and β-structural regions in globular proteins. *J. Mol. Biol.* **88**, 873–894.

381. P. Argos, J. Schwarz, and J. Schwarz (1976), An assessment of protein secondary structure prediction methods based on amino acid sequence. *Biochim. Biophys. Acta* **439**, 261–273.

382. J. A. Lenstra (1977), Evaluation of secondary structure prediction in proteins. *Biochim. Biophys. Acta* **491**, 333–338.

383. G. E. Schulz, C. D. Barry, J. Friedman, P. Y. Chou, G. D. Fasman, A. V. Finkelstein, V. I. Lim, O. B. Ptitsyn, E. A. Kabat, T. T. Wu, M. Levitt, B. Robson, and K. Nagano (1974), Comparison of predicted and experimentally determined secondary structure of adenylate kinase. *Nature* **250**, 140–142.

384. B. W. Matthews (1975), Comparison of the predicted and observed secondary structure of T4 phage lysozyme. *Biochim. Biophys. Acta* **405**, 442–451.

385. J. A. Lenstra, J. Hofsteenge, and J. J. Beintema (1977), Invariant features of the structure of pancreatic ribonuclease. *J. Mol. Biol.* **109**, 185–193.

386. J. C. Wootton (1974), The coenzyme binding domains of glutamate dehydrogenases. *Nature* **252**, 542–546.

387. G. E. Schulz (1977), Recognition of phylogenetic relationships from polypeptide chain fold similarities. *J. Mol. Evol.* **9**, 339–342.

388. P. Argos (1977), Secondary structure predictions of calcium-binding proteins. *Biochemistry* **16**, 665–672.

389. A. F. Heil, G. Müller, L. Noda, T. Pinder, H. Schirmer, I. Schirmer, and I. von Zabern (1974), The amino-acid sequence of porcine adenylate kinase from skeletal muscle. *Eur. J. Biochem.* **43**, 131–144.

390. Protein Data Bank, Department of Chemistry, Brookhaven National Laboratory, Associated Universities, Inc., Upton, L.I., New York 11973.

391. Kendrew–Watson Models, Cambridge Repetition Engineers, Ltd., Green's Road, Cambridge, England.

392. F. M. Richards (1968), The matching of physical models to three-dimensional electron-density maps: A simple optical device. *J. Mol. Biol.* **37**, 225–230.

393. Byron's Bender, Charles Supper Co., 15 Tech Circle, Natick, Massachusetts 01760.

394. Nicholson Models, Labquip Ltd. 18 Rosehill Park Estate, Caversham, Reading RG4 8XE, England.

395. CPK-models, The Ealing Corporation, South Natick, Massachusetts 01760.

396. B. Lee and F. M. Richards (1971), The interpretation of protein structures: Estimation of static accessibility. *J. Mol. Biol.* **55**, 379–400.

397. Evans and Sutherland, 580 Arpeen Drive, Salt Lake City, Utah 84108.

398. R. Diamond (1978), Bilder: A computer graphics program for biopolymers. International Symposium of Biomolecular Structure, Conformation, Function and Evolution, Madras.

399. R. J. Feldmann and T. K. Porter (1978), Div. Computer Research and Technology, National Institutes of Health, Bethesda, Maryland 20014.

400. Stereo viewer, Abrams Instr. Corp., 606 East Shiawassee Street, Lansing, Michigan 48901.

401. AMSOM, Tracor Jitco Inc., 1776 E. Jefferson St., Rockville, Maryland 20852.

402. R. Balasubramanian (1977), New type of representation for mapping chain-folding in protein molecules. *Nature* **266**, 856–857.

403. R. Srinivasan, R. Balasubramanian, and S. S. Rajan (1975), Some new methods and general results of analysis of protein crystallographic structural data. *J. Mol. Biol.* **98**, 739–747.

404. M. Levitt and A. Warshel (1975), Computer simulation of protein folding. *Nature* **253**, 694–698.

405. R. Balasubramanian (1976), A new type of representation of dipeptide conformation. *Biochem. J.* **157**, 769–771.

406. I. D. Kuntz (1975), An approach to the tertiary structure of globular proteins. *J. Amer. Chem. Soc.* **97**, 4362–4366.

407. M. G. Rossmann and A. Liljas (1974), Recognition of structural domains in globular proteins. *J. Mol. Biol.* **85**, 177–181.

408. E. A. Kabat, T. T. Wu, and H. Bilofsky (1976), *Variable Regions of Immunoglobulin Chains*. Bolt Beranck and Newman, Cambridge, Massachusetts.

409. P. G. Behrens, A. M. Spiekerman, and J. R. Brown (1974), Structure of human serum albumin. *Fed. Proc.* **34**, 2106.

410. B. Gutte and R. B. Merrifield (1969), The total synthesis of an enzyme with ribonuclease A activity. *J. Amer. Chem. Soc.* **91**, 501–506.

411. R. Hirschmann, R. F. Nutt, D. F. Veber, R. A. Vitali, S. L. Varga, T. A. Jacob, F. W. Holly, and R. G. Denkewalter (1969), Studies on the total synthesis of an enzyme. V. The preparation of enzymically active material. *J. Amer. Chem. Soc.* **91**, 507–508.

412. C. B. Anfinsen, E. Haber, M. Sela, and F. H. White, Jr. (1961), The kinetics of formation of native ribonuclease during oxidation of the reduced polypeptide chain. *Proc. Nat. Acad. Sci. USA* **47**, 1309–1314.

413. C. N. Pace (1975), The stability of globular proteins. *CRC Crit. Rev. Biochem.* **3**, 1–43.

414. A. Hvidt and S. O. Nielsen (1966), Hydrogen exchange in proteins. *Adv. Prot. Chem.* **21**, 287–386.

415. M. Ottesen (1971), Methods for measurement of hydrogen isotope exchange in globular proteins. *Meth. Biochem. Anal.* **20**, 135–160.

416. M. Nakanishi, M. Tsuboi, and A. Ikegami (1972), Fluctuation of the lysozyme structure. *J. Mol. Biol.* **70**, 351–361.

417. P. L. Privalov and N. N. Khechinashvili (1974), A thermodynamic approach to the problem of stabilization of globular protein structure: A calorimetric study. *J. Mol. Biol.* **86**, 665–684.

418. D. H. Sachs, A. N. Schechter, A. Eastlake, and C. B. Anfinsen (1972), An immunologic approach to the conformational equilibria of polypeptides. *Proc. Nat. Acad. Sci. USA* **69**, 3790–3794.

419. A. Cooper (1976), Thermodynamic fluctuations in protein molecules. *Proc. Nat. Acad. Sci. USA* **73**, 2740–2741.

420. H. B. Callen (1960), *Thermodynamics*. John Wiley, New York.

421. P. L. Privalov (1974), Thermal investigations of biopolymer solutions and scanning microcalorimetry. *FEBS Lett. Suppl.* **40**, 140–153.

422. J. Suurkuusk (1974), Specific heat measurements on lysozyme, chymotrypsinogen, and ovalbumin in aqueous solution and in solid state. *Acta Chem. Scand.* **B28**, 409–417.

423. R. Hetzel, K. Wüthrich, J. Deisenhofer, and R. Huber (1976), Dynamics of the aromatic amino acid residues in the globular conformation of the basic pancreatic trypsin inhibitor (BPTI). II. Semi-empirical energy calculations. *Biophys. Struct. Mech.* **2**, 159–180.

424. G. Wagner, A. DeMarco, and K. Wüthrich (1976), Dynamics of the aromatic amino acid residues in the globular conformation of the basic pancreatic trypsin inhibitor (BPTI). I. ^{1}H NMR studies. *Biophys. Struct. Mech.* **2**, 139–158.

425. I. D. Campbell, C. M. Dobson, and R. J. P. Williams (1975), Proton magnetic resonance studies of the tyrosine residues of hen lysozyme - assignment and detection of conformational mobility. *Proc. R. Soc. London* Ser. **B. 189**, 503–509.

426. A. Allerhand, D. Doddrell, V. Glushko, D. Cochran, E. Wenkert, P. Lawson, and F. Gurd (1971), Conformation and segmental motion of native and denatured ribonuclease A in solution. Application of natural-abundance carbon-13 partially relaxed Fourier transform nuclear magnetic resonance. *J. Amer. Chem. Soc.* **93**, 544–546.

427. O. Oster, G. W. Neireiter, A. O. Clouse, and F. R. N. Gurd (1975), Deuterium nuclear magnetic resonance of deuterium-labeled diacetyldeuterohemin incorporated into sperm whale myoglobin. *J. Biol. Chem.* **250**, 7990–7996.

428. J. R. Lakowicz and G. Weber (1973), Quenching of protein fluorescence by oxygen. Detection of structural fluctuations in proteins on the nanosecond time scale. *Biochemistry* **12**, 4171–4179.

429. M. R. Eftink and C. A. Ghiron (1975), Dynamics of a protein matrix revealed by fluorescence quenching. *Proc. Nat. Acad. Sci. USA* **72**, 3290–3294.

430. A. Grinvald and I. Z. Steinberg (1974), Fast relaxation processes in a protein revealed by the decay kinetics of tryptophan fluorescence. *Biochemistry* **13**, 5170–5178.

431. M. L. Saviotti and W. C. Galley (1974), Room temperature phosphorescence and the dynamic aspects of protein structure. *Proc. Nat. Acad. Sci. USA* **71**, 4154–4158.

432. G. Careri, P. Fasella, and E. Gratton (1975), Statistical time events in enzymes: A physical assessment. *CRC Crit. Rev. Biochem.* **3**, 141–164.

433. C. J. Epstein, R. F. Goldberger, and C. B. Anfinsen (1963), The genetic control of tertiary protein structure: Studies with model systems. *Cold Spring Harbor Symp. Quant. Biol.* **27**, 439–449.

434. D. B. Wetlaufer and S. Ristow (1973), Aquisition of three-dimensional structure of proteins. *Annu. Rev. Biochem.* **42**, 135–158.

435. E. Haber and C. B. Anfinsen (1962), Side-chain interactions governing the pairing of half-cystine residues in ribonuclease. *J. Biol. Chem.* **237**, 1839–1844.

436. M. Sela and S. Lifson (1959), On the reformation of disulphide bridges in proteins. *Biochim. Biophys. Acta* **36**, 471–478.

437. W. Kauzmann (1959), In: *Sulfur in Proteins* (R. Benesch, P. D. Boyer, I. M. Klotz, W. R. Middlebrook, A. Szent-Györgyi, and Schwartz, Eds.), p. 93. Academic Press, New York.

438. H. F. Epstein, A. N. Schechter, R. F. Chen, and C. B. Anfinsen (1971), Folding of staphylococcal nuclease: Kinetic studies of two processes in acid renaturation. *J. Mol. Biol.* **60**, 499–508.

439. L. L. Shen and J. Hermans, Jr. (1972), Kinetics of conformation change of sperm-whale myoglobin. I. Folding and unfolding of metmyoglobin following pH jump. *Biochemistry* **11**, 1836–1841.

440. P. J. Hagerman and R. L. Baldwin (1976), A quantitative treatment of the kinetics of the folding transition of ribonuclease A. *Biochemistry* **15**, 1462–1473.

441. M. L. Anson (1945), Protein denaturation and the properties of protein groups. *Adv. Prot. Chem.* **2**, 361–386.

442. D. F. Steiner and P. Oyer (1967), The biosynthesis of insulin and a probable precursor of insulin by a human islet cell adenoma. *Proc. Nat. Acad. Sci. USA* **57**, 473–480.

443. D. F. Steiner and J. L. Clark (1968), The spontaneous reoxidation of reduced beef and rat proinsulins. *Proc. Nat. Acad. Sci. USA* **60**, 622–629.

444. P. Varandani and L. A. Shroyer (1973), Insulin degradation. *Biochim. Biophys. Acta* **320**, 249–257.

445. M. E. Friedmann, H. A. Scheraga, and R. F. Goldberger (1966), Structural studies of ribonuclease. XXVI. The role of tyrosine 115 in the refolding of ribonuclease. *Biochemistry* **5**, 3770–3778.

446. I. Zabin and M. R. Villarejo (1975), Protein complementation. *Annu. Rev. Biochem.* **44**, 295–313.

447. R. L. Baldwin (1975), Intermediates in protein folding reactions and the mechanism of protein folding. *Annu. Rev. Biochem.* **44**, 453–475.

448. J. R. Garel and R. L. Baldwin (1973), Both the fast and slow refolding reactions of ribonuclease A yield native enzyme. *Proc. Nat. Acad. Sci. USA* **70**, 3347–3351.

449. J.-R. Garel, B. T. Nall, and R. L. Baldwin (1976), Guanidine-unfolded state of ribonuclease A contains both fast- and slow-refolding species. *Proc. Nat. Acad. Sci. USA* **73**, 1853–1857.

450. J. R. Garel and R. L. Baldwin (1975), The heat-unfolded state of ribonuclease A is an equilibrium mixture of fast and slow refolding species. *J. Mol. Biol.* **94**, 611–620.

451. T. E. Creighton (1977), Energetics of folding and unfolding of pancreatic trypsin inhibitor. *J. Mol. Biol.* **113**, 295–312. (See also earlier papers in the same journal.)

452. J. F. Brandts, H. R. Halvorson, and M. Brennan (1975), Consideration of the possibility that the slow step in protein denaturation reactions is due to *cis–trans* isomerism of proline residues. *Biochemistry* **14**, 4953–4963.

453. B. Furie, A. N. Schechter, D. H. Sachs, and C. B. Anfinsen (1975), An immunological approach to the conformational equilibrium of staphylococcal nuclease. *J. Mol. Biol.* **92**, 497–506.

454. J. G. Curd, A. N. Schechter, and C. B. Anfinsen (1975), Isolation of antibodies specific for a helical region of staphylococcal nuclease by affinity chromatography with synthetic polypeptides. *Fed. Proc.* **34**, 550.

455. P. J. Flory (1956), Theory of elastic mechanisms in fibrous proteins. *J. Amer. Chem. Soc.* **78**, 5222–5235.

456. A. Nakajima and H. A. Scheraga (1961), Thermodynamic study of shrinkage and of phase equilibrium under stress in films made from ribonuclease. *J. Amer. Chem. Soc.* **83**, 1575–1584.

457. T. E. Creighton (1974), The single-disulphide intermediates in the refolding of reduced pancreatic trypsin inhibitor. *J. Mol. Biol.* **87**, 603–624.

458. T. E. Creighton (1977), Kinetics of refolding of reduced ribonuclease. *J. Mol. Biol.* **113**, 329–341.

459. R. R. Hantgan, G. G. Hammes, and H. A. Scheraga (1974), Pathways of folding of reduced bovine pancreatic ribonuclease. *Biochemistry* **13**, 3421–3431.

460. W. L. Anderson and D. B. Wetlaufer (1976), The folding pathway of reduced lysozyme. *J. Biol. Chem.* **251**, 3147–3153.

461. A. N. Schechter and C. J. Epstein (1968), Spectral studies on the denaturation of myoglobin. *J. Mol. Biol.* **35**, 567–589.

462. E. Stellwagen, R. Rysavy, and G. Babul (1972), The conformation of horse heart apocytochrome *c. J. Biol. Chem.* **247**, 8074–8077.

463. W. Fischer, H. Taniuchi, and C. B. Anfinsen (1973), On the role of heme in the formation of the structure of cytochrome *c. J. Biol. Chem.* **248**, 3188–3195.

464. A. Light, H. Taniuchi, and R. F. Chen (1974), A kinetic study of the complementation of fragments of staphylococcal nuclease. *J. Biol. Chem.* **249**, 2285–2293.

465. H. Taniuchi, J. L. Bohnert, and D. S. Parker (1974), The strengthening of cooperative interactions in the structure of nuclease T' by binding of ligands. *Fed. Proc.* **33**, 1589A.

466. L. Bornmann, B. Hess, and H. Zimmermann-Telschow (1974), Mechanism of renaturation of pyruvate kinase of *S. carlsbergensis:* Activation by L-valine and magnesium and mangenese ions. *Proc. Nat. Acad. Sci. USA* **71**, 1525–1529.

467. B. Hess and L. Bornmann (1975), Control of enzyme activity through regulation of three-dimensional structure of proteins: Studies on pyruvate kinase. *Adv. Enz. Reg.* **13**, 235–245.

468. G. W. Hatfield and R. O. Burns (1970), Ligand-induced maturation of threonine deaminase. *Science* **167**, 75–76.

469. J. B. Alpers, H. Paulus, and G. A. Bazylewicz (1971), ATP-catalyzed preconditioning of phosphofructokinase. *Proc. Nat. Acad. Sci. USA* **68**, 2937–2940.

470. O. B. Ptitsyn and A. A. Rashin (1975), A model of myoglobin self-organization. *Biophys. Chem.* **3**, 1–20.

471. A. Warshel and M. Levitt (1976), Folding and stability of helical proteins: Carp myogen. *J. Mol. Biol.* **106**, 421–437.

472. H. Paulus and J. B. Alpers (1971), Preconditioning: An obligatory step in the biosynthesis of oligomeric enzymes and its promotion by allosteric ligands. *Enzyme* **12**, 385–401.

473. R. E. Dickerson (1977), Energy and evolution in the folding of proteins. In: *Molecular evolution and Polymorphism* (M. Kimura, Ed.). National Institute of Genetics, Mishima, Japan.

474. M. O. Dayhoff (1976), The origin and evolution of protein superfamilies. *Fed. Proc.* **35**, 2132–2138.

475. R. J. Britten and E. H. Davidson (1976), DNA sequence arrangement and preliminary evidence on its evolution. *Fed. Proc.* **35**, 2151–2157.

476. D. E. Kohne (1970), Evolution of higher-organism DNA. *Quart. Rev. Biophys.* **3**, 327–375

477. N. J. Cowan, D. S. Secher, and C. Milstein (1976), Purification and sequence analysis of the mRNA coding for an Ig heavy chain. *Eur. J. Biochem.* **61**, 355–368.

478. M. Kimura (1968), Evolutionary rate at the molecular level. *Nature* **217**, 624–626.

479. M. Kimura (1969), The rate of molecular evolution considered from the standpoint of population genetics. *Proc. Nat. Acad. Sci. USA* **63**, 1181–1188.

480. M. Kimura and T. Ohta (1974), On some principles governing molecular evolution. *Proc. Nat. Acad. Sci. USA* **71**, 2848–2852.

481. H. Harris (1976), Molecular evolution: The neutralist–selectionist controversy. *Fed. Proc.* **35**, 2079–2082.

482. J. M. Thoday (1975), Non-Darwinian evolution and biological progress. *Nature* **255**, 675–677.

483. W. M. Fitch and C. H. Langley (1976), Protein evolution and the molecular clock. *Fed. Proc.* **35**, 2092–2097.

484. W. Salser, S. Bowen, D. Browne, F. El Adli, N. Fedoroff, K. Fry, H. Heindell, G. Paddock, R. Poon, B. Wallace, and P. Whitcome (1976), Investigation of the organisation of mammalian chromosomes at the DNA sequence level. *Fed. Proc.* **35**, 23–35.

485. C. A. Marotta, B. G. Forget, S. M. Weissmann, I. M. Verma, R. P. McCaffrey, and D. Baltimore (1974), Nucleotide sequences of human globin messenger RNA. *Proc. Nat. Acad. Sci. USA* **71**, 2300–2304.

486. M. Grunstein, P. Schedl, and L. Kedes (1976), Sequence analysis and evolution of sea urchin (*Lytechinus pictus* and *Strongylocentrotus purpuratus*) histone H4 messenger RNAs. *J. Mol. Biol.* **104**, 351–369.

487. E. S. Weinberg, M. L. Birnstiel, I. F. Purdom, and R. Williamson (1972), Genes coding for polysomal 9S RNA of sea urchins: Conservation and divergence. *Nature* **240**, 225–228.

488. E. A. Barnard, M. S. Cohen, M. H. Gold, and K. Jae-Kyoung (1972), Evolution of ribonuclease in relation to polypeptide folding mechanism. *Nature* **240**, 395–398.

489. J. L. King (1976), Progress in the neutral mutation - random drift controversy. *Fed. Proc.* **35**, 2087–2091.

490. B. S. Hartley (1970), Homologies in serine proteases. *Phil. Trans. Roy. Soc. London,* Ser. **B 257**, 77–87.

491. P. L. Levy, M. K. Pangburn, Y. Burstein, L. H. Ericsson, H. Neurath, and K. A. Walsh (1975), Evidence of homologous relationship between thermolysin and neutral protease A of *Bacillus subtilis*. *Proc. Nat. Acad. Sci. USA* **72**, 4341–4345.

492. P. W. Hochachka (1975), Why study proteins of abyssal organisms? *Comp. Biochem. Physiol.* **52**, B supplement.

493. R. W. Carrell, H. Lehmann, P. A. Lorkin, E. Raik, and E. Hunter (1967), Haemoglobin Sydney: β67 valin $\rightarrow$ alanin: An emerging pattern of unstable haemoglobins. *Nature* **215**, 626–628.

494. M. F. Perutz and H. Lehmann (1968), Molecular pathology of human haemoglobin. *Nature* **219**, 902–909.

495. T. Takano, B. L. Trus, N. Mandel, G. Mandel, O. B. Kallai, R. Swanson, and R. E. Dickerson (1977), Tuna cytochrome c at 2.0 Å resolution. *J. Biol. Chem.* **252**, 776–785.

496. S. Ohno (1973), Ancient linkage groups and frozen accidents. *Nature* **244**, 259–262.

497. D. G. Smyth (1975), Constructing semisynthetic polypeptides. (editorial) *Nature* **256**, 699–700.

498. A. R. Rees and R. E. Offord (1976), The semisynthesis of portions of hen's egg lysozyme by fragment condensation. *Biochem. J.* **159**, 487–493.

499. B. Gutte (1976), Shortened ribonuclease A: Conformational and mechanistic aspects. Xth Int. Congr. Biochem. (Hamburg), Proceedings. Abstr. 04-6-376.

500. O. B. Ptitsyn (1974), Invariant features of globin primary structure and coding of their secondary structure. *J. Mol. Biol.* **88**, 287–300.

501. P. K. Warme, F. A. Momany, S. V. Rumball, R. W. Tuttle, and H. A.

Scheraga (1974), Computation of structures of homologous proteins. Alpha-lactalbumin from lysozyme. *Biochemistry* **13**, 768–782.

502. E. Margoliash (1963), Primary structure and evolution of cytochrome *c*. *Proc. Nat. Acad. Sci. USA* **50**, 672–679.

503. E. L. Smith and E. Margoliash (1964), Evolution of cytochrome *c*. *Fed. Proc.* **23**, 1243–1247.

504. W. M. Fitch and E. Margoliash (1967), Construction of phylogenetic trees. *Science* **155**, 279–284.

505. C. Nolan and E. Margoliash (1968), Comparative aspects of primary structures of proteins. *Annu. Rev. Biochem.* **37**, 727–790.

506. E. L. Smith (1970), Evolution of enzymes. *The Enzymes* **1**, 267–339.

507. A. M. Reiner (1975), Genes for ribitol and D-arabitol catabolism in *Escherichia coli:* Their loci in C strains and absence in K-12 and B strains. *J. Bacteriol.* **123**, 530–536.

508. B. S. Hartley, I. Altosaar, J. M. Dothie, and M. S. Neuberger (1976), Experimental evolution of a xylitol dehydrogenase. In: *Structure–Function Relationships of Proteins* (R. Markham and R. W. Horne, Eds.) pp. 191–200. North Holland, Amsterdam, New York, Oxford.

509. R. E. Dickerson and R. Timkovich (1975), Cytochromes *c. The Enzymes* **11**, 397–547.

510. L. T. Hunt and M. O. Dayhoff (1976), Globins. *Atlas of Protein Sequences* **5**, suppl. 2, 191–224.

511. W. Garstang (1894), Preliminary note on a new theory of the phylogeny of the chordata. *Zool. Anz.* **17**, 122–125.

512. D. C. Watts (1971), Evolution of phosphagen kinases. In: *Molecular Evolution* (E. Schoffeniels, Ed.), Vol. 2, pp. 150–173. North Holland, Amsterdam, New York, Oxford.

513. M. O. Dayhoff (1969), Computer analysis of protein evolution. *Sci. Amer.* July, 86–95.

514. M.-C. King and A. C. Wilson (1975), Evolution at two levels in humans and chimpanzees. *Science* **188**, 107–116.

515. A. C. Wilson, L. R. Maxson, and V. M. Sarich (1974), Two types of molecular evolution. Evidence from studies of interspecific hybridization. *Proc. Nat. Acad. Sci. USA* **71**, 2843–2847.

516. A. C. Wilson (1975), Evolutionary importance of gene regulation. *Stadler Symp.* **7**, 117–134.

517. A. C. Wilson, V. M. Sarich, and L. R. Maxson (1974), The importance of gene rearrangement in evolution: Evidence from studies on rates of chromosomal, protein, and anatomical evolution. *Proc. Nat. Acad. Sci. USA* **71**, 3028–3030.

518. C. C. F. Blake, L. N. Johnson, G. A. Mair, A. C. T. North, D. C. Phillips, and V. R. Sarma (1967), Crystallographic studies of the activity of hen egg-white lysozyme. *Proc. Roy. Soc.* Ser. **B 167**, 378–388. (Other aspects of lysozyme structure and function are also discussed in Vol. 167.)

519. P. H. Clarke (1976), Genes and enzymes. *FEBS Lett.* **62**, E37–E46.

520. C. Wills (1976), Production of yeast alcohol dehydrogenase isoenzymes by selection. *Nature* **261**, 26–29.

521. J. L. Betz, P. R. Brown, M. J. Smyth, and P. H. Clarke (1974), Evolution in action. *Nature* **247**, 261–264.

522. P. W. J. Rigby, B. D. Burleigh, Jr., and B. S. Hartley (1974), Gene duplication in experimental enzyme evolution. *Nature* **251**, 200–204.

523. B. S. Hartley, B. D. Burleigh, G. G. Midwinter, C. H. Moore, H. R. Morris, P. W. J. Rigby, M. J. Smith, and S. S. Taylor (1966), Where do new enzymes come from? *FEBS Proc.* **29**, 151–176.

524. B. G. Hall (1976), Experimental evolution of a new enzymatic function. Kinetic analysis of the ancestral ebg⁰ and evolved ebg⁺ enzymes. *J. Mol. Biol.* **107**, 71–84.

525. V. M. Ingram (1961), Gene evolution and the haemoglobins. *Nature* **189**, 704–708.

526. R. Acher (1971), The neurohypophyseal hormones: An example of molecular evolution. In: *Molecular Evolution* (E. Schoffeniels, Ed.), Vol. 2, pp. 43–51. North Holland, Amsterdam, New York, Oxford.

527. A. Wichmann (1899), Ueber die Krystallformen der Albumine (Krystallisation von Lactalbumin). *Z. Phys. Chem.* **27**, 575–593.

528. R. L. Hill and K. Brew (1975), Lactose synthetase. *Adv. Enzymol.* **43**, 411–490.

529. A. Fleming (1922), On a remarkable bacteriolytic element found in tissues and secretions. *Proc. Roy. Soc.* **B 93**, 306–317.

530. E. P. Abraham and R. Robinson (1937), Crystallization of lysozyme. *Nature* **140**, 24. "Miss D. Crowfoot has kindly undertaken their crystallographic examination."

531. D. C. Phillips (1967), The hen egg-white lysozyme molecule. *Proc. Nat. Acad. Sci. USA* **57**, 484–495.

532. K. Brew, T. C. Vanaman, and R. L. Hill (1967), Comparison of the amino-acid sequence of bovine α-lactalbumin and hen's egg-white lysozyme. *J. Biol. Chem.* **242**, 3747–3749.

533. K. Brew, F. J. Castellino, T. C. Vanaman, and R. L. Hill (1970), The complete amino acid sequence of bovine α-lactalbumin. *J. Biol. Chem.* **245**, 4570–4582.

534. W. J. Browne, A. C. T. North, D. C. Phillips, K. Brew, T. C. Vanaman, and R. L. Hill (1969), A possible three-dimensional structure of bovine α-lactalbumin based on that of hen's egg-white lysozyme. *J. Mol. Biol.* **42**, 65–86.

535. P. K. Warme and H. A. Scheraga (1974), Refinement of the X-ray structure of lysozyme by complete energy minimization. *Biochemistry* **13**, 757–767.

536. R. M. Stroud (1974), A family of protein-cutting proteins. *Sci. Amer.* July, 74–88.

537. D. M. Blow (1976), Structure and mechanism of chymotrypsin. *Acc. Chem. Res.* **9**, 145–152.

538. Y. Beppu and S. Yomosa (1977), Molecular orbital studies on the enzymatic reaction mechanism of serine proteases. *J. Physic. Soc. Japan* **42**, 1694–1700.

539. D. C. Phillips, D. M. Blow, B. S. Hartley, and G. Lowe (Eds.) (1970), A discussion on the structures and functions of proteolytic enzymes. *Phil. Trans. Roy. Soc. London, Ser.* **B 257**, 65–266.

540. R. J. Poljak (1975), Three-dimensional structure, function and genetic control of immunoglobulins. *Nature* **256**, 373–376.

541. W. C. Barker and M. O. Dayhoff (1976), Immunoglobulins and related proteins. *Atlas of Protein Sequence and Structure* **5**, suppl. 2, 165–190.

542. J. D. Watson (1976), *Molecular Biology of the Gene,* 3rd ed. W. A. Benjamin, Menlo Park, California.

543. R. Huber (1976), Antibody structure. *Trends Biochem. Sci.* **1**, 174–178.

544. D. R. Davies, E. A. Padlan, and D. M. Segal (1975), Three-dimensional structure of immunoglobulins. *Annu. Rev. Biochem.* **44**, 639–667.

545. R. J. Poljak, L. M. Amzel, H. P. Avey, B. L. Chen, R. P. Phizackerley, and F. Saul (1973), Three-dimensional structure of the Fab' fragment of a human immunoglobulin at 2.8 Å-resolution. *Proc. Nat. Acad. Sci. USA* **70**, 3305–3310.

546. I. Fridovich (1974), Superoxide dismutases. *Adv. Enzymol.* **41**, 35–97.

547. J. L. Strominger (1979), Histocompatibility antigens. *Annu. Rev. Biochem.* **48**, in press.

548. J. S. Richardson, D. C. Richardson, K. A. Thomas, E. W. Silverton, and D. R. Davies (1976), Similarity of three-dimensional structure between the immunoglobulin domain and the copper, zinc superoxide dismutase subunit. *J. Mol. Biol.* **102**, 221–235.

549. A. Riggs (1976), Factors in the evolution of hemoglobin function. *Fed. Proc.* **35**, 2115–2118.

550. M. F. Perutz (1979), Haemoglobin. *Annu. Rev. Biochem.* **48**, in press.

551. W. Kühne (1865), Ueber den Farbstoff der Muskeln. *Virchows Arch.* **33**, 79–94.

552. F. Hoppe-Seyler (1864), Ueber die chemischen und optischen Eigenschaften des Blutfarbstoffs. *Virchows Arch.* **29**, 233–235.

553. M. F. Perutz, M. G. Rossmann, A. F. Cullis, H. Muirhead, G. Will, and A. C. T. North (1960), Structure of haemoglobin. *Nature* **185**, 416–422.

554. A. B. Edmundson (1965), Amino-acid sequence of sperm whale myoglobin. *Nature* **205**, 883–887.

555. G. Braunitzer, R. Gehring-Müller, N. Hilschmann, K. Hilse, G. Hobom, V. Rudloff, and B. Wittmann-Liebold (1961), Die Konstitution des normalen adulten Humanhämoglobins. *Z. Physiol. Chem.* **325**, 283–286.

556. R. Keller, O. Groudinsky, and K. Wüthrich (1973), Proton magnetic resonances in cytochrome b_2 core. Structural similarities with cytochrome b_5. *Biochim. Biophys. Acta* **328**, 233–238.

557. B. Guiard, O. Groudinsky, and F. Lederer (1974), Homology between bakers' yeast cytochrome b_2 and liver microsomal cytochrome b_5. *Proc. Nat. Acad. Sci. USA* **71**, 2539–2543.

558. P. K. Warme and L. P. Hager (1970), Mesoheme sulfuric anhydride as a heme protein structure probe. Reaction with cytochrome b_{562}. *Biochemistry* **9**, 4244–4251.

559. J. Ozols and P. Strittmatter (1967), The homology between cytochrome b_5, hemoglobin and myoglobin. *Proc. Nat. Acad. Sci. USA* **58**, 264–267.

560. M. G. Rossmann and P. Argos (1975), A comparison of the heme binding pocket in globins and cytochrome b_5. *J. Biol. Chem.* **250**, 7525–7532.

561. B. Hagihara, N. Sato, and T. Yamanaka (1975), Type b cytochromes. *The Enzymes* **11**, 549–593.

562. F. S. Mathews and F. Lederer (1976), Crystallographic study of bakers' yeast cytochrome b_2. *J. Mol. Biol.* **102**, 853–857.

563. E. W. Czerwinski, F. S. Mathews, P. Hollenberg, K. Drickamer, and L. P. Hager (1972), Crystallographic study of cytochrome b_{562} from *E. coli*. *J. Mol. Biol.* **71**, 819–821.

564. T. Takano, R. Swanson, O. B. Kallai, and R. E. Dickerson (1971), Conforma-

tional changes upon reduction of cytochrome *c*. *Cold Spring Harbor Symp.* **36**, 397–403.

565. D. W. Seybert and K. Moffat (1977), Structure of haemoglobin reconstituted with mesoheme. *J. Mol. Biol.* **113**, 419–430.

566. P. Argos and F. S. Mathews (1975), The structure of ferrocytochrome b_5 at 2.8 Å resolution. *J. Biol. Chem.* **250**, 747–751.

567. P. Strittmatter (1960), The nature of the heme binding in microsomal cytochrome b_5. *J. Biol. Chem.* **235**, 2492–2497.

568. J. Ozols and P. Strittmatter (1964), The interaction of porphyrins and metalloporphyrins with apocytochrome b_5. *J. Biol. Chem.* **239**, 1018–1023.

569. M. G. Rossmann and P. Argos (1976), Exploring structural homology of proteins. *J. Mol. Biol.* **105**, 75–96.

570. P. John and F. R. Whatley (1975), *Paracoccus denitrificans* and the evolutionary origin of the mitochondrion. *Nature* **254**, 495–498.

571. R. E. Dickerson, R. Timkovich, and R. J. Almassy (1976), The cytochrome fold and the evolution of bacterial energy metabolism. *J. Mol. Biol.* **100**, 473–491.

572. J. L. Connelly, O. T. G. Jones, V. A. Saunders, and D. W. Yates (1973), Kinetic and thermodynamic properties of membrane-bound cytochromes of aerobically and photosynthetically grown *Rhodopseudomonas spheroides*. *Biochim. Biophys. Acta* **292**, 644–653.

573. N. H. Horowitz (1945), On the evolution of biochemical syntheses. *Proc. Nat. Acad. Sci. USA* **31**, 153–157.

574. W. T. Astbury (1950), Adventures in molecular biology. *Harvey Lect.* **46**, 3–44.

575. H. P. Vosberg (1977), Molecular cloning of DNA. *Human Gen.* **40**, 1–72.

576. G. G. Brownlee, E. M. Cartwright, N. J. Cowan, M. Jarvis, and C. Milstein (1973), Purification and sequence of messenger RNA for immunoglobulin light chains. *Nature New Biol.* **244**, 236–240.

577. S. Tonegawa, N. Hozumi, G. Matthyssens, and R. Schuller (1977), Somatic changes in the content and context of immunoglobulin genes. *Cold Spring Harbor Symp.* **41**, 877–889.

578. P. Truffa-Bachi, M. Veron, and G. N. Cohen (1974), Structure, function, and possible origin of a bifunctional allosteric enzyme, *Escherichia coli* aspartokinase I- homoserine dehydrogenase I. *Crit. Rev. Biochem.* **2**, 379–415.

579. I. P. Crawford (1975), Gene rearrangements in the evolution of the tryptophan synthetic pathway. *Bacteriol. Rev.* **39**, 87–120.

580. E. Schweizer, A. Knobling, G. Manger, and G. Dietlein (1975), Gene fusion in the fatty acid synthetase system of *Saccharomyces cerevisiae*. *FEBS Proc.* **32**, 133–146.

581. J. Yourno, T. Kohno, and J. R. Roth (1970), Enzyme evolution: Generation of a bifunctional enzyme by fusion of adjacent genes. *Nature* **228**, 820–824.

582. K. D. Tartof (1975), Redundant genes. *Annu. Rev. Gen.* **9**, 355–385.

583. L. Hood, J. H. Campbell, and S. C. R. Elgin (1975), The organisation, expression and evolution of antibodies and other multigene families. *Annu. Rev. Genet.* **9**, 305–353.

584. C. B. Bridges (1936), The BAR "gene", a duplication. *Science* **83**, 210–211.

585. L. Hood (1976), Antibody genes and other multigene families. *Fed. Proc.* **35**, 2158–2167.

586. E. Zuckerkandl (1975), The appearance of new structures and functions in proteins during evolution. *J. Mol. Evol.* **7**, 1–57.

587. A. D. McLachlan (1972), Repeating sequences and gene duplications in proteins. *J. Mol. Biol.* **64**, 417–437.

588. J. A. Black and G. H. Dixon (1968), Amino-acid sequence of alpha chains of human haptoglobins. *Nature* **218**, 736–741.

589. A. D. McLachlan and J. E. Walker (1977), Evolution of serum albumin. *J. Mol. Biol.* **112**, 543–558.

590. J. H. Collins (1976), Homology of myosin DTNB light chain with alkali light chains, troponin C and parvalbumin. *Nature* **259**, 699–700.

591. C. R. Cantor and T. H. Jukes (1966), The repetition of homologous sequences in the polypeptide chains of certain cytochromes and globins. *Proc. Nat. Acad. Sci. USA* **56**, 177–184.

592. R. L. Russell, J. N. Abelson, A. Landy, M. L. Gefter, S. Brenner, and J. D. Smith (1970), Duplicate genes for tyrosine transfer RNA in *Escherichia coli*. *J. Mol. Biol.* **47**, 1–14.

593. M. Ycas (1976), Origin of periodic proteins. *Fed. Proc.* **35**, 2139–2140.

594. W. A. Hendrickson (1977), The molecular architecture of oxygen-carrying proteins. *Trends Biochem. Sci.* **2**, 108–110.

595. J. Bridgen, J. I. Harris, and F. Northrop (1975), Evolutionary relationships in superoxide dismutase. *FEBS Lett.* **49**, 392–395.

596. E. Subramanian (1978), Molecular structure of acid-proteases. *Trends Biochem. Sci.* **3**, 1–3.

597. A. B. Edmundson, K. R. Ely, E. E. Abola, M. Schiffer, N. Panagiotopoulos, and H. F. Deutsch (1976), Conformational isomerism, rotational allomerism, and divergent evolution in immunoglobulin light chains. *Fed. Proc.* **35**, 2119–2123.

598. A. D. McLachlan (1971), Tests for comparing related amino acid sequences. Cytochrome c and cytochrome c_{551}. *J. Mol. Biol.* **61**, 409–424.

599. A. D. McLachlan and M. Stewart (1976), The 14-fold periodicity in α-tropomyosin and the interaction with actin. *J. Mol. Biol.* **103**, 271–298.

600. A. D. McLachlan (1972), Gene duplication in carp muscle calcium binding protein. *Nature New Biol.* **240**, 83–85.

601. M. G. Rossmann and P. Argos (1977), The taxonomy of protein structure. *J. Mol. Biol.* **109**, 99–129.

602. R. H. Schirmer and I. Schirmer (1976), The first enzyme—Kühne's trypsin—a hundred years old favourite. *Pharma Int.* **5/6**, 10–16.

603. B. A. Cunningham, J. L. Wang, I. Berggård, and P. A. Peterson (1973), The complete amino acid sequence of β_2-microglobulin. *Biochemistry* **12**, 4811–4822.

604. H. J. Evans, H. M. Steinmann, and R. L. Hill (1974), Bovine erythrocyte superoxide dismutase II. *J. Biol. Chem.* **249**, 7315–7325.

605. H. M. Steinmann, V. R. Naik, J. L. Abernethy, and R. L. Hill (1974), Bovine erythrocyte superoxide dismutase I. *J. Biol. Chem.* **249**, 7326–7338.

606. J. L. Abernethy, H. M. Steinmann, and R. L. Hill (1974), Bovine erythrocyte superoxide dismutase III. *J. Biol. Chem.* **249**, 7339–7347.

607. D. M. Segal, E. A. Padlan, G. H. Cohen, S. Rudikoff, M. Potter, and D. R. Davies (1974), The three-dimensional structure of a phosphorylcholine-binding mouse immunoglobulin Fab and the nature of the antigen binding site. *Proc. Nat. Acad. Sci. USA* **71**, 4298–4302.

608. P. M. Colman, J. Deisenhofer, R. Huber, and W. Palm (1976), Structure of the human antibody molecule Kol (Immunoglobulin G1). An electron density map at 5 Å resolution. *J. Mol. Biol.* **100**, 257–282.

609. E. Itagaki and L. P. Hager (1966), Studies on cytochrome b_{562} of *Escherichia coli. J. Biol. Chem.* **241**, 3687–3695.

610. N. Oshino, Y. Imai, and R. Sato (1971), A function of cytochrome b_5 in fatty acid desaturation by rat liver microsomes. *J. Biochem. (Tokyo)* **69**, 155–167.

611. J. Ozols and P. Strittmatter (1969), Correction of the amino acid sequence of calf liver microsomal cytochrome b_5. *J. Biol. Chem.* **244**, 6617–6618.

612. A. Tsugita, M. Kobyashi, S. Tani, S. Kyo, M. A. Rashid, Y. Yoshida, T. Kajihara, and B. Hagihara (1970), Comparative study of the primary structures of cytochrome b_5 from four species. *Proc. Nat. Acad. Sci. USA* **67**, 442–447.

613. E. Itagaki and L. P. Hager (1968), The amino acid sequence of cytochrome b_{562} of *Escherichia coli. Biochem. Biophys. Res. Commun.* **32**, 1013–1019.

614. J. C. Cavadore (1971), Polycondensation d'α-amino acides en milieu aqueux. Thèse doctorat ès sciences physiques, Université des Sciences et Techniques du Languedoc, Montpellier.

615. R. H. Kretsinger (1976), Calcium-binding proteins. *Annu. Rev. Biochem.* **45**, 239–266.

616. M. Nozaki (1976), Structure and function of metalloenzymes (review). *J. Syn. Org. Chem.* **34** (NM), 805–817.

617. M. Potter, S. Rudikoff, M. Vrana, D. N. Rao, and E. B. Mushinksi (1976), Primary structural differences in myeloma proteins that bind the same haptens. *Cold Spring Harbor Symp.* **41**, 661–666.

618. L. M. Amzel, R. J. Poljak, F. Saul, B. M. Varga, and F. F. Richards (1974), The three-dimensional structure of a combining region–ligand complex of immunoglobulin NEW at 3.5 Å resolution. *Proc. Nat. Acad. Sci. USA* **71**, 1427–1430.

619. T. T. Wu and E. A. Kabat (1970), An analysis of the sequences of the variable regions of Bence–Jones proteins and myeloma light chains and their implications for antibody complementarity. *J. Exp. Med.* **132**, 211–250.

620. E. A. Padlan, D. R. Davies, I. Pecht, D. Givol, and C. Wright (1976), Model-building studies of antigen-binding sites: The hapten-binding site of MOPC-315. *Cold Spring Harbor Symp.* **41**, 627–637.

621. R. A. Dwek, S. Wain-Hobson, S. Dower, P. Gettins, B. Sutton, S. J. Perkins, and D. Givol (1977), Structure of an antibody combining site by magnetic resonance. *Nature* **266**, 31–37.

622. M. W. Hunkapiller, A. M. Goetze, J. H. Richards, and L. E. Hood (1977), Structure–function correlates for DNP and PC binding myeloma proteins. *Immunogenetics* **4**, 424–425.

623. J. Klotz and D. L. Hunstron (1975), Protein-interaction with small molecules. *J. Biol. Chem.* **250**, 3001–3009.

624. M. E. Jolley, S. Rudikoff, M. Potter, and C. P. Glaudemans (1973), Spectral changes on binding of oligosaccharides to murine IgA myeloma immunoglobulins. *Biochemistry* **12**, 3039–3044.

625. E. A. Kabat (1961), In: *Experimental Immunochemistry* (E. A. Kabat and M. M. Mayer, Eds.), 2nd ed. Charles C Thomas, Springfield, Illinois.

626. N. Citri (1973), Conformational adaptability in enzymes. *Adv. Enzymol.* **37**, 397–647.

627. J. D. Robertus, R. A. Alden, J. J. Birktoft, J. Kraut, J. C. Powers, and P. E. Wilcox (1972), An X-ray crystallographic study of the binding of peptide chloromethyl ketone inhibitors to subtilisin BPN'. *Biochemistry* **11**, 2439–2449.

628. D. M. Blow, J. J. Birktoft, and B. S. Hartley (1969), Role of a buried acid group in the mechanism of action of chymotrypsin. *Nature* **221**, 337–340.

629. D. W. Ingles and J. R. Knowles (1967), Specificity and stereo-specificity of α-chymotrypsin. *Biochem. J.* **104**, 369–377.

630. J. B. S. Haldane (1930), *Enzymes*. Longmans, Green, London.

631. W. P. Jencks (1975), Binding energy, specificity and enzymic catalysis: The Circe effect. *Adv. Enzymol.* **43**, 220–410.

632. R. M. Sweet, H. T. Wright, J. Janin, C. H. Chothia, and D. M. Blow (1974), Crystal structure of the complex of porcine trypsin with soybean trypsin inhibitor (Kunitz) at 2.6 Å resolution. *Biochemistry* **13**, 4212–4228.

633. H. Fischer and K. Zeile (1929), Synthese des Hämatoporphyrins, Protoporphyrins und Hämins. *Ann. Chem.* **468**, 98–116.

634. J. M. Baldwin (1975), Structure and function of haemoglobin. *Progr. Biophys. Mol. Biol.* **29**, 225–320.

635. H. C. Watson and C. L. Nobbs (1968), *The Structure of Oxygenated and Deoxygenated Myoglobin,* Colloq. Ges. Biol. Chem. (Mosbach/Baden), pp. 37–48. Springer-Verlag, Heidelberg.

636. H. C. Watson (1969), The stereochemistry of the protein myoglobin. *Progr. Stereochem.* **4**, 299–333.

637. J. E. Falk (1964), *Porphyrins and Metalloporphyrins,* BBA Library 2, Elsevier, Amsterdam–London–New York.

638. J. Le Gall and J. R. Postgate (1973), Bacterial cytochromes *c, Adv. Microbiol. Physiol.* **10**, 81–133.

639. J. H. Wang (1970), Synthetic biochemical models. *Accounts Chem. Res.* **3**, 90–97.

640. F. R. Salemme (1976), A hypothetical structure for an intermolecular electron transfer complex of cytochromes *c* and b_5. *J. Mol. Biol.* **102**, 563–568.

641. A. C. Maehley (1961), Haemin-protein binding in peroxidase and methaemalbumin. *Nature* **192**, 630–632.

642. H. Schichi, D. P. Hackett, and G. Funatsu (1963), Studies on the *b*-type cytochromes from mung bean seedlings. *J. Biol. Chem.* **238**, 1156–1161.

643. C. L. Nobbs, H. C. Watson, and J. C. Kendrew (1966), Structure of deoxymyoglobin: A crystallographic study. *Nature* **209**, 339–341.

644. M. M. Maltempo, T. H. Moss, and M. A. Cusanovich (1974), Magnetic studies on the changes in the iron environment in *chromatium* ferricytochrome c^1. *Biochim. Biophys. Acta* **342**, 290–305.

645. S. C. Harrison and E. R. Blout (1965), Reversible conformational changes of myoglobin and apomyoglobin. *J. Biol. Chem.* **240**, 299–303.

646. V. I. Lim and A. V. Efimov (1977), The folding pathway for globins. *FEBS Lett.* **78**, 279–283.

647. A. H. Corwin and J. G. Erdman (1946), A synthetic ferroporphyrin complex that is passive to oxygen. *J. Amer. Chem. Soc.* **68**, 2473–2478.

648. J. P. Collmann, J. J. Brauman, T. R. Halbert, and K. S. Suslick (1976), Nature of O_2 and CO binding to metalloporphyrins and haem proteins. *Proc. Nat. Acad. Sci. USA* **73**, 3333–3337.

649. M. Rougee and D. Brault (1975), Influence of trans weak or strong field ligands upon the affinity of deuteroheme for carbon monoxide. Monoimidazoleheme as a reference for unconstrained five-coordinate hemoproteins. *Biochemistry* **14**, 4100–4105.

650. L. Pauling (1964), Nature of the iron–oxygen bond in oxyhaemoglobin. *Nature* **203**, 182–183.

651. J. P. Collmann, R. R. Gagne, C. A. Reed, W. T. Robinson, and G. A. Rodley (1974), Structure of an iron II dioxygen complex; a model for oxygen-carrying hemeproteins. *Proc. Nat. Acad. Sci. USA* **71**, 1326–1329.

652. J. J. Weiss (1964), Nature of the iron–oxygen bond in oxyhaemoglobin. *Nature* **202**, 83–84.

653. A. J. Thomson (1977), Ligand binding properties of the haem group. *Nature* **265**, 15–16.

654. J. L. Hoard (1975), In: *Porphyrins and Metalloporphyrins* (K. M. Smith, Ed.) pp. 356–358. Elsevier, New York.

655. R. Huber, O. Epp, and H. Formanek (1970), Structures of deoxy- and carbonmonoxy-erythrocruorin. *J. Mol. Biol.* **52**, 349–354.

656. E. A. Padlan and W. E. Love (1975), Three-dimensional structure of hemoglobin from the polychaete annelid, *Glycera dibranchiata,* at 2.5 Å resolution. *J. Biol. Chem.* **249**, 4067–4078.

657. J. C. Norvell, A. C. Nunes, and B. P. Schoenborn (1975), Neutron diffraction analysis of myoglobin: Structure of the carbon monoxide derivative. *Science* **190**, 568–570.

658. L. S. Kaminsky and V. J. Miller (1972), The ascorbate reduction of denatured ferricytochrome *c. Biochem. Biophys. Res. Commun.* **49**, 252–256.

659. G. R. Moore and R. J. P. Williams (1977), Structural basis for the variation in redox potential of cytochromes. *FEBS Lett.* **79**, 229–232.

660. R. J. Kassner (1972), Effects of nonpolar environments on the redox potentials of heme complexes. *Proc. Nat. Acad. Sci. USA* **69**, 2263–2268.

661. H. C. Freeman (1967), Crystal structures of metal-peptide complexes. *Adv. Prot. Chem.* **22**, 257–424.

662. M. F. Perutz (1969), The haemoglobin molecule. *Proc. Roy. Soc. (B)* **173**, 113–140.

663. L. Pauling and C. D. Coryell (1936), The magnetic properties and structure of the hemochromogens and related substances. *Proc. Nat. Acad. Sci. USA* **22**, 159–164.

664. J. L. Hoard (1966), In: *Hemes and Hemoproteins* (B. Chance, R. W. Estabrook, and T. Yonetani, Eds.). Academic Press, New York and London.

665. E. F. Pai, W. Sachsenheimer, R. H. Schirmer, and G. E. Schulz (1977), Substrate positions and induced-fit in crystalline adenylate kinase. *J. Mol. Biol.* **114**, 37–45.

666. M. F. Perutz (1976), Structure and mechanism of haemoglobin. *Br. Med. Bull.* **32**, 195–208.

667. J. T. Edsall (1972), Blood hemoglobin. *J. Hist. Biol.* **5**, 205–257.

668. J. V. Kilmartin and L. Rossi-Bernardi (1973), Interaction of haemoglobin with hydrogen ions, carbon dioxide, and organic phosphates. *Physiol. Rev.* **53**, 836–890.

669. J. Baggot (1978), The magnitudes of the Bohr and Haldane effects. *Trends Biochem. Sci.* **3**, N30–N32.

670. R. Benesch and R. E. Benesch (1969), Intracellular organic phosphates as regulators of oxygen release by haemoglobin. *Nature* **221**, 618–622.

671. I. Tyuma and K. Shimizu (1970), Effect of organic phosphates on the difference in oxygen affinity between fetal and adult human haemoglobin. *Fed. Proc.* **29**, 1112–1114.

672. A. Arnone (1972), X-ray diffraction study of binding of 2,3-diphosphoglycerate to human deoxyhaemoglobin. *Nature* **237**, 146–149.

673. C. Lenfant, J. D. Torrance, R. D. Woodson, P. Jacobs, and C. A. Finch (1970), Role of organic phosphates in the adaptation of man to hypoxia. *Fed. Proc.* **29**, 1115–1117.

674. C. Bohr (1903), Theoretische Behandlung der quantitativen Verhältnisse bei der Sauerstoffaufnahme des Hämoglobins. *Zentr. Physiol.* **17**, 682–688.

675. C. Bohr, K. Hasselbalch, and A. Krogh (1904), Über einen in biologischer Beziehung wichtigen Einfluss, den die Kohlensäurespannung des Blutes auf dessen Sauerstoffbindung übt. *Skand. Arch. Physiol.* **16**, 402–412.

676. J. Christiansen, C. G. Douglas, and J. S. Haldane (1914), The absorption and dissociation of carbon dioxide by human blood. *J. Physiol.* **48**, 245–271.

677. A. V. Hill (1910), The possible effects of the aggregation of the molecules of haemoglobin on its dissociation curves. *J. Physiol.* **40**, iv–vii.

678. G. S. Adair (1925), The haemoglobin system. *J. Biol. Chem.* **63**, 493–545.

679. J. Wyman and D. W. Allen (1951), The problem of the heme interactions in hemoglobin and the basis of the Bohr effect. *J. Polymer Sci.* **7**, 499–518.

680. J. Wyman (1948), Heme proteins. *Adv. Prot. Chem.* **4**, 407–531.

681. J. Monod, J. P. Changeux, and F. Jacob (1963), Allosteric proteins and cellular control systems. *J. Mol. Biol.* **6**, 306–329.

682. D. E. Koshland, Jr., G. Nemethy, and D. Filmer (1966), Comparison of experimental binding data and theoretical models in proteins containing subunits. *Biochemistry* **5**, 365–385.

683. W. Eventoff and M. G. Rossmann (1975), The evolution of dehydrogenases and kinases. *CRC Crit. Rev. Biochem.* **3**, 112–140.

684. N. Katunuma, K. Kito, and E. Kominami (1971), A new enzyme that specifically inactivates apo-protein of NAD-dependent dehydrogenases. *Biochem. Biophys. Res. Commun.* **45**, 76–81.

685. R. Einarsson, H. Eklund, E. Zeppezauer, T. Boiwe, and C.-I. Brändén (1974), Binding of salicylate in the adenosine-binding pocket of dehydrogenases. *Eur. J. Biochem.* **49**, 41–47.

686. D. E. Koshland, Jr. (1958), Application of a theory of enzyme specificity to protein synthesis. *Proc. Nat. Acad. Sci. USA* **44**, 98–104.

687. D. E. Koshland (1973), Protein shape and biological control. *Sci. Amer.* October, 52–64.

688. W. Sachsenheimer and G. E. Schulz (1977), Two conformations of crystalline adenylate kinase. *J. Mol. Biol.* **114**, 23–36.

689. F. M. Richards and H. W. Wyckoff (1971), Bovine pancreatic ribonuclease. *The Enzymes* **4**, 647–806.

690. S. E. V. Phillips (1978), Structure of oxymyoglobin. *Nature* **273**, 247–248.

691. L. E. Webb, E. J. Hill, and L. J. Banaszak (1973), Conformation of nicotinamide adenine dinucleotide bound to cytoplasmic malate dehydrogenase. *Biochemistry* **12**, 5101–5109.

692. I. Ohlsson, B. Nordström, and C.-I. Brändén (1974), Structural and functional

similarities within the coenzyme binding domains of dehydrogenases. *J. Mol. Biol.* **89**, 339–354.

693. R. E. Williams (1976), Phosphorylated sites in substrates of intracellular protein kinases: A common feature in amino acid sequences. *Science* **192**, 473–474.

694. A. M. Scanu (1977), Plasma lipoprotein structure (editorial). *Nature* **270**, 209–210.

695. C. F. Fox (1972), The structure of cell membranes. *Sci. Amer.* **226**, 30–38.

696. S. J. Singer and G. L. Nicolson (1972), The fluid mosaic model of the structure of cell membranes. *Science* **175**, 720–731.

697. D. E. Metzler (1977), *Biochemistry*. Academic Press, New York.

698. J. Ozols and C. Gerard (1977), Covalent structure of the membranous segment of horse cytochrome b_5. *J. Biol. Chem.* **252**, 8549–8553.

699. R. Henderson and P. N. T. Unwin (1975), Three-dimensional model of purple membrane obtained by electron microscopy. *Nature* **257**, 28–32.

700. A. E. Blaurock (1975), Bacteriorhodopsin: a trans-membrane pump containing α-helix. *J. Mol. Biol.* **93**, 139–158.

701. W. Hasselbach (1977), The sarcoplasmic calcium pump—a most efficient ion translocating system. *Biophys. Struct. Mech.* **3**, 43–54.

702. D. H. MacLennan and P. C. Holland (1975), Calcium transport in sarcoplasmic reticulum. *Annu. Rev. Biophys. Bioeng.* **4**, 377–404.

703. W. Hasselbach and M. Makinose (1961), Die Calciumpumpe der Erschlaffungsgrana des Muskels und ihre Abhängigkeit von der ATP-Spaltung. *Biochem. Z.* **333**, 518–528.

704. M. Makinose and W. Hasselbach (1971), ATP synthesis by the reverse of the sarcoplasmic calcium pump. *FEBS Lett.* **12**, 271.

705. L. Rothfield, D. Romeo, and A. Hinckley (1972), Reassembly of purified bacterial membrane components. *Fed. Proc.* **31**, 12–17.

706. G. S. Levey (1971), Restoration of norepinephrine responsiveness of solubilised myocardial adenylate cyclase by phosphatidylinositol. *J. Biol. Chem.* **246**, 7405–7410.

707. A. N. Martonosi (1975), In: *Calcium Transport in Secretion and Contraction* (E. Carafoli, F. Clementi, W. Drabikowski, and A. Margreth, Eds.), pp. 77–86. North Holland, Amsterdam.

708. P. K. Brown (1972), Rhodopsin rotates in the visual receptor membrane. *Nature New Biol.* **236**, 35–38.

709. G. L. Nicolson (1976), Transmembrane control of the receptors on normal and tumor cells. *Biochim. Biophys. Acta* **457**, 57–108.

710. S. J. Singer (1974), The molecular organisation of membranes. *Annu. Rev. Biochem.* **43**, 805–833.

711. R. Kornfeld and S. Kornfeld (1976), Comparative aspects of glycoprotein structure. *Annu. Rev. Biochem.* **45**, 217–237.

712. R. D. Marshall (1972), Glycoproteins. *Annu. Rev. Biochem.* **41**, 673–702.

713. G. M. Edelman (1977), Summary: Understanding selective molecular recognition. *Cold Spring Harbor Symp.* **39**, 891–902.

714. H. G. Wittmann (1976), Structure, function and evolution of ribosomes. *Eur. J. Biochem.* **61**, 1–13.

715. C. G. Kurland (1977), Structure and function of the bacterial ribosome. *Annu. Rev. Biochem.* **46**, 173–200.

716. A. Liljas, S. Eriksson, D. Donner, and C. G. Kurland (1978), Isolation and crystallization of stable domains of the protein L7/L12 from *Escherichia coli* ribosomes. *FEBS-Letters* **88**, 300–304.

717. R. Leberman, A. Wittinghofer, and G. E. Schulz (1976), Polymorphism in crystalline elongation factor Tu·GDP from *E. coli. J. Mol. Biol.* **106**, 951–961.

718. W. Kabsch, W. H. Gast, G. E. Schulz, and R. Leberman (1977), Low resolution structure of partially trypsin-degraded polypeptide elongation factor, EF-Tu, from *E. coli. J. Mol. Biol.* **117**, 999–1012.

719. T. A. Steitz, R. E. Stenkamp, N. Geisler, K. Weber, and J. Finch (1978), X-ray and electron microscopic studies of crystals of native and proteolytically cleaved lac repressor protein. Int. Symp. Biomol. Struct., Conform., Funct. and Evol., Madras.

720. K. Beyreuther, H. Raufuss, O. Schrecker, and W. Hengstenberg (1977), The phosphoenolpyruvate-dependent phosphotransferase system of *Staphylococcus aureus. Eur. J. Biochem.* **75**, 275–286.

721. R. J. Winzler (1970), Glycoproteins. In: *Handbook of Biochemistry,* pp. C42–C49. Publ. Chemical Rubber Co., Cleveland, Ohio.

722. J. T. Gallagher and A. P. Corfield (1978), Mucin-type glycoproteins—new perspectives on their structure and synthesis. *Trends Biochem. Sci.* **3**, 38–41.

723. A. L. DeVries (1971), Glycoproteins as biological antifreeze agents in Antarctic fish. *Science* **172**, 1152–1155.

724. R. E. Feeney and D. T. Osuga (1977), Polar fish proteins. *Trends Biochem. Sci.* **2**, 269–271.

725. G. L. Nelsestuen and J. W. Suttie (1971), Properties of asialo- and aglycoprothrombin. *Biochem. Biophys. Res. Commun.* **45**, 198–203.

726. G. M. Edelman, J. Yahara, and J. L. Wang (1973), Receptor mobility and receptor cytoplasmic interactions in lymphocytes. *Proc. Nat. Acad. Sci. USA* **70**, 1442–1446.

727. E. Bayer and G. Holzbach (1977), Synthetische Hämopolymere zur reversiblen Anlagerung von molekularem Sauerstoff. *Angew. Chem.* **89**, 120–122.

728. F. O. Schmidt, D. M. Schneider, and D. M. Crothers (Eds.) (1975), *Functional Linkage in Biomolecular Systems.* Raven Press, New York.

729. L. v. Bertalanffy (1968), *General System Theory.* George Braziller, New York.

730. D. E. Koshland (1962), The comparison of non-enzymic and enzymic reaction velocities. *J. Theoret. Biol.* **2**, 75–83.

731. W. P. Jencks (1969), *Catalysis in Chemistry and Enzymology,* chap. 5. McGraw–Hill, New York.

732. G. G. Hammes (1964), Mechanism of enzyme catalysis. *Nature* **204**, 342–343.

733. C. L. Hamilton, C. Niemann, and G. S. Hammond (1966), A quantitative analysis of the binding of N-acyl derivatives of α-aminoamides by α-chymotrypsin. *Proc. Nat. Acad. Sci. USA* **55**, 664–669.

734. M. L. Bender and F. J. Kezdy (1964), The current status of the α-chymotrypsin mechanism. *J. Amer. Chem. Soc.* **86**, 3704–3714.

735. H. B. Bürgi, E. Shefter, and J. D. Dunitz (1973), Geometrical reaction coordinates. II. Nucleophilic addition to a carbonyl group. *J. Amer. Chem. Soc.* **95**, 5065–5067.

736. M. Hunkapiller, S. H. Smallcombe, D. R. Whitaker, and J. H. Richards (1973), Ionization behaviour of the histidine residue in the catalytic triad of serine proteases. *J. Biol. Chem.* **248**, 8306–8308.

737. R. M. Garavito, M. G. Rossmann, P. Argos, and W. Eventoff (1977), Convergence of active center geometries. *Biochemistry* **16**, 5065–5071.

738. L. Pauling (1946), Molecular architecture and biological reactions. *Chem. Engineer. News* **24**, 1375–1377.

739. R. Wolfenden (1972), Analog approaches to the structure of the transition state in enzyme reactions. *Acc. Chem. Res.* **5**, 10–18.

740. G. E. Lienhard (1973), Enzymatic catalysis and transition state theory. *Science* **180**, 149–154.

741. M. I. Page and W. P. Jencks (1971), Entropic contributions to rate accelerations in enzymic and intramolecular reactions and the chelate effect. *Proc. Nat. Acad. Sci. USA* **68**, 1678–1683.

742. T. C. B. Bruice (1970), Proximity effects and enzyme catalysis. *The Enzymes* **2**, 217–279.

743. D. R. Storm and D. E. Koshland, Jr. (1970), A source for the special catalytic power of enzymes: Orbital steering. *Proc. Nat. Acad. Sci. USA* **66**, 445–452.

744. D. R. Storm and D. E. Koshland, Jr. (1972), An indication of the magnitude of orientation factors in esterification. *J. Amer. Chem. Soc.* **94**, 5805–5814; Effect of small changes in orientation on reaction rate. *J. Amer. Chem. Soc.* **94**, 5815–5825.

745. G. C. K. Roberts (1978), What's special about enzymes? (editorial). *Nature* **271**, 409–410.

746. J. Crosby, R. Stone, and G. E. Lienhard (1970), Mechanisms of thiamine-catalyzed reactions. Decarboxylation of 2-(1-carboxy-1-hydroxyethyl)-3,4-dimethylthiazolium chloride. *J. Amer. Chem. Soc.* **92**, 2891–2900.

747. W. N. Lipscomb, J. A. Hartsuck, F. A. Quiocho, and G. N. Reeke, Jr. (1969), The structure of carboxypeptidase A. *Proc. Nat. Acad. Sci. USA* **64**, 28–35.

748. L. O. Ford, L. N. Johnson, P. A. Machin, D. C. Phillips, and R. Tjian (1974), Crystal structure of a lysozyme-tetrasaccharide lactone complex. *J. Mol. Biol.* **88**, 349–371.

749. C. H. Chothia and J. Janin (1976), Stability and specificity of protein–protein interactions: The case of the trypsin–trypsin inhibitor complexes. *J. Mol. Biol.* **100**, 197–212.

750. G. Wulff (1977), Synthetische Polymere mit chiralen Hohlräumen. *Nachr. Chem. Techn. Lab.* **25**, 239–243.

751. D. C. Watts (1973), Creatine kinase. *The Enzymes* **8**, 383–455.

752. D. C. Turner, T. Wallimann, and H. M. Eppenberger (1973), A protein that binds specifically to the M-line of skeletal muscle is identified as the muscle form of creatine kinase. *Proc. Nat. Acad. Sci. USA* **70**, 702–705.

753. D. E. Atkinson (1975), Functional linkage as a basic principle of biology. In: *Functional Linkage in Biomolecular Systems* (F. O. Schmidt, D. M. Schneider, and D. M. Crothers, Eds.), p. 43, Raven Press, New York.

754. K. C. Holmes (1975), Selbstorganisation biologischer Strukturen. *Klin. Wochenschr.* **53**, 997–1005.

755. H. E. Huxley (1973), Muscular contraction and cell motility. *Nature* **243**, 445–449.

756. M. Clarke and J. A. Spudich (1977), Nonmuscle contractile proteins: The role of actin and myosin in cell motility and shape determination. *Annu. Rev. Biochem.* **46**, 797–822.

757. D. DeRosier, E. Mandelkow, A. Silliman, L. Tilney, and R. Kane (1977),

Structure of actin-containing filaments from two types of nonmuscle cells. *J. Mol. Biol.* **113**, 679–695.

758. H. G. Mannherz and R. S. Goody (1976), Proteins of contractile systems. *Annu. Rev. Biochem.* **45**, 427–465.

759. A. F. Huxley and R. Niedergerke (1954), Structural changes in muscle during contraction. *Nature* **173**, 971–973.

760. H. E. Huxley and J. Hanson (1954), Changes in the cross-striations of muscle during contraction and stretch and their structural interpretation. *Nature* **173**, 973–976.

761. D. E. Koshland, Jr. and K. E. Neet (1968), The catalytic and regulatory properties of enzymes. *Annu. Rev. Biochem.* **37**, 359–410.

762. H. Diebler, M. Eigen, and G. G. Hammes (1960), Relaxations-spektrometrische Untersuchungen schneller Reaktionen von ATP in wässriger Lösung. *Z. Naturf.* **15b**, 554–560.

763. B. Chance (1960), Catalysis in biochemical reactions. *Z. Elektrochem.* **64**, 7–13.

764. M. Eigen and L. De Maeyer (1963), Relaxation methods. In: *Investigations of Rates and Mechanisms of Reaction* (S. L. Friess, E. S. Lewis, and A. Weissberger, Eds.), pp. 895–1054. *Technique of Organic Chemistry,* Vol. 8, Part II.

765. A. Einstein (1920), Schallausbreitung in teilweise dissoziierten Gasen. *Sitz. ber. Preuss. Akad. Wiss., Phys. Math. Kl., Berlin*, pp. 382–385.

766. H. Hartridge and F. J. W. Roughton (1923), A method of measuring the velocity of very rapid chemical reactions. *Proc. Roy. Soc. Ser.* **A 104**, 376–394.

767. H. E. Huxley (1972), Structural changes in the actin- and myosin-containing filaments during contraction. *Cold Spring Harbor Symp. Quant. Biol.* **37**, 361–376.

768. J. A. Spudich, H. E. Huxley, and J. T. Finch (1972), Regulation of skeletal muscle contraction. II. Structural studies of the interaction of the tropomyosin–troponin complex with actin. *J. Mol. Biol.* **72**, 619–632.

769. D. A. D. Parry and J. M. Squire (1973), Structural role of tropomyosin in muscle regulation: Analysis of the X-ray diffraction patterns from relaxed and contracting muscles. *J. Mol. Biol.* **75**, 33–55.

770. W. Hasselbach (1976), The sarcoplasmic calcium transport, a basic mechanism of the regulation of muscle activity. *Verh. Dtsch. Zool. Ges.* **1976**, 100–110.

771. S. Ebashi and M. Endo (1968), Calcium ion and muscle contraction. *Progr. Biophys. Mol. Biol.* **18**, 125–183.

772. C. Cohen (1975), The protein switch of muscle contraction. *Sci. Amer.* November, 36–43.

773. A. Weber (1975), Regulation of muscle contraction through cooperative interaction in the actin filament. In: *Functional Linkage in Biomolecular Systems* (F. O. Schmidt, D. M. Schneider and D. M. Crothers Eds.) pp 312–318, Raven Press, New York.

774. K. Kirschner (1971), Kinetic analysis of allosteric enzymes. *Curr. Top. Cell. Reg.* **4**, 167–210.

775. Z. Dische (1941), Interdependence of various enzymes of the glycolytic system and the automatic regulation of their activity within the cells. I. Inhibition

of the phosphorylation of glucose in red corpuscles by monophosphoglyceric and diphosphoglyceric acids; state of the diphosphoglyceric acid and the phosphorylation of glucose. *Bull. Soc. Chim. Biol.* (Paris) **23**, 1140–1148.

776. H. E. Umbarger (1956), Evidence for a negative-feedback mechanism in the biosynthesis of isoleucine. *Science* **123**, 848.

777. D. A. Walsh and C. D. Ashby (1973), Protein kinases: Aspects of their regulation and diversity. *Recent Progr. Horm. Res.* **29**, 329–359.

778. G. A. Robinson, F. R. W. Butcher, and E. W. Sutherland (1971), *Cyclic AMP.* Academic Press, New York.

779. W. A. Hagins and S. Yoshikami (1974), A role for Ca^{2+} in excitation of retinal rods and cones. *Exp. Eye Res.* **18**, 299–305.

780. D. O'Dell (1977), Egg activation (editorial). *Nature* **270**, 558.

781. M. Eigen and G. Hammes (1963), Elementary steps in enzyme reactions (as studied by relaxation spectrometry). *Adv. Enzymol.* **25**, 1–38.

782. M. Eigen (1975), Construction of control and self-organizing systems. In: *Functional linkage in Biomolecular Systems* (F. O. Schmidt, D. M. Schneider, and D. M. Crothers, Eds.), pp. 38–41. Raven Press, New York.

783. J. C. Rüegg and R. T. Tregear (1966), Mechanical factors affecting the ATPase activity of glycerol-extracted insect fibrillar flight muscle. *Proc. Roy. Soc.* **B. 165**, 497–512.

784. J. W. S. Pringle (1967), The contractile mechanism of insect fibrillar muscle. *Progr. Biophys. Mol. Biol.* **17**, 3–60.

785. A. Weber and J. M. Murray (1973), Molecular control mechanisms in muscle contraction. *Physiol. Rev.* **53**, 612–673.

786. F. M. Pohl (1971), Empirical protein energy maps. *Nature New Biol.* **234**, 277–279.

787. P. Doty, A. M. Holtzer, J. H. Bradbury, and E. R. Blout (1954), Polypeptides. II. The configuration of polymers of γ-benzyl-L-glutamate in solution. *J. Amer. Chem. Soc.* **76**, 4493–4494.

788. E. Ising (1925), Beitrag zur Theorie des Ferromagnetismus. *Zeitschr. Phys.* **31**, 253–258.

789. B. H. Zimm and J. K. Bragg (1959), Theory of the phase transition between helix and random coil in polypeptide chains. *J. Chem. Phys.* **31**, 526–535.

790. H. A. Kramers and G. H. Wannier (1941), Statistics of the two-dimensional ferromagnet. *Phys. Rev.* **60**, 252–276.

791. L. D. Landau and E. M. Lifschitz (1975), *Lehrbuch der Theoretischen Physik,* Vol. V, *Statistische Physik.* Akademie Verlag, Berlin.

792. W. G. J. Hol, P. T. van Duijnen, and H. J. C. Berendsen (1978), The α-helix dipole and the properties of proteins. *Nature* **273**, 443–446.

793. T. E. Creighton (1978), Experimental studies of protein folding and unfolding. *Prog. Biophys. Mol. Biol.* **33**, 231–297.

794. R. E. Dickerson and I. Geis (1969), *The Structure and Action of Proteins.* Harper & Row, New York, Evanston, London.

795. T. Takano (1977), Structure of myoglobin refined at 2 Å resolution. *J. Mol. Biol.* **110**, 537–584.

796. J. J. Marshall (1978), Manipulation of the properties of enzymes by covalent attachment of carbohydrate. *Trends Biochem. Sci.* **3**, 79–83.

797. J. P. Collman, J. I. Brauman, E. Rose, and K. S. Suslick (1978), Cooperativity in O_2 binding to iron porphyrins. *Proc. Nat. Acad. Sci. USA* **75**, 1052–1055.

798. Y. Mitsui, Y. Satow, T. Sakamaki, and Y. Iitaka (1977), Crystal structure of a protein proteinase inhibitor, *Streptomyces* subtilisin inhibitor, at 2.3 Å resolution. *J. Biochem. (Tokyo)* **82**, 295–298.

799. J. H. Ploegman, G. Drent, K. H. Kalk, W. G. J. Hol, R. L. Heinrikson, P. Keim, L. Weng, and J. Russell (1978), The covalent and tertiary structure of bovine liver rhodanese. *Nature* **273**, 124–129.

800. J. S. Richardson, D. C. Richardson, K. A. Thomas, E. W. Silverton, and D. R. Davies (1976), Similarity of three-dimensional structure between the immunoglobulin domain and the copper, zinc superoxide dismutase subunit. *J. Mol. Biol.* **102**, 221–235.

801. A. T. Hagler and B. Honig (1978), On the formation of protein tertiary structure on a computer. *Proc. Nat. Acad. Sci. USA* **75**, 554–558.

802. S. J. Remington and B. W. Matthews (1978), A general method to assess similarity of protein structures, with applications to T4 bacteriophage lysozyme. *Proc. Nat. Acad. Sci. USA* **75**, 2180–2184.

803. S. B. Needleman and C. D. Wunsch (1970), A general method applicable to the search for similarities in the amino acid sequence of two proteins. *J. Mol. Biol.* **48**, 443–453.

804. G. Fermi (1975) Three-dimensional Fourier synthesis of human deoxyhaemoglobin at 2.5 Å resolution: Refinement of the atomic model. *J. Mol. Biol.* **97**, 237–256.

805. H. Gutfreund & D. R. Trentham (1975), Energy changes during the formation and interconversion of enzyme-substrate complexes. *Ciba Foundation Symposium* **31**, 69–86.

Index

F

G

Examines established concepts, current activity and new advances in RNA biochemistry.

The Ribonucleic Acids
Second Edition
Edited by **P. R. Stewart** and **D. S. Letham**
1977. ix, 374p. 55 illus. cloth.

Here is the revised and up-dated second edition of Stewart and Letham's widely acclaimed book. *The Ribonucleic Acids* presents a comprehensive summary of modern concepts concerning the synthesis and function of RNA in living cells. It bridges the gap between the somewhat general approach of biochemistry texts and the highly specialized research reports on the role of RNA in living cells.

The book examines concepts of cell differentiation from the point of view of the processes which occur at the molecular level. It also discusses the role of RNA in the transfer and decoding of genetic information. Numerous illustrations appear throughout, and each chapter ends with a specific list of references.

From the reviews of the first edition:

"This review was months delayed because on its arrival the book was instantaneously swallowed up into constant use by researchers in our lab. Having managed to read it all at last, I think every one of the chapters is really fine, and the usefulness of the discussions is very great indeed. . . . I suggest this book as a 'best buy' for every library and nucleic acid research group."

American Society for Microbiology News

"Considering the breadth of coverage, there appears to be remarkably little repetition of material, perhaps because of the different perspectives in each chapter. . . . There is no question that this monograph is a useful survey of our knowledge of RNA structure and function . . ."

American Scientist

"Readable . . . well organized, well referenced and concerned with a surprising amount of fringe information, this is a very well worthwhile publication . . ."

Quarterly Journal of Experimental Physiology

Bacterial Metabolism

By **Gerhard Gottschalk**

1978. approx. 320p. approx. 160 illus. cloth

(Springer Series in Microbiology)

This concise, yet comprehensive textbook surveys the domain of bacterial metabolism and describes the various facets of this subject in terms useful to students and research workers. Emphasis is on those metabolic reactions which occur only in bacteria or those of particular importance for bacteria. Thus, the book describes in detail the energy metabolism of various groups of bacteria, including aerobic and anaerobic heterotrophs as well as chemolithotrophs and phototrophs. In addition, the book examines the various pathways used by bacteria for the degradation of certain organic compounds, the fixation of molecular nitrogen, the biosynthesis of cellular constituents, and the regulation of bacterial metabolism.

Contents